Interactive Multimedia Systems and Services

Interactive Multimedia Systems and Services

Edited by **Nelly Foreman**

CWILLFORD PRESS

New York

Published by Willford Press,
118-35 Queens Blvd., Suite 400,
Forest Hills, NY 11375, USA
www.willfordpress.com

Interactive Multimedia Systems and Services
Edited by Nelly Foreman

International Standard Book Number: 978-1-68285-041-1 (Hardback)

Printed in the United States of America.

Contents

Preface

In my initial years as a student, I used to run to the library at every possible instance to grab a book and learn something new. Books were my primary source of knowledge and I would not have come such a long way without all that I learnt from them. Thus, when I was approached to edit this book; I became understandably nostalgic. It was an absolute honor to be considered worthy of guiding the current generation as well as those to come. I put all my knowledge and hard work into making this book most beneficial for its readers.

Interactive multimedia systems aim to develop user interactive content by incorporating a combination of text, images, animation, etc. The various concepts such as social networks' user experience, human-computer interaction, design and testing of interactive multimedia systems are comprehensively discussed in this book. It provides the information needed to efficiently translate new research findings and technological advancements into practical applications. It aims to serve as a resource guide for students and professionals alike.

I wish to thank my publisher for supporting me at every step. I would also like to thank all the authors who have contributed their researches in this book. I hope this book will be a valuable contribution to the progress of the field.

Editor

The Digital Economy: Social Interaction Technologies – an Overview

Teófilo Redondo

ZED Wordwide, Department of Research and Innovation, Madrid, Spain

Abstract — Social interaction technologies (SIT) is a very broad field that encompasses a large list of topics: interactive and networked computing, mobile social services and the Social Web, social software and social media, marketing and advertising, various aspects and uses of blogs and podcasting, corporate value and web-based collaboration, e-government and online democracy, virtual volunteering, different aspects and uses of folksonomies, tagging and the social semantic cloud of tags, blog-based knowledge management systems, systems of online learning, with their ePortfolios, blogs and wikis in education and journalism, legal issues and social interaction technology, dataveillance and online fraud, neogeography, social software usability, social software in libraries and nonprofit organizations, and broadband visual communication technology for enhancing social interaction. The fact is that the daily activities of many businesses are being *socialized*, as is the case with Yammer (https://www.yammer.com/), the social enterprise social network. The *leitmotivs* of social software are: create, connect, contribute, and collaborate.

Keywords — blogs, folksonomies, online learning, social interaction technologies, social media, social web, wikis

I. INTRODUCTION AND BACKGROUND

IN recent years, we have been bearing witness to an exponential growth in capabilities to electronically collect, process, store, retrieve and disseminate information and create new knowledge. This has been the case with Internet-based collaboration tools and platforms reaching end-users in unprecedented ways: online social networking, blogs, wikis, podcasts, web feeds, folksonomies, social bookmarking, photo and video sharing, discussion forums, virtual worlds, all intended to advance interaction, collaboration, and sharing online. Social Computing is the generic term used to refer to any type of computing where software serves as an intermediary for a social relation. In social computing the user takes an active role in the process, often creating content or modifying previous content, and the computing experience has extended from the individual to the social.

Social interaction technologies (SIT) and collaboration software touch many fields and they have impacted on many fields, that is to say that Web 2.0 communication converge both socially and technologically. Web science must be interdisciplinary, since it brings together experts from computer science, software engineering, management information systems, business and economics, knowledge management systems, marketing, public relations and advertising, law, journalism and media, communication, psychology, anthropology, social work, design, library and information science, and education. The new emphasis is on user-generated content, creativity, and community-based knowledge building are characteristic of Web 2.0. The term "Web 2.0" suggests a fundamental technological improvement by assigning a version number like typical IT products. Common characterizations of Web 2.0 are mostly based on seven paradigms defined by Tim O'Reilly, in his now famous blog entry *What is Web 2.0* [1]: "The Web as a Platform," "Harnessing Collective Intelligence," "Data is the Next Intel Inside," "End of Software Release Cycle," "Lightweight Programming Models," "Software above the Level of a Single Device," and "Rich User Experiences."

In this "Web 2.0" world, web users have begun publishing their own content on a large scale and started using social software to store and share documents, such as photos, videos or bookmarks. The current trends indicate that a large number of US adult online consumers make daily use of social networking sites, publishing blogs/webpages, uploading visual content, commenting on blogs, posting reviews, or simply consuming user-generated content.

Social media sites, 2012-2014
% of online adults who use the following social media websites, by year

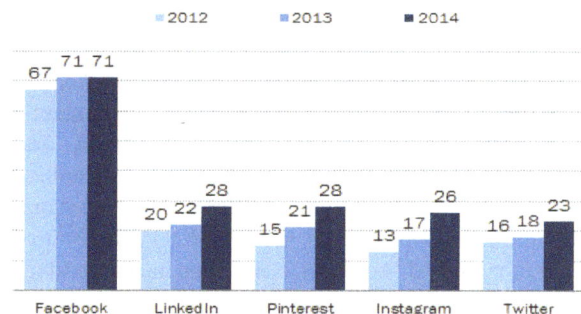

Pew Research Center's Internet Project Surveys, 2012-2014. 2014 data collected September 11-14 & September 18-21, 2014. N=1,597 internet users ages 18+.
PEW RESEARCH CENTER

Fir.1. Social Media Sites (Source: PewReseach Internet Project: http://www.pewinternet.org/2015/01/09/social-media-update-2014/)

The theoretical foundation of the social web can be traced back to J.C.R. Licklider with his seminal article "Man-

Computer Symbiosis," [2] published in 1960, the first of three articles that attempted to redefine the human-computer interaction. Licklider outlined a vision for interactive, networked computing and, ultimately, the Internet that we experience today.

Licklider's career did not originate in computing; his studies began in "physiological psychology," the field known today as neuroscience. Licklider investigated the brain's ability to understand speech in the presence of signal distortion [3]. These early studies helped Licklider understand the workings of the human brain and prepared him to foresee the potential for improved human-computer interactions, both individually and collectively, the very basis of the online communities that conform the Social Web.

The term "Social Web" is often used in everyday language as well as in academic literature as a synonym for "virtual" and "online communities". An online community is seen as a social group that interacts through a web platform over an extended period of time. An online community can be characterized by four elements:

- group of people with shared objectives (e.g., interests, goals)
- interaction over an extended period of time
- closeness due to bonds and relationships
- shared space for interactions governed by certain rules (for example, role definitions)

The Social Web refers to an aggregation of social interaction and collaboration technologies and can be viewed as a concept and a platform that utilizes social software to support some human needs. The Social Web encompasses numerous Internet applications, such as social networking sites, blogs, podcasts, wikis, massively multiplayer online role-playing games, photo and video sharing, online stores and auction houses, simulated 3-D virtual worlds, and wiki collaborations. Various attempts to provide a definition for the Social Web have resulted in three different approaches: technical, social, and economic. The technical approach focuses on the Internet as a medium or platform for a community. The sociological point of view stresses the forming and functioning of communities, whereas the economic perspective examines potential gains and intended profits [4].

MIT's Media Lab and Intel Corporation each developed two early mobile social web applications. *Social Serendipity* was MIT's Bluetooth-based social service meant to harness the power of mobile technology and social information. *Social Serendipity* facilitated social interaction among geographically proximate users by matching user profiles and then exchanging profile information with similar matches. Intel's *Jabberwocky* sought to monitor and broadcast a user's movement to identify "familiar strangers" and encourage a sense of urban community. Both of these technologies relied on the mobility of the devices to evaluate locational information to facilitate social connections among users.

The Social Web is realized through *social software*, which is a combination of various social tools within a growing ecosystem of online data and services, all joined together (aggregated) using common protocols, and Application

Programming Interface (API) methods. Social software is at the center of the so-called API economy, a collective term referring to the economic effects enabled by companies, governments, or individuals providing direct programmable access to their systems and processes through exposing specific APIs for creating larger applications and solutions.

II. SOCIAL SOFTWARE

Several tools are associated with social software:

1. tools allow people to participate by creating, publishing and distributing content, such as video, pictures, music and texts through the Internet.
2. social software allows people with similar interests to find one another and connect through social networking sites, such as Facebook.
3. people can coordinate their activities and collaborate through raising petitions and funds, and planning and conducting mobile campaigns and communities programs.
4. people can create reliable, robust, and complex products such as open source software applications such as Linux (the largest example of community development).

There are three characteristics commonly attributed to *social software*:

- conversational interaction between individuals or groups,
- social feedback that allows a group to rate the contributions of others.
- social networks to explicitly create and manage a digital expression of people's personal relationships.

Social software serves many purposes:

- *Delivery* of communication between groups
- *Enabling* communication between many people
- *Providing* gathering and sharing resources
- *Delivery* of collaborative collecting and indexing of information
- *Providing* new tools for knowledge aggregation and creation of new knowledge
- *Delivery* to different platforms depending on the creator, recipient, and context.

In summary, above all Social Software is about group interaction. For instance, mobile social networks allow users to connect with each other, share information, and create technologically enabled mobile communities. With the introduction of the iPhone in 2007, the public dream of mobile computing was realized. Mobile communication is becoming ubiquitous in many parts of the world today with over 4 billion mobile phone users worldwide.

Although mobile phones may lead to the atomization and privatization among users by discouraging face-to-face communication, the instant accessibility to whatever social app validates the social effects of mobile phone use. As mobile technology advances, new services for mobile phones have been developed, which allow people to create, develop, and strengthen social interaction.

Next we will explore a little bit further the most common Social Interaction Tools.

Number of mobile phone users worldwide from 2012 to 2018 (in billions)

The statistic shows the total number of mobile phone users worldwide from 2012 to 2018. For 2017 the source projects the number of mobile phone users to reach almost 5.3 billion.

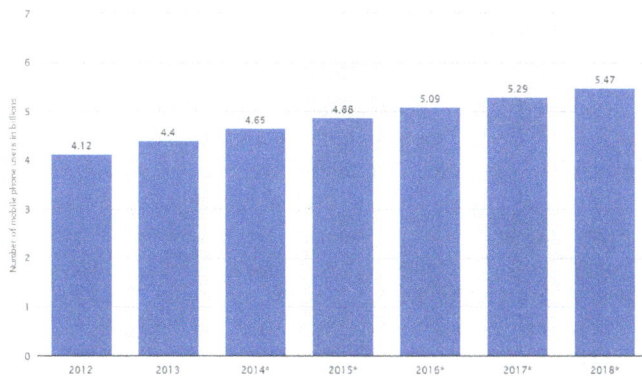

Fig. 2. Number of mobile phone (Source: Statista – The Statistics Portal: http://www.statista.com/statistics/274774/forecast-of-mobile-phone-users-worldwide/)

III. SOCIAL INTERACTION TOOLS

A. Discussion forums

The origin of Internet forums can be traced back to Usenet and its newsgroups, and the most recent form are the comments section at the end of articles in newspapers/magazines web sites. A forum provides an online exchange of information between people about a particular topic. It can be a place for questions and answers, or a comment-centric site and may be monitored to keep the content appropriate. Forums can be either text-only or media-rich (using images or videos to illustrate a point), and sometimes can be like a mini-portal on the topic. Forums can be entirely anonymous or require registration with username and password.

There are many types of forums: educational, professional, political, … for instance, forums have been institutionalized as an integral part of the political communication system.

B. Blogs and Wikis

One of the main results of the Web 2.0 are new modes of online communication and self-expression. The basic form of a blog is generally accepted to be brief posts, collected on one web page, which are chronologically ordered rather than by topic or argument. Blogs have a commentary concentrated style, which may also include links, pictures, video, or any other media forms, as well as reader comments. There are two main categories: (a) *filter blogs* tend to be focused on external events with all kinds of links, and (b) *journal blogs*, a blog created and maintained by an individual where authors write about events in their own lives. Blogs can be viewed as both a technology and a communication channel.

Blogs are sometimes used by organizations as educational or knowledge management tools. Organizations are using blogs for internal information sharing and knowledge management applications. Information shared within corporate blogs may include: industry or company news, strategy brainstorming, activities within a specific department, and the sharing of customer related information. But a number of concerns have arisen, mainly a preoccupation with regards to productivity. These concerns include: a waste of work time, loss of productivity, posting of inappropriate content, reluctance of employees to share knowledge, and increasing bandwidth requirements.

A number of sites provide all the necessary tools to get started in blogging: *Wordpress*, *Typepad* or *Blogger* are just a few of them.

The term Wiki refers to both technology and a concept of how one can create and edit online content. Wikis are editable websites that enable users to build content and collaborate. Wikis allow users to browse through Wiki pages, edit and modify existing pages, or develop new content in a collaborative way. A Wiki is both a website and a database for keeping track of all versions of the site as modified by the users. *Wikipedia* [5] is probably the most well-known of wikis, but there are many, some as singular as the *Intellipedia* [6], a wiki for sharing restricted information among intelligence services in the United States.

Ward Cunningham conceived the first Wiki in 1994 as a series of index cards stacked upon one another, each able to be changed or altered individually without affecting the other. The format allowed for each "card" to be a new alteration to the web page and to the central database.

Wikis were first adopted by businesses as collaboration software to allow a large number of people to work on a single idea in real time, where people separated geographically could all give input into a single database using a common format. Depending on how the moderator of the Wiki sets the parameters, users can edit, add or subtract information, remain anonymous or have to login to use the system.

C. Online social networking

Internet has made possible for people to connect with each other beyond geographical frontiers. Social networks encompass interactions between different people, members of a community or members across different communities. Each person in this social network is represented as node or vertex and the communications represent the links or edges among these nodes. A social network comprises several focused groups or communities that can be treated as subgraphs, in order to study the structural and temporal characteristics of social networks. To separate social networks from other types of social interactions we focus here on friendship networks ("friend" in Facebook, or "follower" in Twitter, show this connectivity). In these social friendship networking sites users explicitly provide trust ratings to other members.

Social networking sites allow individuals to create online profiles providing information about themselves and their interests, create lists of users with whom they wish to share information, and to view information published within the network by their friends. Social networks allow us to share experiences, thoughts, opinions, and ideas.

These social interactions also led researchers to hypothesize "Small World Phenomenon" (also known as "Small World Effect"), everyone in this world can be contacted via a short chain of social acquaintances. The social psychologist Stanley Milgram in a famous experiment in 1967 set out to find the length of this short chain, and resulted in the famous concept, "six degrees of separation". This "connectedness" aspect of social interactions between people have been applied to fields as varied as genealogy studies, mathematics, economy, team sports and even corporate dynamics.

Without providing an exhaustive list of social networks, each one highlighting a special means, we can mention Facebook, Twitter, Instagram, Snapchat, Tumblr, or more specific social networks: business (*LinkedIn*, *Xing*); enterprise (*Yammer*); academic (*ResearchGate*, *Gaudeamus*, *Academia.edu*, *Mendeley*).

D. Virtual worlds

Apart from various social media (Facebook, Blogger, YouTube) Web 2.0 technologies have generated online virtual reality environments (Second Life, World of Warcraft, Sims) that have influenced today's students in many ways. There are good opportunities for immersive experiences within user-constructed environments, communities and quests. Hands-on learning (even virtual hands-on) provide an intense engagement of immersive cognitive responses.

One particular case is Massively Multiplayer Online Role-Playing Games (MMORPGs), where participants move from loose to strong associations forming social networks via structured guidelines and interaction patterns. These virtual world inhabitants create communication conduits, collaborate to attain goals and solve problems, or entertain themselves, to chart the various associations ranging from casual conversations to groups in which role specialization is critical to community success. The basis for gamification lies in using rewards for accepted behavior, then creating a socialization continuum that stimulates players to interact with one another.

The vitality of MMORPGs and MMOGs (Massively Multiplayer Online Games) assure that more MMORPGs will be on the way [8].

E. Folksonomies

Folksonomy is a portmanteau of the words folk and taxonomy. Folksonomies are a type of annotation usually referred to as social tagging, with the purpose of knowledge representation and knowledge management. Social tagging is a multidisciplinary linking knowledge representation and classification that creates an open domain network. Many recent tools and techniques focus on exploring aspects of the connection between social tagging and the underlying community, in particular the role of tagging as a means of shared informal annotation. These tags are based on user motivation and function.

Folksonomies are a relatively novel way of indexing documents and locating information based on user generated keywords. This type of grass-roots community classification (similar to other social networking approaches, such as blogs and wikis) is a good example of collective intelligence. While taxonomies are hierarchical classifications defined by formal methods that do not necessarily include user-generated tags, folksonomies structure content via user tags and the vocabulary is not preassigned. Hierarchical taxonomies attempt to organize information and give context to data through a branching structure while folksonomies allow for a multiplicity of contexts.

A folksonomy can be seen as an indexing method open for users to apply freely chosen index terms. Peter Merholz [9] entitles this method "metadata for the masses". The term "folksonomy", was introduced in 2005 by Thomas Vander Wal [10], who defines folksonomy as the outcome of individual free tagging of online content and resources in a social environment for one's own retrieval, collaboratively assigning keywords to resources or items, and sometimes have been used synonymously with the terms social classification, social indexing, or social tagging.

Large-scale social bookmarking sites (such as *Del.icio.us* or *Reddit*) have been among the earliest adopters of using folksonomies to organize information. These sites are effective tools for storing, finding, and sharing Internet-based resources, a form of social knowledge management. Much of the success of social bookmarking is attributed to its loosely structured approach to organizing data and the ease with which consumers can learn and integrate tags into a folksonomy.

F. Podcasts and webcasts

Podcasting, a portmanteau word created out of the brand name iPod and the term broadcasting is a distinctive area within social interaction technology. Content is often listened to or viewed within the world of a personal audio/video device, this more so with the ubiquitous mobile device. The upload or the download is the interaction. Podcasts are used for entertainment, education, instruction, profit and just to enjoy some time.

The first podcast took place in 2003 by automatically streaming a single audio file half way around the world. Months later Apple Computer, Inc. proved that its personal listening device, the iPod, could synchronize with a new program called iTunes and download files using the same technology. The broadcast media have begun to use podcasting as a method of time shifting programming, to allow its audience to listen to content claiming the audience consume "what they want, when they want". Educators are using podcasting for reaching out to students. Businesses are using podcasting as a marketing tool. The commercial future of podcasting appears to be in the area of advertising and broadcasting.

The user listens to or views the file, deletes it, and waits for the next episode. Because a podcast audio or video file is digital and in a format common to many devices (normally MP3), that file can go viral being reposted, edited, linked to through social sites (such as YouTube, or Facebook), or moved around through email or by some other social interaction means.

Webcasts in the form of broadband visual communication (BVC) technologies (such as videoconferencing and video sharing) allow for the exchange of rich simultaneous or pre-recorded visual and audio data over broadband networks. BVC involves both simple and complex social and technical interactions. The complexities arise as the interaction grows from communication between two individuals in the same location to communication between multiple individuals in multiple locations, working for multiple organizations located in different communities.

G. Photo and video sharing

Photo sharing can be said to be one of the first social engaging uses of the early Web 2.0 days, even before the term was coined. At a time when most cameras were not digital, users started uploading (publishing their digital photos online) and providing links for friends to share and comment, using websites like *Picasa*, *Flickr*, or *Instagram*.

Similarly, video sharing sites allow users to upload and share their video clips with their friends or connections (private) or with public at large (public). The best known examples are *YouTube*, *Vimeo*, and *Dailymotion*, as well as *Netflix*, *Hulu*, or *Vine*. Apart from traditional video some of these sites provide webcasting capabilities too. For instance, YouTube personal channels can be used for streaming content, which is particularly useful in the case of online learning.

H. Geotagging

Neogeography refers to geography in the Web 2.0 style, a collaborative technology from the public rather than from those in the profession, that is, a group of people (many unknown to one another) who volunteer collectively to contribute data about a topic, in this case, mapping. The practice of neogeography shares the characteristics of other social interactive technologies. Volunteer-supplied geographic tags may assume informational value beyond entertainment. Neogeography-related websites provide different ways for people to contribute photographs, locations, tags, and comments.

Neogeography might be considered a subset of cybercartography or interactive, web-based spatially referenced data, and can trace its origins to the Geospatial Web or the GeoWeb. Any sort of data that conveys place can qualify as geographical data, including for example, zip codes, area codes, images of a place, census data or place names, using a specific XML format for geographic data known as GML (Geography Markup Language).

Geosocial networking is the result of combining geotagging with social networking by including geographic services and capabilities to enable additional social dynamics, and this is the case by inserting location coordinates assigned to pictures, and then adding those pictures to maps with many applications: *Flickr* or *Instagram* for photographs, or *Panoramio* for Google Earth / Google Maps. Users of geosocial applications like *Yelp*, *Facebook Places* and *Foursquare* (*Swarm*) share their locations as well as ranked recommendations for locations or 'venues'.

IV. Some Special Use Cases

A. Advertising / marketing

Marketing is historically considered an activity that businesses performs to direct the flow of goods and services from producers to consumers. The growth of the Internet and the development of social software have changed this top-down process in the age of citizen marketing, that is, consumers voluntarily posting product information based on their knowledge and experience. Citizen marketers were envisioned by futurist Alvin Toffler [11] who coined the term "prosumers," blending the words producer and consumer, consumers who educated themselves and became involved in the design and manufacture of products. Product here refers to goods, services, brands, companies, organizations, or people, such as political candidates. The product information may take the form of opinions, reviews, videos, to be found on forums, blogs, ratings or opinion sites, social networking sites, video sharing sites, or even on mainstream marketers' websites as consumer reviews or discussion boards.

Consumers seek product information provided by citizen marketers, who are eager to share their experiences and their knowledge of a product. Citizen marketers are not on the company payroll and are not trying to sell anything. They volunteer their time as writers, animators, designers, and videographers to express their opinions about products.

B. Social capital

The social capital framework is applied to illustrate how Web 2.0 tools and techniques can support effective information and knowledge management in organizations. Managing social capital for effective knowledge sharing is a complex process, and Web 2.0 helps by creating a new culture of voluntary, contributive, and collaborative participation.

Research on social capital has been carried out in different disciplines and at different levels depending on the chosen perspective (e.g., Putnam [12]; Fukuyama [13]). Social capital includes the individual and the social aspects and is defined as the sum of actual and potential resources embedded within, available through, and derived from the network of relationships possessed by an individual or social unit. Social capital encompasses both the network and the assets that may be mobilized through that network.

Social capital is also often described in three dimensions: a structural dimension (network character), a relational dimension (trust and social identity), and a content dimension (communication to facilitate social capital). People generate economic, emotional, spiritual, and social value by engaging in social relationships. Social capital is the glue holding communities together with the power of cooperative actions. Social capital is dictated by how networks of individuals in a community create conditions where people are inclined to do things for one another.

C. Virtual Teams

Social interaction technologies have made it possible for teams to exist in a virtual reality. Team members can create, maintain, transmit and influence their competitiveness and effectiveness. The key requirements for the functioning of successful virtual teams and online culture are building trust, consolidating authentic communication flows and thinking critically.

Traditionally, a team is viewed as a group of people who bring balanced competencies to shared purposes, approaches and performance targets. There is usually synergy between the individual members of the team, which means that when the individual efforts and actions are harmonized, something different and unique is created that could not be produced by any single individual of the team. The concept of a team is expressed through seven vital elements:

1. size linked to the scope of the task
2. members' skills are balanced (basis for interdependence)
3. mutual accountability (members are synergistic and trust each other)
4. synergies of purpose
5. approach
6. performance targets
7. distance between members

D. Online Learning

Online behavior, distributed collaboration, and social interaction are already having a transformative effect on education, triggering changes in how teachers and students communicate and learn. Learners can engage in creative authorship by producing and manipulating digital content and making it available for consumption and critique by classmates, teachers, and a wider audience on the web.

Informal education (or learning) sits outside the traditional educational context and is voluntary, self-directed, lifelong, and motivated mainly by intrinsic interests, curiosity, exploration, and social interaction, and typically lacks the presence of an instructor. Informal learning often is self-paced and visual- or object-oriented. It provides an experiential base and motivation for further activity and learning.

The benefits and challenges of hybrid courses, which blend face-to-face instruction with online learning, leverage opportunities provided by the introduction of web-based social interaction technologies. Hybrid courses continue to evolve to meet the needs of students, instructors, and institutions of higher learning.

Today, more classrooms are equipped with various types of technology including Internet access, integrated projectors for computers, digital board, or audio and video devices. Online education management systems (*Blackboard*, *Sakai*, or *Moodle*), sometimes called learning management systems (LMS), are becoming more commonplace and are enabling communication, learning materials, assignments, and grading to occur online.

Early online learning environments were not engaging, and limited in supporting the interaction, coordination and cooperation between students and instructors, with low levels of confidence while learning at a distance, and low satisfaction levels resulted as a consequence. At earlier times dropout rates were relatively high. In online learning the motivation, and the

sense of shared social experience are greatly constrained. New social interaction technologies can improve the social experience and social support of online learning. The members experience the online environment as a social place for learning and not in isolation. Opinion and preference of online versus classroom has turned around and now online instruction is the preferred way when a choice is offered.

	Total Enrollment	Annual Growth Rate Total Enrollment	Students Taking at Least One Online Course	Online Enrollment Increase over Previous Year	Annual Growth Rate Online Enrollment	Online Enrollment as a Percent of Total Enrollment
Fall 2002	16,611,710	NA	1,602,970	NA	NA	9.6%
Fall 2003	16,911,481	1.8%	1,971,397	368,427	23.0%	11.7%
Fall 2004	17,272,043	2.1%	2,329,783	358,386	18.2%	13.5%
Fall 2005	17,487,481	1.2%	3,180,050	850,267	36.5%	18.2%
Fall 2006	17,758,872	1.6%	3,488,381	308,331	9.7%	19.6%
Fall 2007	18,248,133	2.8%	3,938,111	449,730	12.9%	21.6%
Fall 2008	19,102,811	4.7%	4,606,353	668,242	16.9%	24.1%
Fall 2009	20,427,711	6.9%	5,579,022	972,669	21.1%	27.3%
Fall 2010	21,016,126	2.9%	6,142,280	563,258	10.1%	29.2%
Fall 2011	20,994,113	-0.1%	6,714,792	572,512	9.3%	32.0%
Fall 2012	21,253,086	1.2%	7,126,549	411,757	6.1%	33.5%

TOTAL AND ONLINE ENROLLMENT IN DEGREE-GRANTING POSTSECONDARY INSTITUTIONS – FALL 2002 THROUGH FALL 2012

Fig. 3. Online Enrollment (Source: Edudemic – Connecting education and technology: http://www.edudemic.com/2013-survey-online-learning/)

Institutional use of e-learning, respondent percentages

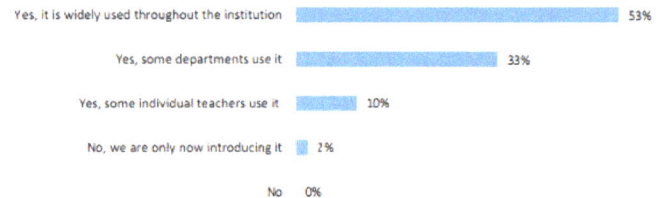

Yes, it is widely used throughout the institution	53%
Yes, some departments use it	33%
Yes, some individual teachers use it	10%
No, we are only now introducing it	2%
No	0%

Fig. 4. (Source: European University Association : http://www.eua.be/Libraries/Publication/e-learning_survey.sflb.ashx)

	Percent of Respondents		
Previous Enrollment in Online Study	All	Undergraduate	Graduate
Yes, I took individual online courses	45%	41%	45%
No	43	51	37
Yes, I completed another fully online program	17	11	25

	Percent of Respondents		
Compare Online to Classroom	All	Undergraduate	Graduate
Better	47%	50%	43%
About the same	43	41	48
Not as good	10	9	9

Fig. 5. (Source: Online College Students 2014: http://www.learninghouse.com/wp-content/uploads/2014/06/2014-Online-College-Students-Final.pdf)

Some examples of applying social interaction tools to education follow:

a) Wikis in education

In recent years, the field of education has discovered the educational value of the instructional use of wiki-based classroom technologies. Certain kinds of wiki-related activities correspond to certain levels of classroom interactions: social interaction, general discussion, topic-focused discussion, and collaborative/cooperative activities. Students use Wikis for collaborative writing exercises, or completing group assignments. [14]

b) Videogames

Squeak Etoys is a free software program and media-rich authoring system with a user-friendly visual interface. The software is designed to help six to twelve year-old children learn through interaction and collaboration. Etoys environment was created to increase the capacity for creative learning, exploration, interaction, and collaboration. Children interact, work together on projects, and engage in computer simulations and games while learning mathematics, physics, chemistry, and geometry. A similar objective is pursued by *Pocket Code*, strongly inspired by MIT's *Scratch*, the free programming language and online community that helps creating interactive stories, games, and animations.

c) Webcasting

Webcasting refers to the delivery of audio and video content over the web. The web is used as a delivery medium for informational, instructional, marketing, and entertainment purposes. Webcasting incorporates social media elements that can assist in the development of a science-oriented educational website.

E. Public libraries

Libraries have started employing social software applications (such as blogs, tagging, social networking, and wikis) to engage readers, encourage user-contributed content, and connect with user populations. User-centered philosophies are at the heart of libraries' service and have been in practice long before the emergence of Web 2.0. However, libraries have seen a radical shift as they are now faced with web-users' expectations. These expectations may not be met with less interactive computer technologies, such as library online public access catalogs (OPACs).

V. CURRENT AND FUTURE TREND

A. Semantic Web

Social tagging has become an essential element for Web 2.0 and the emerging Semantic Web applications. Web 2.0 sites express their structure, features, and relations in different ways. The model, termed the Social Semantic Cloud of Tags (SCOT), allows for the exchange of semantic tag metadata and the reuse of tags in various social software applications.

The initial purpose of tagging is to help users organize and manage their own resources, and collective tagging of common resources can be used to organize information via informal distributed groups of users. The power of semantic social tagging lies in the aggregation of information, which involves social cohesion by reinforcing social connections and providing social search mechanisms. A community built around tagging activities can be considered a social network with an insight into relations between topics and users. Semantic Web techniques and approaches help social tagging systems to eliminate tagging ambiguities.

B. eGovernment and privacy in social media

Globalization has brought a special emphasis on knowledge creation and transfer as the primary driver of economic growth and competitiveness with information technologies playing an ever-increasing role. The economic, social and political landscape in which development is taking place has changed completely.

E-government initiatives are aimed at modernizing governmental agencies in their dealings with the public and extending services into online environments. These initiatives have begun in various countries, which have allowed citizens easy access to public services and lobbying opportunities at policy level decision-making.

A number of legal issues around privacy preservation may arise from the increasing use of social interaction technologies: prospective employers searching the Internet to discover information from candidates' blogs, personal web pages, or social networking profiles; or employees being fired because of blog comments. These situations present challenges to legal systems which historically have been slow to adapt to new technologies. As a result, many of these legal issues remain unsettled.

C. Social enterprise

Social software is assuming a significant role in business, and has been utilized recently on a growing scale by companies in customer relationship management (CRM). A firm needs to identify the optimal level of social software deployment when planning to maximize its transactional benefits through the management of a customer knowledge base. The optimal level of social software depends on a range of factors: the initial volume of knowledge base, transaction benefits, and the estimates of the positive and negative effects of social software use.

Only recently have companies started to apply social software for managing customer knowledge, maintaining good customer relationships, and enhancing customer satisfaction, sometimes even reaching to customers in a very personalized manner. Although social software is gradually assuming a more essential role in e-business, it is still unclear at what level firms should implement it. Social software dynamically influences customer knowledge bases with direct and indirect effects of social software implementation on businesses. Customers who are dissatisfied with their shopping experiences may impact the current knowledge base affecting the transactions of future potential customers.

D. Social Web of Things

The relationship between Social Networks and Internet of Things (IoT) was introduced as "Social Web of Things". This idea is a redefinition of the IoT paradigm, in which things leverage social networks and specifically social standards to communicate, assigning a specific social identity to things ("smart objects") at the same level than people. Social Networks can help in that sense in "elevating" the semantics of IoT interactions to the user level and thus fostering the adoption of connected objects such as wearables, home automation, connected cars, to interconnect different devices with users.

"Smart social objects" consists in creating a network of "trusted" friends between humans and objects. Objects can post information to the social network, show their availability and discover new "friends", interacting with other objects or humans. In that sense the social component adds a user-friendly interaction (dialogue) paradigm for people to interact with their surrounding "Smart Social Space" environment.

Smart Social Spaces could be public, such as a local business, or private, such as a smart office or home, in which appliances and sensors communicate with one another and post their behaviors on the social wall. Users could receive (multimedia) notifications or alerts about sensors and are able to send commands remotely, for example, to their home security cam.

Some current examples of the Social Web of Things are: *Toyota Friend*, *Nike+*, *Xively* or *Evrythng*.

E. Artificial Intelligence in social networks

Facebook, *Twitter*, *LinkedIn*, and others are beginning to use artificial intelligence techniques to build their "deep learning" capacities. They are starting to process all the activity occurring over their networks, from conversations, to facial recognition, to gaming activity. Advances in cutting-edge artificial intelligence research, which program machines to perform high-level thought and abstractions, are helping social networks and their advertisers get insights from unstructured consumer data.

And one area demanding further progress is the Human-Computer symbiosis as represented by such experiments as IBM's Watson [15] and IpSoft's Amelia [16], whose stated purpose is extending a human's capabilities by applying intelligent artificial systems techniques, such as deep learning and social and interpersonal communication.

F. Crowdfunding

Crowdfunding, which can be likened to donations, is a capital collection method where common people, and not necessarily professional investors, could fund small personal or business projects by putting their own money into a kind of collective account. Originating from the evolution of Social Web technologies, crowdfunding has gained a followship because of its simplicity, and by removing more formal and traditional forms of loans, like those provided by banks, out of the picture. The most well-known crowdsourcing sites are: *Kickstarter*, *Indiegogo*, *RocketHub* or *Razoo*.

G. Crowdsourcing software development

Voluntary contribution to the creation of new software products, and amelioration of existing versions, is also a recent phenomenon originating from Social Web interaction. Crowdsourcing of software development implies the participation of large numbers of what could be termed a multidisciplinary team involving from designers, to IT architects, to code developers, to relational and documental databases administrators and developers. It is a paradigm shift from industrial mode to peer production mode with a clear impact on both time and money needed for the implementation of an IT product. By having a crowd of volunteers available, testing is much more thorough, and possible issues are detected earlier and then corrected, to the overall benefit of the community of interested parties, and consequently quality is enhanced. This collaborative software development model has now a very widespread use and all of the Free and Open Source Software (FOSS) initiatives pursue the same objectives.

One good example of crowdsourcing of software development are mashups. Loosely defined as mixing and matching content from more than one source to create a single new service displayed in a single graphical interface, the earlier uses for mashups were maps on which geolocating pictures and videos. Mashups have been made possible by the common availability of APIs to make a developer's life a lot easier.

VI. CONCLUSION

Social Interaction Technologies (SIT) have had a transformational effect in many aspects of our lives, since they touch many fields and they have impacted on so many fields in a clear convergence both socially and technologically. Daily activities of many businesses are being socialized, incorporating the central topics of social software (create, connect, contribute, and collaborate) into a multidisciplinary ecosystem of interactive and networked computing.

We have reviewed a number of social interaction tools and some special usages where they have shown a greater effect and impact. The economic results of the so-called social media economy have yet to be produced, but only in terms of productivity increase, and employees and customers satisfaction, the value is certainly remarkable.

REFERENCES

[1] T. O'Reilly (2005). Available: http://www.oreilly.com/pub/a/web2/archive/what-is-web-20.html

[2] J.C.R. Licklider. "Man-Computer Symbiosis" (1960), *IRE Transactions on Human Factors in Electronics* (a copy can be accessed at http://worrydream.com/refs/Licklider%20-%20Man-Computer%20Symbiosis.pdf)

[3] M. Mitchell Waldrop (2001). *The Dream Machine: J. C. R. Licklider and the Revolution That Made Computing Personal*. New York: Viking Penguin

[4] J. Hummel, *Online-Gemeinschaften als Geschäftsmodell: Eine Analyse aus sozio-ökonomischer Perspektive*, 2005, Deutscher Universitäts-Verlag.

[5] Mochón, F., and M. Rojas, "IJIMAI Editor's Note - Vol. 2 Issue 5", International Journal of Interactive Multimedia and Artificial Intelligence, vol. 2, issue Special Issue on AI Techniques to Evaluate Economics and Happines, no. 5, pp. 4-5, 03/2014

[6] Calderón JCP, et al., Informe sobre la situación del Sector TIC 2010 a 2013: Mirada global y de España, Sociedad y Utopía. Revista de Ciencias Sociales 43, 84-101

[7] S. Milgram (1967) "The Small World Problem", *Psychology Today*, Vol. 2, 60–67

[8] MMORPG gamelist: http://www.mmorpg.com/gamelist.cfm

[9] P. Merholz (2004) *Metadata for the masses*. (http://www.adaptivepath.com/ideas/e000361/)

[10] T. Vander Wal (2004). Folksonomies (http://vanderwal.net/folksonomy.html)

[11] A. Toffler (1980). *The Third Wave*. New York: Bantam Books.

[12] R. Putnam (2000). *Bowling Alone: The Collapse and Revival of American Community*. New York: Simon & Schuster

[13] F. Fukuyama (1995) *Trust: The Social Virtues and the Creation of Prosperity*. Cambridge University Press.

[14] W. Richardson. (2006). *Blogs, wikis, podcasts, and other powerful web tools for classrooms*. Thousand Oaks, CA: Corwin Press.

[15] IBM?s Watson: http://www.ibm.com/smarterplanet/us/en/ibmwatson/

[16] IpSoft's Amelia: http://www.ipsoft.com/what-we-do/amelia/

2

Kuruma: The Vehicle Automatic Data Capture for Urban Computing Collaborative Systems

Guillermo Cueva-Fernandez, Jordán Pascual Espada, Vicente García-Díaz, and Martin Gonzalez-Rodriguez

University of Oviedo, Department of Computer Science

Abstract — **Smartphones can provide coverage in large areas all around the world and with the availability of powerful operating systems they can become solid sensing infrastructures. In fact, static sensors are hard to deploy and maintain while modern mobile devices include many sensors that can be used to sense and benefit from collaborative communities. This project tries to improve urban computing by developing a framework able to create monitoring applications for mobile devices, focusing on obtaining the highest degree of interoperability between sensors. A prototype application has been developed to demonstrate the feasibility of creating multidisciplinary applications with several different approaches. The application developed consists of a Road Roughness Information System that measures smoothness and detects irregularities on the roads.**

Keywords — **OBD, Internet of Things, Urban computing, Vehicle**

I. INTRODUCTION

Nowadays passenger vehicles are filled with technology; some modern cars have numerous different sensors that are continuously sensing. This converts vehicles into a great platform for urban sensing networks. All the amount of information that can be obtained from vehicles can be very valuable if it is shared and compared with other readings. This data can be used for numerous practical applications. Our intention is to collect as much information as possible and upload it to the cloud. We expect this information to be useful for many applications.

The motivation is originated with the generalization of the use of mobile phones that combined with the use of a cheap OBD (On Board Diagnosis) connector can be a very powerful tool, providing almost complete engine control as well as monitoring parts of the chassis, body and accessory devices, including the diagnostic control network of the carrier. Since these sensors are connected to vehicles that continuously move, we can achieve a much larger sensing areas than with a static sensor network. Static sensors can be deployed in main roads but it would be very expensive to install them in the road network. Covering wide areas of roads is part of the Nericell [1] project that consists on monitoring road and traffic conditions by using the sensors in the users smartphones.

Using the built-in accelerometers, microphone and GPS they are able to detect potholes, bumps, breaking and honking, the use of the Smartphone sensors in other mobile areas [2-3].

The idea of using a vehicle as distributed mobile sensors is not new. CarTel [4] has established a system that can be installed into vehicles. Each vehicle, when installed with an embedded computer, is considered a node that gathers and processes sensor readings locally. Data is sent to a central portal, where it is analyzed and visualized.

There are other researches that have shown the possible benefits of studying data collected by sensors in cars. Traffic can be efficiently monitored as proved by several researches [5,6] by recording the speed of several vehicles and comparing the information.

Even commercial products like TomTom HD Traffic record the speed of their users to make traffic estimations. TomTom claims that if 10% of vehicles in the road would drive with their system they would have a real time accurate description of the traffic conditions.

Not only traffic can be monitored. Some researches such as ParkNet [7] and Fueoogle [8] have already demonstrated that sensing different parameters in vehicles can have many utilities. They established that by collecting data from sensors they can create the optimum routes for fuel efficiency or locate street parking spaces in an accurate way.

In addition, complex simulations on distributed systems, for instance, SignalGuru [9] consists on generating traffic light patterns that provide the optimum speed for vehicles to reduce fuel consumption and reduce the environmental impact.

All these collaborative systems aim for different objectives, but they all have a common base, they all sense from devices mounted in moving vehicles. We aim to create a common platform that can gather all kinds of different localized data that can be compared with data from other users to obtain some kind of common benefit.

The contributions of this work are the following:

1. The creation of a framework capable to help in the implementation and maintenance of sensor based applications. The framework is called NIKKO.
2. The creation of a user focused application that can sense and upload data by the use of Smartphones connected to the vehicles OBD port. This application is called Kuruma.

3. The discussion of four possible uses that can be based on the Kuruma application, including black box applications, road hazard detections systems, automatic suspension setting recommendations and a study of the relationship between road roughness and fuel consumption.

The remainder of this paper is structured as follows: first, in Section 2 we present an overview of the NIKKO framework. In Section 3 we describe the main concepts related to the Kuruma application. In Section 4 we show a practical use of Kuruma through a road roughness information system. Section 5 underlines other possible applications for Kuruma. Finally, in Section 6 we indicate our conclusions and future work to be done.

II. THE NIKKO FRAMEWORK

In order to satisfy the objectives of the project, we designed and created a new framework that helps us to generate a system in which sensors can be easily managed to create many kinds of applications that use different types of sensors (Fig. 1).

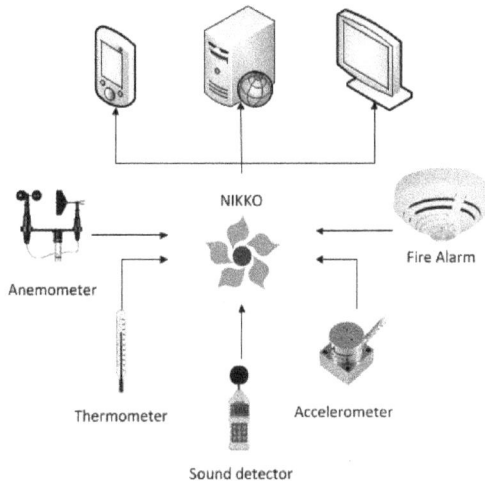

Fig. 1. NIKKO Overview

The framework was designed to obtain as much flexibility as possible with sensors. In order to enable a quick and efficient way to change the destination of the sensor notifications, a configuration file has been kept separated from the source code so that it can be easily modified without the need of recompiling. NIKKO has the ability of adding, deleting and interchanging sensor notifications with a simple file that can change the complete behavior of the applications. That way, the applications that use the framework can be very flexible and valuable tools that are easily maintainable.

The core function is to monitor the events that the sensors generate and to create the specific response actions included in the configuration file (Fig. 2).

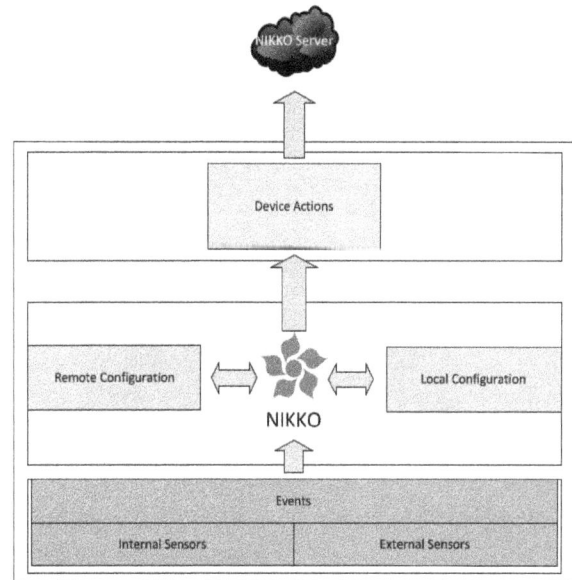

Fig. 2. NIKKO Modules

When the module receives a new event, it checks if the received event is one of the events that it has to handle. If it is included on the list, it calls the method of the actions it has assigned.

NIKKO is conFig.d by an XML file. The file provides the relationships between the sensors events and the actions they have to trigger. The configuration file can be obtained from a local file in the device or a remote location in a web server.

Internal and external sensors are encapsulated in events that are triggered when sensors reading reach a specific measure.

The Device Actions module is used to send the actions when an event is triggered. This module can be highly conFig.d and can create all sorts of actions, including uploading to servers or notifications in applications.

III. KURUMA

The Kuruma application senses the information from vehicle surroundings. The application is based on the NIKKO framework developed for Android devices. It uses an OBD Bluetooth port to connect to the sensors of the vehicle that is monitored.

Taking in consideration all the sensors that can be connected to a phone with the OBD port, it points out the need to collect data and upload it to share it with other users. There could be hundreds of possible applications.

Our intention is to create the base for an application that gathers localized data from vehicles to show the feasibility of these kinds of systems.

The developed application can monitor any standard parameter such as speed or engine load (called PIDs) from an OBD port of a vehicle. Likewise, any nonstandard PIDs could be easily added and monitored (Fig. 3).

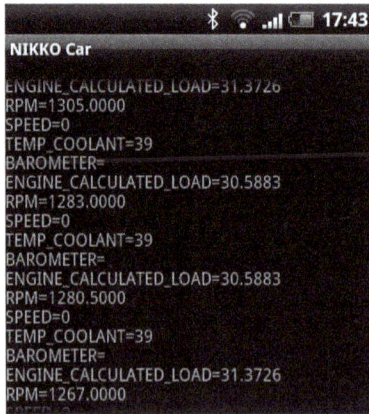

Fig. 3. Kuruma Application

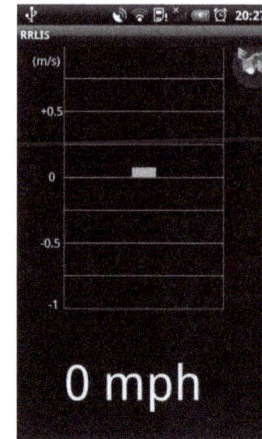

Fig. 4. Road Roughness Information System

Once the application is started it will begin searching for paired Bluetooth devices. If it finds an ELM327 Compatible Bluetooth OBD adapter it will start sending requests to obtain the sensors indicated in the configuration file. Then, the application will show in the device screen the values obtained from the vehicle's OBD port.

The application uses the NIKKO framework to manage the different sensors. With NIKKO, the actions that the application has to perform when there is a specific event are easily configurable. By changing the XML configuration file the application can show in the application screen or upload to a server the monitored sensors.

IV. PRACTICAL APPLICATION

To demonstrate the Kuruma's potential we have developed a naïve system that could lead to future research lines of investigation.

A. Road Roughness Information System

The application consists on developing a cheap system to create and maintain information of the roughness in roads. Pothole Patrole [10] focuses on detecting potholes and bumps, but does not measure the state of the roads. In a deteriorated road there could be no potholes at all, so this system would not find any irregularities. Potholes appear in weakened roads, so the ideal system would help detecting irregular roads and potholes. Our objective is to not only focus in detecting potholes but also adding the ability to track the state and smoothness of the roads.

To achieve these objectives the system uses Smartphones with built-in accelerometers and GPS receivers. The mobile phone users will record with the accelerometers the alterations of the roads and will report them to a centralized server. This information is useful to generate a detailed map of cracking, rutting, patching and potholes without the need of special expensive vehicles that measure roughness of the roads.

The motivation is to create a system that monitors in real time the smoothness of the roads by the help of phone accelerometers. Fig. 4 shows an example.

Phone users obtain data that will be able to generate a more detailed and realistic map of the actual state of the roads. Thus, by increasing the accuracy of the information on the roads, maintenance can be done in a much more efficient way, selecting only the roads that really need a service. In addition, after the conservation works the road evolution can also be monitored.

B. Implementation

To implement the system we have used an application that runs in an Android phone that will collect data from the accelerometers using our special algorithm and emit it to the system data server.

The data is sent to the server and must be collected while driving a vehicle. The Android application will be in standby mode without transmitting data; only when the accelerometer sensors detect a perturbation, the data will be sent. The data that the application sends will be the intensity of the acceleration, the speed of the vehicle in the acceleration moment and the location provided by the GPS.

Fig. 5. Road Roughness Information System Output

Providing the speed is crucial because the same bump at different speeds can cause different accelerations. For example, a big bump at a very low speed could be inappreciable for the accelerometer sensors. That is why speed data is necessary to ensure that the data is valid. The user can forget to stop the application and while he is walking, the application will be aware of this and discard this data without sending it. After the data is classified it is shown in a map available through an Internet browser.

To test the application, a small deployment of the system with one vehicle was done. The preliminary results obtained are shown in the Fig. 5. They were obtained on the University of South Florida campus and show the bumps and smoothness of some roads

V. OTHER POSSIBLE APPLICATIONS OF KURUMA

A. Relation between Consumption and Road Roughness

Working with Kuruma application, research could be developed to determine how irregular roads affect the consumption of vehicles. To generate this study, vehicle consumption in addition to the irregularities of roads should be monitored at all times.

To determine a correct relationship between the data many additional factors have to be considered. Not all drivers will drive with a manner of the same efficiency. Also different cars have different consumptions and their suspension differ in order to record the bumps. Hills have to be reflected, in terms of consumption is different to go uphill than to go downhill. Also external factors like weather conditions or traffic state have to be measured to ensure the correctness of the data.

B. Black Box

The idea is create a Black Box similar to those used in aircraft or railways locomotives. Its function is to capture all the possible information from vehicles and record it in case of an accident. Depending on the intensity of the crash the system will contact automatically with the emergency services providing relevant existing information that can collect from the accident. Also all the other data will be captured for it to be examined with more detail.

Depending on the vehicle it can record different parameters from the vehicle and its surroundings like video, audio, speed, accelerations, vehicle position, throttle position, DTC's (Trouble Codes) and other parameters obtained from the OBD port.

We have already done a first prototype that detects crashes and reports them by calling the emergency services. Fig. 6 shows the augmented reality application based on the Wikitude API that is used to find the crashed vehicles.

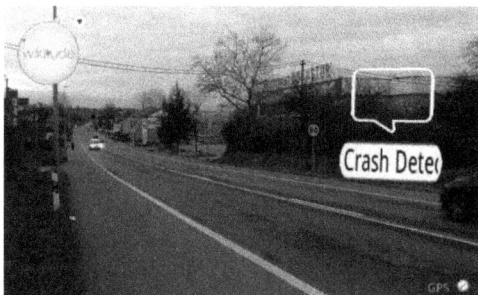

Fig. 6. AR Vehicle Crash Detector Application

C. Road Hazard Detector

This system is destined to detect risks in the road like for example, slippery roads. To detect risky conditions of the road the system must be installed in modern vehicles that are equipped with some kind of Electronic Control System (ESP). The ESP improves the safety of a vehicle's stability by detecting and minimizing skids.

The idea to monitor the activity of the ESP and when it starts working determine the situation that made it trigger. Thanks to other parameters like speed, throttle position and the kind of road the vehicle is going though, it could be determined if it was an inadequate speed of the driver or if there was actually a danger in the road. This information would be sent to a server. So if other drivers report the same situation in the same place of the road, a notification would automatically be sent to alert other drivers around the areas. This information could be also very useful and effective for authorities not only to respond to a specific alert, but to determine in which parts of the roads it would be more likely to generate an accident.

D. Automatic Suspension Setting

Based on the information of the accelerometers provided by users a model of the irregularities of the road can be obtained. This is useful for vehicles that can set up their suspension settings changing its firm. If the road is very bumpy you would like to change your settings to a softer suspension, and if the road is very smooth you can change the setting for a stiffen suspension that increases safety.

VI. CONCLUSIONS AND FUTURE WORK

In this paper, we have presented a framework for urban computing monitoring that is aimed to develop collaborative user focused applications. The framework settles the base for future applications which creates a series of benefits such as standardization, integration, flexibility and maintainability.

Through a unified architecture we have been able to develop a base application and full working prototype that classifies the state of the firm of the roads. Road Roughness Information System is a naïve application that was created just as an example. Despite that, it could have great future potentials for different kinds of applications.

The framework is still in development, but with the developed applications and the examples provided we have demonstrated the feasibility of the framework and discovered new directions and features that need to be implemented for future versions. Additionally, because of the collaborative user focused nature of the framework, issues involving trust, privacy and security still have to be resolved in a near future.

In addition, we have observed that all sensors and events are treated equally. In real-world applications, not all the information has the same importance. Some events could be mere statistic Fig.s while other could be vital emergency information that has to be immediately sent. The obvious solution would be to make all the information important and send it as soon as possible; however this has terrible consequences in real-world applications viability. For these reasons we have designed, but not yet implemented, an improved version of NIKKO with priorities (Fig. 7).

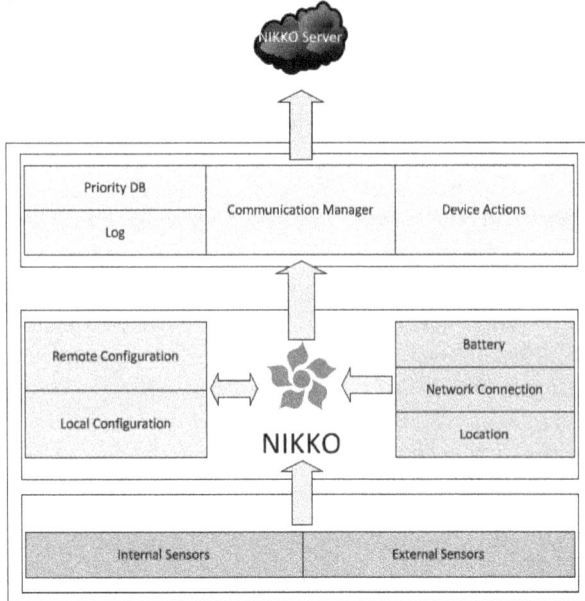

Fig. 7. NIKKO Diagram with Priorities

There are important because real users of mobile devices have data restriction limitations. Added to this problem is the issue of power consumption, that it could be mitigated when the Smartphone is plugged in a vehicle but still could be an interest to take into consideration for person centered applications.

The principal idea of flexibility is maintained, having the same basic scheme. The same three modules are preserved, but new components needs to be added to them.

The center module is still in charge of monitoring the events that the sensors generate and to create the specific respond actions included in the configuration file. Likewise, the configuration file or files are still managed by it and can be imported from the local device or from a remote location.

The main change is that the configuration file will include extra information concerning priorities. NIKKO will obtain information from the device such as, data connection availability, battery status and location information that will help decide the action that has to perform.

A Communication Manager and a new database will be added to store the obtained data. The communications manager would be in charge of making the connections with the NIKKO server. Depending on a 5 level priority scale it will upload the information when the requirements are met. Gathered information could be discarded, logged, saved, uploaded or urgently uploaded.

ACKNOWLEDGMENT

Kuruma is part of the SHUBAI Project: Augmented Accessibility for Handicapped Users in Ambient Intelligence and in Urban computing environments (TIN2009-12132) is developed with the support of the MCYT (Spanish Ministry of Science and Technology).

REFERENCES

[1] P. Mohan, V. N. Padmanabhan, R. Ramjee, "Nericell: Rich Monitoring of Road and Traffic Conditions using Mobile Smartphones", *ACM Sensys*, Nov. 2008.
[2] J.P. Espada, "Service Orchestration on the Internet of Things", International Journal of Interactive Multimedia and Artificial Intelligence, vol. 1(7), pp. 76-77, 2012.
[3] A.G. García, M.A. Álvarez, J.P. Espada, O.S Martínez, J.M.C. Lovelle, C.P. G-Bustelo, "Introduction to devices orchestration in Internet of things using SBPMN", International Journal of Interactive Multimedia and Artificial Intelligence, vol. 1(4), pp. 16-18, 2011.
[4] B. Hull, V. Bychkovsky, Y., Kevin Chen, M. Goraczko, A. Miu, E. Shih, H. Balakrishna, S. Madden, "CarTel: A Distributed Mobile Sensor Computing System", *ACM Sensys*, Nov. 2006.
[5] J. Yoon, B. N. Mingyan, Liu. "Surface Street Traffic Estimation", *ACM MobiSys*, Jun. 2007.
[6] A. Thiagarajan, L. Ravindranath, K. LaCurts, S. M. H. Balakrishnan, S. Toledo, J. Eriksson. "VTrack: accurate energy-aware road traffic delay estimation using mobile phones", *ACM Sensys*, Nov. 2009.
[7] S. Mathur, T. Jin, N. Kasturirangan, J. Chandrashekharan,W. Xue, M. Gruteser, W. Trappe. "ParkNet: Drive-by Sensing of Road-Side Parking Statistics", *ACM MobiSys*, Jun. 2010.
[8] N. Pham, R. Ganti, S. Nangia, T. Pongthawornkamol, S. Ahmed, T. Abdelzaher, J. Heo, M. Khan, and H. Ahmadi. "Fueoogle: A Participatory Sensing Fuel-Efficient Maps Application", *Computer Science Research and Tech Reports*. Oct. 2009.
[9] E. Koukoumidis, L. Peh, M. Martonosi. "SignalGuru: Leveraging Mobile Phones for CollaborativeTraffic Signal Schedule Advisory", *ACM MobiSys*, Jun. 2011.
[10] J. Eriksson, L. Girod, B. Hull, R. Newton, S. Madden, H. Balakrishnan. "The Pothole Patrol: Using a Mobile Sensor Network for Road Surface Monitoring", *ACM MobiSys*, Jun. 2008.

3

Using rules to adapt applications for business models with high evolutionary rates

Juan Fuente, A. A.[1], López Pérez, B.[1], Infante Hernández, G.[2], Cases Fernández, L. J.[2]

[1]*Computer Science Department, University of Oviedo, Asturias, Spain*
[2]*Laboratory of Software Architecture, University of Oviedo, Asturias, Spain*

Abstract — Nowadays, business models are in permanent evolution since the requirements belongs to a rapidly evolving world. In a context where communications all around the world travel so fast the business models need to be adapted permanently to the information the managers receive. In such world, traditional software development, needed for adapting software to changes, do not work properly since business changes need to be in exploitation in shorter times. In that situation, it is needed to go quicker from the business idea to the exploitation environment. This issue can be solved accelerating the development speed: from the expert to the customer, with no —or few, technical intervention. This paper proposes an approach to empower domain experts in developing adaptability solutions by using automated sets of production rules in a friendly way. Furthermore, a use case that implements this kind of development was used in a real problem prototype.

Keywords — business rules, domain experts, software adaptability, software architecture

I. INTRODUCTION

BUSINESS environments and business needs are changing rapidly, thus a progressive change and adaptation of the systems development is unavoidable in order to maintain the customer satisfaction. Even though, it is an expensive and difficult task for software engineers and developers to align the changing business requirements with actual software systems to keep them working properly [1]. Software adaptability must therefore be taken into account throughout the full software life cycle. Systems adaptation may be undertaken using two different levels in most of the cases: simple adaptations usually performed by using configuration files and complex adaptations where solutions are commonly structural ones. This paper focuses on the latter type of system adaptation, the complex or logical systems. These systems can be modified by using rules that solves first order logical issues over the predicates in order to assist decision making process [2].

The difficulty in ensuring systems adaptability is highly related with the software development lifecycle. Usually, human knowledge is transformed into software systems by the mediation of requirements documents and design models. These documents and models provide a high level view of the system and guide developers in producing running systems from the specification. Even though, the original requirements textual descriptions of system functions are separated from the developed design models, which lack the capability to capture the exact behavioral semantics from what is stated in the functional requirements [3].

Hence, these behavioral semantics need to be expressed in a more flexible and abstract manner to avoid coupling with actual developed systems and at the same time ease the adaptability process. The domain expert's role in defining the behavior of the systems expressed in comprehensible business rules is then a matter to take into account since their business knowledge can be transformed in adaptability solutions. There are a number of proposals that include tools with interfaces for non-skilled users, more than personalization, they either allow rapid development of prototypes [4], [5] or provide for the visual expression of simple rules, which, although powerful enough in certain cases, is somewhat limited in the application domain. Therefore, it would be beneficial to explicitly involve the final users and allow them to provide part of the desired configuration for system adaptation, since they will be familiar with their own environment and their requirements.

II. BACKGROUND

There are some methods used usually to adapt applications to existent business models. One of such methods include the Model-Driven Architecture (MDA) [3], [6] and [7] which promotes the production of business models with sufficient detail so that they can be used to generate or be transformed into executable software, running on target systems [8].

MDA proposes a Platform-Independent Model (PIM), a highly abstracted model, independent of any implementation technology. This is translated to one or more Platform-specific Models (PSM). The translation is based on a particular technological implementation including specific constructs and features of the implementation [9]. PSM is translated into code in a similar pattern.

The transformation process of PIM to PSM and finally code starts from the design products rather than requirements models. Hence, it requires highly creative work [6] to build a PIM from narrative requirements documents. This results in high costs in requirements change because of the need of skilled software engineers. Furthermore, as stated in [13], UML alone is not able to capture some semantics in its diagrams and a combination of UML and OCL [7] is used in MDA. However, OCL constraints are static and used in the design stages rather than the requirements stages. Moreover, MDA relies heavily on the tools which are supposed to have strong transformation capabilities from PIM to PSM and then to code.

MDA can reproduce object oriented OO systems despite the intrinsic static nature of object structure and behavior, code being regenerated from models. However, changes cannot be made to systems at runtime without interruption. Another important issue is that some business representation cannot be directly formed as objects, such as business rules. Additional maintenance problems would be otherwise added to systems if business rules were hard-coded [10]. These weaknesses MDA have led to the exploration of an alternative component technology at a higher level abstraction, being capable to retrieve, understand, as well as interpret business knowledge directly and dynamically.

There are a number of different technologies that may be used to express this sort of information. Almost any language that supports some form of rule-based inference can be used; this includes rule engines such as Drools [11], Jena [12] and Jess [13]. The Java specification request JSR-94 [14] covers the definition of a Java rule engine API, and most commercial rule engines are implementations of this standard. Drools, is an open source business rule management system and inference rule engine implemented in Java [11]. Inference rules are evaluated using an enhanced implementation of the Rete algorithm [15]. Drools natively provides an expressive textual language for defining inference rules, but also supports the integration of a custom rule DSL to improve the productivity of defining rules within certain domains. The underlying model that Drools operates within is simple plain old java objects (POJOs), making it easy to integrate into an existing Java-based software system. The structure of inserted POJOs does not need to be defined as part of the rule base; this means that all metamodel properties and operations are always accessible to a Drools rule. These are the main reasons why to choose Drools in this approach in order to build and execute the rule sets.

III. RELATED WORK

There are some significant studies aimed at providing a mechanism for not skilled users to specify the rules needed by the system in order to be better adapted to their needs. Authors in [4] present an application prototyping tool which does not require coding and instead uses a graphical interface based on controls, which allows context and devices to be collected and rules to be constructed from them by taking only logical–relational operators and restrictions on types of complex conditions. The technique might not be considered suitable for domain users to modify the applications since is actually a prototyping support intended for developers.

In [16] the authors present a programming prototyping environment intended for domain users. The system provides a series of data flows from different inputs where users can select the input flows required for the behavior they want to express. It also specifies the actions to be executed. This proposal makes use of machine learning algorithms to interpret the annotated flows in order to determine the user's intention. Since domain users are familiar with their own activities and environments they are able to tell the environment how it should behave, but they might have

The work described in [17] sketches a visual interface that specifically targets non-expert users based on a drag-and-drop metaphor. It relies on a rule grammar for expressing conditions and rule alternatives. This tool is intended to be implemented in future with an emphasis on providing visual hints and suggestions to facilitate incremental rule construction by end-users, but has not yet been tested with real non-skilled users.

Although these systems are intended to be used by domain experts, they fail in the way to represent the information in a comprehensible way for not skilled domain experts. Some of them use programming languages or domain specific languages that are more suitable for developers in order to express the adaptation rules. The solution proposed in this paper can be more suitable for domain experts since the representation of the rules is done graphically with no special knowledge of the technology used. Moreover, the rules predicates are expressed with a very simple way close to natural language, avoiding complex logical structures.

IV. MATERIALS AND METHODS

A target application developed to be in permanent adaptation by the use of rules needs a previous architecture design where the modules that will be affected by the adaptation process along with invariant ones need to be defined. Moreover, these rules need to be edited and manipulated by domain experts instead of undertake the development of new features by software engineers.

In order to develop such applications using rules, an engineering model able to support integration [18] have been used. The method is composed by the following steps:

Problem statement

The study of the business issues and those exposed to high evolutionary rates is addressed in this step. This step represents an important task within the whole process since the accuracy in identifying these issues will impact future developments or avoid them if possible, saving money and gaining efficiency in a long term.

Domain experts' knowledge represents a valuable source of information in this step. This information describes the general business features and the most frequently scenarios to take into account by software engineers. The business knowledge is taken by software engineers to identify the main core of the system (which is more immutable and thus less subject to change), along with the more dynamic elements that may vary the most in exploitation time.

From the identified elements, those which bring the possibility to be adapted with simpler techniques (e.g. configuration files) are separated. The remaining dynamic elements are classified whether they are sensitive to be adapted with rule-based systems or require new developments.

The final number of dynamic elements sensitive to be adapted by rules, resultant from this classification is big enough to justify the use of rule-based systems.

The elements that require new development in order to be adapted are studied apart. These elements must be designed with architectural patterns that minimize the interdependencies among them and simplify the systems evolution.

Architecture design

This step follows the attribute driven design (ADD) used in [18]. The most dynamic modules are incorporated with a "modifiability" quality attribute and quality scenarios are designed to check this feature. Each of the modules that are adapted by rules receives the classification of rule-based architectural style modules within the architecture context [2].

Rule-based system design

This step describes the design of an architecture module that will be adapted by rules. The rules to be applied along with their attributes and predicates are identified. The component that enables domain experts to interact with the rules' management in an intuitive way is also designed. There are four basic elements needed to address systems adaptability with rule-based systems:

Attributes. Bring the possibility to query any object feature in the system.

Predicates. Code elements that perform complex queries to system elements, evaluates them and return a value that can be processed by rules.

Actions. Situated in the rules consequent, they have effects over the system since they can modify its behavior, reason why it is important not to extra limit their scope. To properly establish actions scope, a set of services are defined (i.e. façade, web services, etc.) this way actions can only trigger these services and do not affect other parts of the system.

Rules. These are the most dynamic aspect of the system. They are intended to be dynamically inserted in the system. Domain experts use the set of rules to design new actions to be executed by the system.

In order to ease domain expert in designing new rules, a rule editor is constructed following the business vocabulary. It is only necessary that the expert have a little notion of logic to interact with the editor. Even though, the editor is intuitive enough to assist the expert in creating syntactically correct rules.

The rules' structural modifiability is restricted only to the set of attributes, predicates and possible actions. In the case that this modifiability has a broader scope, then its structural significance gets bigger than the logical adaptation and the module can be classified as complex system where the adaptations which solutions are commonly structural ones, therefore, they out of this paper scope. In order to solve future structural issues, the intervention of technological teams aided by domain experts is necessary. This roles' combination enables the construction of modules (which insertion in the system is previously set) typed as: new attributes, new rules and new actions. All modules must be created with the same previously defined constraints.

V. USE CASE

This work presents a design model to adapt a traceability management that is able to handle, in sufficiently short times, the alarms triggered by user actions that do not follow established procedures. The use case address the establishment of a traceability system integrated in the enterprise applications of a transportations company. This company has several quality procedures that require certain records in some specific moments over time within its activities. In particular, the activity studied is the personnel hiring for driver positions.

The process under investigation is composed by the activities performed by the actors involved in hiring a new driver. Every activity generates a corresponding record that is

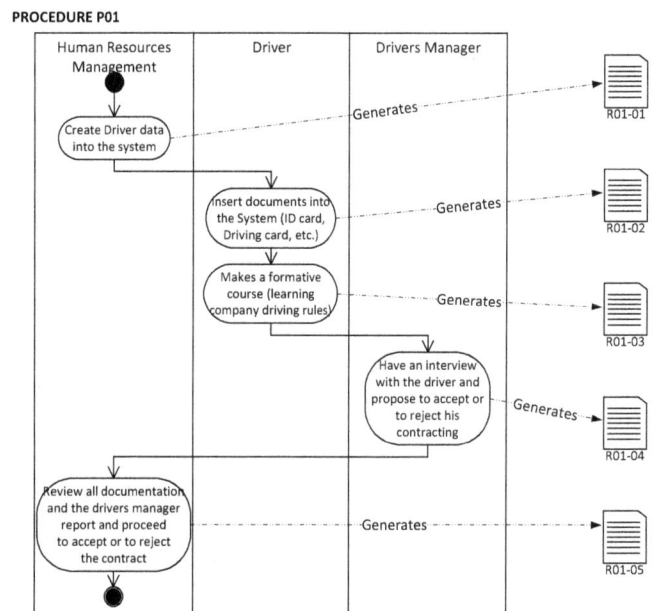

Fig. 1. Description of the procedure for contracting drivers

stored in a record control system (see Fig. 1)

The company's product manager (PM) needs to know whether the procedures are performed properly or not and most important if the drivers' interviews are carried out. Furthermore, he/she requires that the information queries be recorded in order to later be aware of the time spent between the interviews and the actual hiring. It is worth to mention that these queries have a very short lifetime, e.g. the PM might need to have this data during a month with a special increment in accidents, while the next month this information is no longer required. For this reason, to undertake and ad-hoc development in order to achieve the adaptation of queries' information needs results expensive and repetitive. The solution proposed in this work has a reasonable low cost to make viable these kinds of adaptations.

This kind of queries can only be executed against the record database. This represents an issue for the PM since he/she needs this information available all the time without expressly perform the query. Hence, the system needs to be proactive and inform the PM in an autonomous way. This situation cannot be solved with a traditional development technique since every query needs to be adapted to fulfill PM information needs. This context fosters the implementation of rule-based architectural style for this module.

VI. PROPOSED SOLUTION

There are several reasons to decouple traceability systems from business modeling systems, the following are some of the most significant:

- The procedures, in general, are items subject to changes which cannot be executed in monolithic systems that require new developments constantly.

- Actual legacy systems are not integrated with traceability systems. This situation hinders to perform changes in their behavior in order to avoid unordered activities.

- Highly dynamic systems may violate the strict path predetermined for the activities execution within a procedure.

The evolution of business models brings the adaptation to new standards which demand a change in the rules that handle the traceability of the products in a reasonable time. This scenario may not be suitable for traditional software development.

Once identified the main changing points, the system proposed was enriched with a rule-based adaptable module easy to modify by domain experts with no technical skills. This enables the performance of business adaptations in a very short period of time and with a very low cost. In order to achieve such adaptation, a graphical rule editor was implemented.

General system context

The proposed system is formed by three main components:

1. Legacy systems component (green) to model the business without traceability integrated.

2. Record systems component (blue) to interact with legacy systems in order to produce records at the right moment.

3. Information exploitation systems component (orange) to apply the rule-based architectural style in the architecture.

Fig 2. shows the general context of the proposed architecture.

The Information exploitation systems component retrieves the information from the records warehouse and processes it to obtain reports, indicators, etc. that enables the responsible actors to have data about the events and situations being gathered by the traceability system.

As can be shown in Fig. 2, the Information Exploitation Component has two basic modules:

- **Rules Manager**. Manages the rules that respond to the events and explores the available system information.

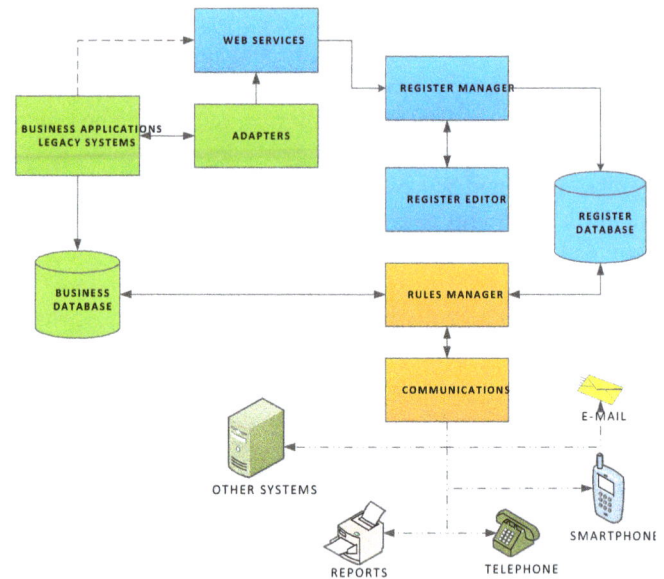

Fig. 2. Architecture context for the entire system

- **Communications.** Manages the users' information communication systems.

A more detailed description of the Information Exploitation Component is depicted in Fig. 3.

Basically, the process consists of the application of a set of rules that identifies different situations that need to be documented in some specific way. These rules are executed in two different ways:

- **On-line rules**. Executed in the precise moment of recording the registration in the database.

- **On demand rules**. Executed on demand by the users with privileges.

Once the rules are executed, if the conditions are met the associated actions are triggered.

The data mining system retrieves the business information needed to fire a rule or to complement the communication of an alarm action. There are four types of actions:

- **Just in time information (Alarms)**. Information that detects situations where some users wants to take just in time information from. These alarms can be configured by non-technical users.

- **Configurable reports**. Generic reports that can be configured to multiple purposes by non-technical personnel.

- **On demand reports**. Specialized reports that require technical personnel intervention.

- **On demand alarms**. Alarms version that cannot be handled by the rule configurable system and requires a technical service intervention.

On demand components perform solutions to more complex design that could not be designed at the development time.

Reports are sent to specific people or to another system by using different information channels; this requires the information adaptation depending on the channel used to send the data. This adaptation task is performed by the Communications Component.

Rules

The rules used follow the predicates logic format. The

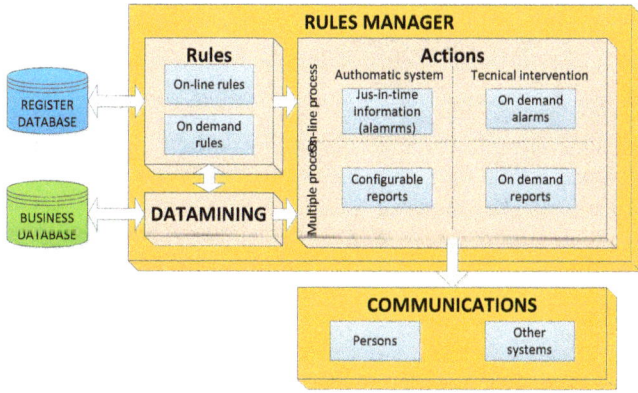

Fig. 3. Basic architecture for the Information Exploitation Component

quantifiers are eliminated since they can be included in the predicates. The rules are fired by triggered events; the basic functioning can be described as (1):

event: evaluation(rules_set)   (1)

For each rule, if its evaluation is true, the associated actions are triggered as in (2):

IF Evaluation (Predicates) = TRUE →Execute (Actions)   (2)

The general appearance of the rules used is the following (3):

P_1 (x,...) ∧ P_2 (y,...)...→Action (Info, Stakeholders)
(3)

Where *Info* represents the information required for the report, which will be composed by the *Action* itself along with the data mining component that seeks the information in the database. *Stakeholder* is the user that will receive this

information. The parentheses are allowed in order to establish the priorities and associations when evaluating the predicates. The connectors are ∧: Connective "AND", ∨: Connective "OR", ¬: Denial and →: Implication.

As a further constraint every predicate must be an object that allows a Boolean evaluation as in (4):

Record_written_at_DB_event:
IsProcedure(P1)∧IsRegistry(R1.5)∧¬(ExistsRegistry(R1.3)∨ExistsRegistry(R1.4)→Alarm(Info,Stakeholders)   (4)

Previous rule formulates the following predicate: IF the procedure is P1, the registration in course is registry R1.5 from the mentioned procedure and the registries R1.3 or el R1.4 does not exist then the alarm is raised.

Every time an alarm or report is created, rules that triggers actions related to them are generated. These rules are in a rule set that has different subsets which are evaluated against the rule engine depending on events detected. For example, there is a subset of temporal events that contains all the rules to be executed when a time-out is reached. When this event takes place, the rules related with it are evaluated. Another important subset is the one that is evaluated every time the system receives a new registry. The objective is to optimize system responses in order to enable the growing of rules number since not all of them are evaluated in every event.

The identified predicates list by default is depicted in Table II.

All the predicates should be in context. The possible contexts are:

- **Default Procedure**. It is the procedure that is generating

TABLE II
PREDICATES CLASSIFICATION

Predicate	Description
IsProcedure(string)	Returns TRUE if the procedure name **matches** with the given string.
IsRegistry(string)	Returns TRUE if the registry name **matches** with the given string.
ExistsRegistry(string)	Returns TRUE if the registry named by string **exists** in the execution of the procedure
RegistryNumberGreaterEqual(string, num)	Returns TRUE if during the execution of the current procedure, the registry name (string) has been repeated **equal or more** times than the number in num.
RegistryNumberGreater(string, num)	Returns TRUE if during the execution of the current procedure, the registry name (string) has been repeated **more** times than the number in num.
RegistryNumberEqual(string, num)	Returns TRUE if during the execution of the current procedure, the registry name (string) has been repeated **equal** times than the number in num.
RegistryNumberLess (string, num)	Returns TRUE if during the execution of the current procedure, the registry name (string) has been repeated **less** times than the number in num.
RegistryNumberLessEqual (string, num)	Returns TRUE if during the execution of the current procedure, the registry name (string) has been repeated **equal or less** times than the number in num.
TimeBetweenTwoRegistriesGreaterEqual (string1, string2, num)	Returns TRUE if the time consumed between two registries named as string1 and string2 is greater or **equal** to num in milliseconds.
TimeBetweenTwoRegistriesGreater (string1, string2, num)	Returns TRUE if the time consumed between string1 and string2 is **greater** than the num in milliseconds.
TimeBetweenTwoRegistriesEqual(string1, string2, num)	Returns TRUE if the time consumed between string1 and string2 is **equal** to the num in milliseconds.
TimeBetweenTwoRegistriesLess(string1, string2, num)	Returns TRUE if the time consumed between string1 and string2 is **less** than the num in milliseconds.
TimeBetweenTwoRegistriesLessEqual(string1, string2, num)	Returns TRUE if the time consumed between string1 and string2 is **equal or less** than the num in milliseconds.
LastRegistryTimeGreaterEqual(num)	Returns TRUE if the time consumed from the last registry of the studied procedure is **greater or equal** to num in milliseconds.

the current register to be stored. It refers to the procedure template.

- **Instance procedure**. It refers to the current execution of the procedure.

The predicates could be added to the system by means of new development processes, for improving the configuration possibilities.

Actions identification

Actions associated with system are basically those that generate reports. Generally, actions' information processing goes across three stages:

- Information recovery
- Information treatment
- Presentation to users

These stages can be executed on-line, just like alarms, or require elements to store the information temporally (recovered and/or treated) in order to compose the report when a period ends.

Alarms

Alarms are information from specific identified events usually on-line. When an event or an unexpected situation that has been programmed as an alarm appears, the traceability system sends the configured information to the defined users.

For example, when registering the information in the database, if a rule detects that some procedure step has been ignored and that step is important enough to inform some person then an alarm is programmed as in (5).

$$IF\ R1.3\ \wedge \neg R1.2 \rightarrow Alarm\ (Info,\ list\ (Stakeholder)) \qquad (5)$$

A more complex example can be the following: when R1.5 (last registry that closes the procedure) arrives and some of the previous registries are missing (i.e. R1.0, R1.1, R1.2, R1.3 or R1.4), then the corresponding alarm is raised as in (6).

$$IF\ R1.5\ \wedge \neg\ (R1.0\ \vee R1.1\ \vee R1.2\ \vee R1.3\ \vee R1.4) \rightarrow Alarm$$
$$(Info,\ list\ (Stakeholder)) \qquad (6)$$

Configurable reports

These reports can be configured by domain experts with no technical skills and may be used in different scenarios. A report can be received periodically containing a set of completed procedures, i.e. those which has completed the last registry. These reports may also contain information about procedures opened but not completed, or lacking of some step registration. Generic reports like these enable to obtain information about some processes execution, this information represent a business report for domain experts. When adapting these reports to be sent by email, e.g. to the personnel manager, they contain the last month hiring processes summaries, the hiring that did not followed the process correctly, the people hired and the ongoing hiring.

On demand reports

This is the component that tries to solve the structural modifications the system allows. These reports or alarms are adapted to a specific situation and created with software development processes by software engineers. They are integrated in the system by means of plug-ins and behave like basic actions. On the other hand the events the system responds to are:

- **New registry event.** Launched when a new registry is stored in the database.
- **Error event.** Launched when some of the next errors is generated: *incomplete registry error*, *repeated registry error* or *login error*.

Rules editor

The last step is the construction of a rules editor in order to simplify domain expert's work. The editor enables experts to model the conditions as a predicates tree and the actions as lists that can be configured. Fig. 4 depicts the rules editor interface where the left panel shows the available predicates

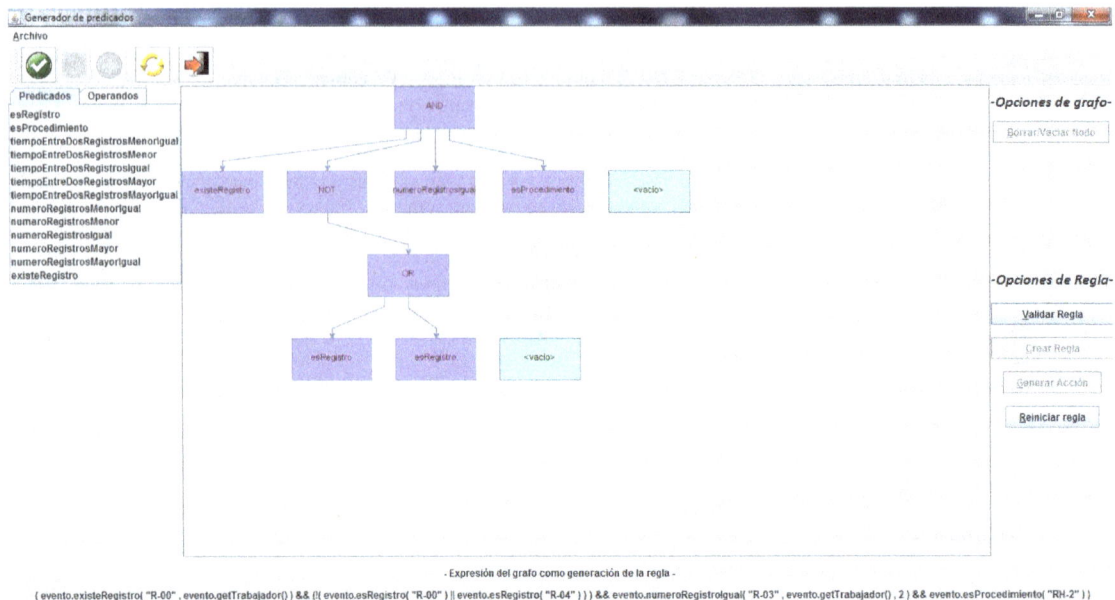

Fig. 4. Rules editor interface

and the logical operators. These options allow the graphical composition of the rules in the central panel with a tree style.

This prototype enables the configuration of the events that fires the rules' evaluation along with the configuration of each predicate and action. It also brings the visualization of the rule list defined for the system. In Fig. 4, the composed rule is being performed at the bottom of the window editor.

VII. CONCLUSIONS

This work proposed a solution to software adaptability reducing the development time with no, or few, technical intervention. A set of automated production rules has been used to achieve software adaptability. Furthermore, a use case that implements this kind of development was used in a real scenario and a prototype developed. It can be said the development time needed to adapt new software solutions to business model is reduced by using the approach presented in this work. Also, this proposal increases the possibilities of domain experts in modeling the most frequent adaptations by using the graphical rules editor developed. As far as the concern of the authors, evaluating the results of this work, rule systems have shown a high suitability for the adaptation of very dynamic systems in reducing the time and cost of putting those systems into exploitation. Furthermore, the rule-based architectural style has been implemented in order achieve systems adaptation to frequent changes with minimal effort.

REFERENCES

[1] M. Fayad and M. P. Cline, "Aspects of software adaptability," Communications of the ACM, vol. 39, no. 10, pp. 58–59, 1996.

[2] R. . Gamble, P. . Stiger, and R. . Plant, "Rule-based systems formalized within a software architectural style," Knowledge-Based Systems, vol. 12, no. 1–2, pp. 13–26, Apr. 1999.

[3] M. Fowler, UML Distilled. Addison-Wesley, 2004, p. 192.

[4] T. Sohn and A. Dey, "iCAP: an informal tool for interactive prototyping of context-aware applications," in Human-Computer Interaction, 2003, vol. 2, pp. 974–975.

[5] Y. Li, J. I. Hong, and J. A. Landay, "Topiary: a tool for prototyping location-enhanced applications," in Proceedings of the 17th annual ACM symposium on User interface software and technology, 2004, vol. 6, no. 2, pp. 217–226.

[6] A. G. Kleppe, J. Warmer, and W. Bast, MDA Explained: The Model Driven Architecture: Practice and Promise. 2003.

[7] Omg, "OMG Model Driven Architecture," Object Management Group, vol. 2009, no. Marh 13th. pp. 1–5, 2009.

[8] S. J. Mellor and M. J. Balcer, Executable UML: A Foundation for Model-Driven Architecture. Addison-Wesley Professional, 2002, p. 416.

[9] J. Huamonte and K. Smith, The use of roles to model agent behaviors for model driven architecture. IEEE, 2005, pp. 594–598.

[10] T. Morgan, Brought to you by Team-Fly Business Rules and Information Systems : Aligning IT with Business Goals. Addison-Wesley Professional, 2002, p. 384.

[11] "Drools Expert," 2013. [Online]. Available: http://www.jboss.org/drools/drools-expert.html. [Accessed: 21-Mar-2013].

[12] B. McBride, Jena: a semantic Web toolkit, vol. 6, no. 6. IEEE Computer Society, 2002, pp. 55–59.

[13] E. F. Hill, Jess in Action: Java Rule-Based Systems. Manning Publications Co., 2003.

[14] D. Selman, "JSR 94: Java Rule Engine API," Java Community Process, 2004. [Online]. Available: http://www.jcp.org/en/jsr/detail?id=94. [Accessed: 13-Apr-2013].

[15] D. Sottara, P. Mello, and M. Proctor, A Configurable Rete-OO Engine for Reasoning with Different Types of Imperfect Information, vol. 22, no. 11. 2010, pp. 1535–1548.

[16] A. K. Dey, R. Hamid, C. Beckmann, I. Li, and D. Hsu, "a CAPpella: programming by demonstration of context-aware applications," Proceedings of the 2004 conference on Human factors in computing systems CHI 04, vol. 6, no. 1, pp. 33–40, 2004.

[17] L. De Russis, F. Corno, and D. Bonino, "A User-Friendly Interface for Rules Composition in Intelligent Environments," in Ambient Intelligence Software and Applications, 2011, vol. 92, pp. 213–217.

[18] R. Hilliard, "IEEE-Std-1471-2000 Recommended Practice for Architectural Description of Software-Intensive Systems," IEEE httpstandards ieee org, no. IEEE-Std-1471–2000, 2000.

4

A Repository of Semantic Open EHR Archetypes

Fernando Sánchez, Samuel Benavides, Fernando Moreno, Guillermo Garzón, Maria del Mar Roldan-Garcia, Ismael Navas-Delgado, Jose F. Aldana-Montes

University of Málaga, Malaga, Spain

Abstract — **This paper describes a repository of openEHR archetypes that have been translated to OWL. In the work presented here, five different CKMs (Clinical Knowledge Managers) have been downloaded and the archetypes have been translated to OWL. This translation is based on an existing translator that has been improved to solve programming problems with certain structures. As part of the repository a tool has been developed to keep it always up-to-date. So, any change in one of the CKMs (addition, elimination or even change of an archetype) will involve translating the changed archetypes once more. The repository is accessible through a Web interface (http://www.openehr.es/).**

Keywords — **openEHR, Semantic Web, CKM**

I. INTRODUCTION

AS stated in [1] EHRs (Electronic Health Records) and ePrescribing have a real impact in the healthcare at a service level and also at the economical level. However the economic impact is reflected in the net benefits only in an average period of 7 years. The use of standards for the establishment of EHRs in healthcare systems would reduce this latency period.

The development of health information systems has been guided by the need for health systems to manage the huge amounts of information that make the use of physical methods unfeasible. However, these systems are not usually constrained to standards. Thus, different hospitals working together or even different services within the same hospital cannot share information about their patients.

Most advanced EHR architectures and standards are based on the dual model-based architecture, which defines two conceptual levels [2].

OpenEHR has at its core the aim of providing the necessary elements for managing electronic health records, providing ways of modelling all the agents implied in a health environment. The openEHR Foundation provides specifications which define a health information reference model together with a language for developing archetypes (clinical models). This language is not part of any software or query language by default. This architecture, based on archetypes, enables the use of external health terminologies (SNOMED CT, LOINC and ICD). OpenEHR uses the dual-model architecture, which has also influenced HL7 CDA. In dual model approaches, archetypes constitute a tool for building clinical consensus and this enables interoperability between different health information systems.

In this approach we are working towards extending how the models are published by providing new perspectives in the use of OWL as a language to provide semantically rich clinical models. Using a translator, we have built a repository of OWL models derived from public ADL models. Ongoing work is helping this proposal to provide ways of improving this semantics by aligning archetypes and health records with ICD-10 and SNOMED-CT. However, because the structure of the EHR is annotated with such terminologies, the information contained in an EHR is mostly composed of text descriptions without terminology annotations on the patient data.

Section 2 presents some related work. Section 3 describes the archetype translation process. Section 4 presents the current version of the repository and its user interface, to conclude with Section 5 explaining the main conclusions and ongoing work.

II. RELATED WORK

Archetypes are considered an important element in the achievement of the semantic interoperability between EHR systems. So, the design of methods to manage them is fundamental [3]. The translation of openEHR archetypes to OWL is not a novel proposal. [4] presents the first proposal of an ontology for representing archetypes in OWL. This ontology is divided into seven integrated ontologies:

- EHR EXTRACT Reference Model. It defines the semantics shared by all kinds of Extract requests and Extracts from openEHR data.
- EHR Reference Model. It contains a representation in OWL of the information model of the openEHR EHR.
- Data Structures Reference Model. It represents the shared data structures used in openEHR reference model, including lists, tables, trees, and history, together with one possible data representation (hierarchical).

- Support Reference Model. It defines identifiers, assumed types, and terminology interface specification used by openEHR reference model.
- Common Reference Model. It contains shared concepts, including the archetype-enabling LOCATABLE class, party references, audits and attestations, change control, and authored resources.
- Demographic Reference Model. It describes the architecture of the openEHR Demographic Information
- Data Types Reference Model. It represents data types, including quantities, date/times, plain and coded text, time specification, multimedia and URIs.

Figure 1 shows a part of this ontology. As can be observed, the design of this ontology is directly driven by the syntactic structure of the archetypes, including their main types, without taking into account compressibility or reusability. From a semantics point of view, this is an inconsistent ontology (tested using the Pellet reasoner in Protégé 4.3), so it cannot be used for reasoning purposes. However, the positive aspect of this ontology is that it is complemented by translation software [5] for obtaining OWL versions of ADL archetypes. This translator is based on the ADL API and the Archetype Object Model (AOM). The OWL model is built using Jena to construct the ontology model in memory while the ADL archetype is simultaneously parsed. A negative aspect of this translator is that it only includes the translation of 2 of the 4 archetype types, and many of the archetypes in these two types cannot be translated due to programming errors.

In this paper we present the roadmap from this approach to reach some goals:

- A comprehensible and reusable consistent OWL ontology.
- A complete translator for any ADL archetypes to consistent OWL ontologies.
- A repository of archetypes and translations able to trace the evolution of the archetypes.
- Software able to automatically align clinical records with external vocabularies.

III. Archetype Translation

An archetype constrains the entities of the reference model. The constraints are applied to the attributes defined for each entity: range, cardinality, etc. In this way, each constrained entity is defined by means of an OWL class in which the corresponding constraints are defined [6].Using the existing translator we have taken several steps to improve it.

A. Error detection.

In this step we have tested the translator using public archetypes in the openEHR CKM (http://www.openehr.org/ckm/). The automatic execution of these archetypes showed the following errors that were solved on the translator provided in our portal (http://www.openehr.es/):

- Non-existing nodes. Some ADL nodes were not expected at certain parsing steps, and this lead the software to an error, stopping the translation process. These nodes were analysed and the translator extended to deal with them properly.
- Repeated class names. The names of the classes in the translation directly rep-resent ADL nodes. ADL does not prevent us from using the same name for different nodes, but OWL does not allow the use of the same name in different classes. In order to solve this problem, the names for these classes were automatically detected and changed to a new name using the parent class name as a prefix.

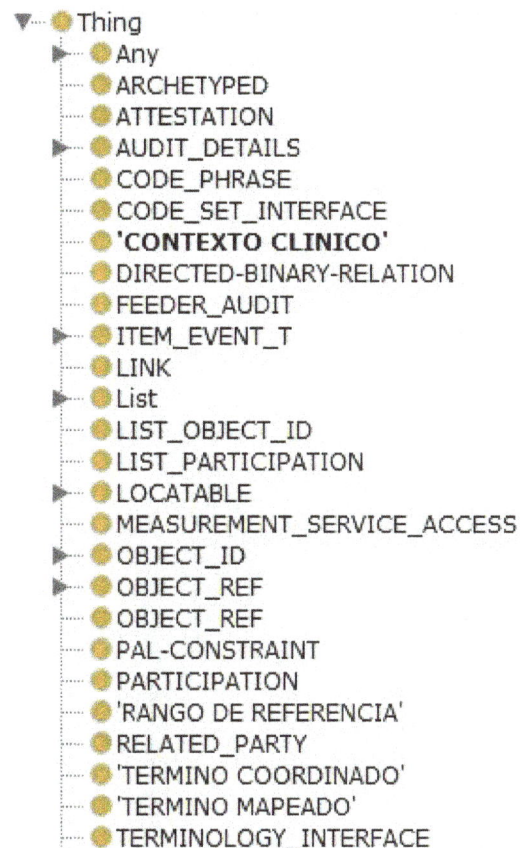

Fig. 1. Main structure of the archetype ontology

B. Incompleteness.

The translator has been shown to be incomplete and so unable to deal with over half of the public archetypes we wished to translate automatically.

- Archetype types that are not translated. The original translator does not trans-late ACTION AND INSTRUCTION archetypes. INSTRUCTION and ACTION have been added to the reference ontology and now the archetypes in this cate-gory are properly translated.
- External vocabulary annotations (SNOMED, ICD, etc.) are not translated. The first step in solving this issue has been to add a new concept to the resulting ontology:

ONTOLOGY_CONCEPT. Thus, the translator has been extended with a component for detecting and dealing with external vocabulary annotations. When the ADL parser detects these annotations this component is activated to add a new instance of the new ONTOLOGY_CONCEPT indicating the external vocabulary used (SNOMED, ICD, etc.) and the term is referenced in the ADL annotation. These annotations will be of help when trying to align clinical data with external vocabularies as this will provide a context to be used by the text mining process.

C. Improve the resulting OWL ontology.

The translator is being modified to eliminate the generation of unnecessary nodes. Some of the concepts added to the OWL ontology were direct translations from the ADL language and are not needed to rep-resent the information of the archetypes. This part of the translator is being modified to use a different structure of the OWL ontology without using these intermediate class names, reducing the complexity of the resulting OWL ontology. This modification which will lead to a totally different translator is still ongoing work which will describe in the following sections.

D. Test case generation.

The translation of archetypes to OWL enables the possibility of using RDF Database Management Systems to deal with clinical data represented as instances (individuals) of these OWL ontologies. However, there are no examples of how clinical data should be represented in these ontologies. Thus, we have developed an instance generator to provide test cases for the data management. Our instance generator asserts individuals in a given ontology in two different ways: inserting individuals according to certain data or inserting individuals randomly generated in a given range.

In order to insert individuals by given data we should follow these steps:

- Instance the reference ontology using "columnX" where "X" is the number of the column from which the program should take the data. We should keep in mind that the first column is "column0".
- The name of each instance in the reference ontology has to be given, with its version at the end, e.g. "example.1", "example.2", "example.1.1", "example.2.3.4", [...].
- This algorithm can be configured to take input files, and decide where to write the results. The separator of data by default is tabulator.
- The program will insert as many individuals as there are lines in the input file.

In order to insert randomly generated individuals in a given range we should follow these steps:

- The input file must have a first line with the type of value that we would like to use separated by spaces (being I=Integer, D=Double and S=String.
- The input file must include one line for each of the types we put in the first line.
- The reference ontology is instanced in the same way as in the previous case, but the data is not collected from the input file, rather it is randomly generated, taking the data types indicated.
- The output is a file that can be used as input for the previous case. Thus, it is possible to create a workflow that uses both cases together, although their maintenance is independent. So, the changes in the reference ontology will only affect the first case, but not the second.

E. Translation examples

Current translation implies that the result is an ontology with a similar structure as an ADL file. Thus, a simple ADL file (Figure 2) will produce a complex structure based on subsumption and object properties.

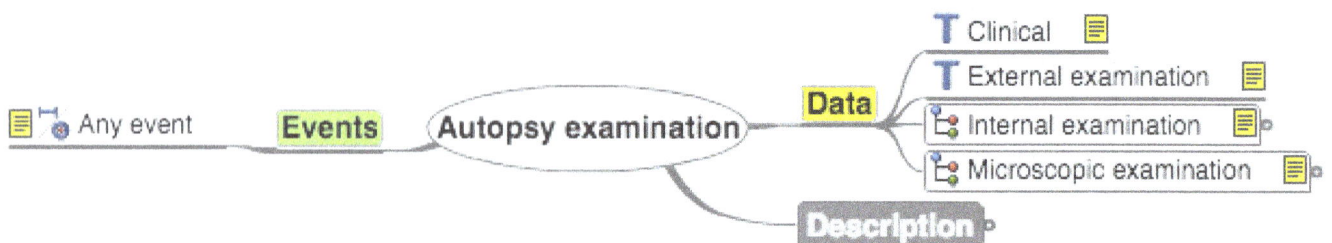

Fig. 2. Autopsy observation archetype

The generation of instances will produce instances for the whole structure of the given archetype ontology. This means generating a lot of instances for intermediate concepts that will serve only as the connection between the archetype and the given data. For example, the following input file (for Blood Pressure archetype) will generate a complex structure of instances as shown in Figure 3:

```
13.10        0 mm[hg] 8.100 mm[hg]  9.50  0
             mm[hg] 75

13.20        0 mm[hg] 8.200 mm[hg]  9.60  0
             mm[hg] 85

13.30        0 mm[hg] 8.300 mm[hg]  9.70  0
             mm[hg] 95

13.40        0 mm[hg] 8.400 mm[hg]  9.80  0
             mm[hg] 105

13.50        0 mm[hg] 8.500 mm[hg]  9.90  0
             mm[hg] 115

13.60        0 mm[hg] 8.600 mm[hg]  9.00  0
             mm[hg] 125
```

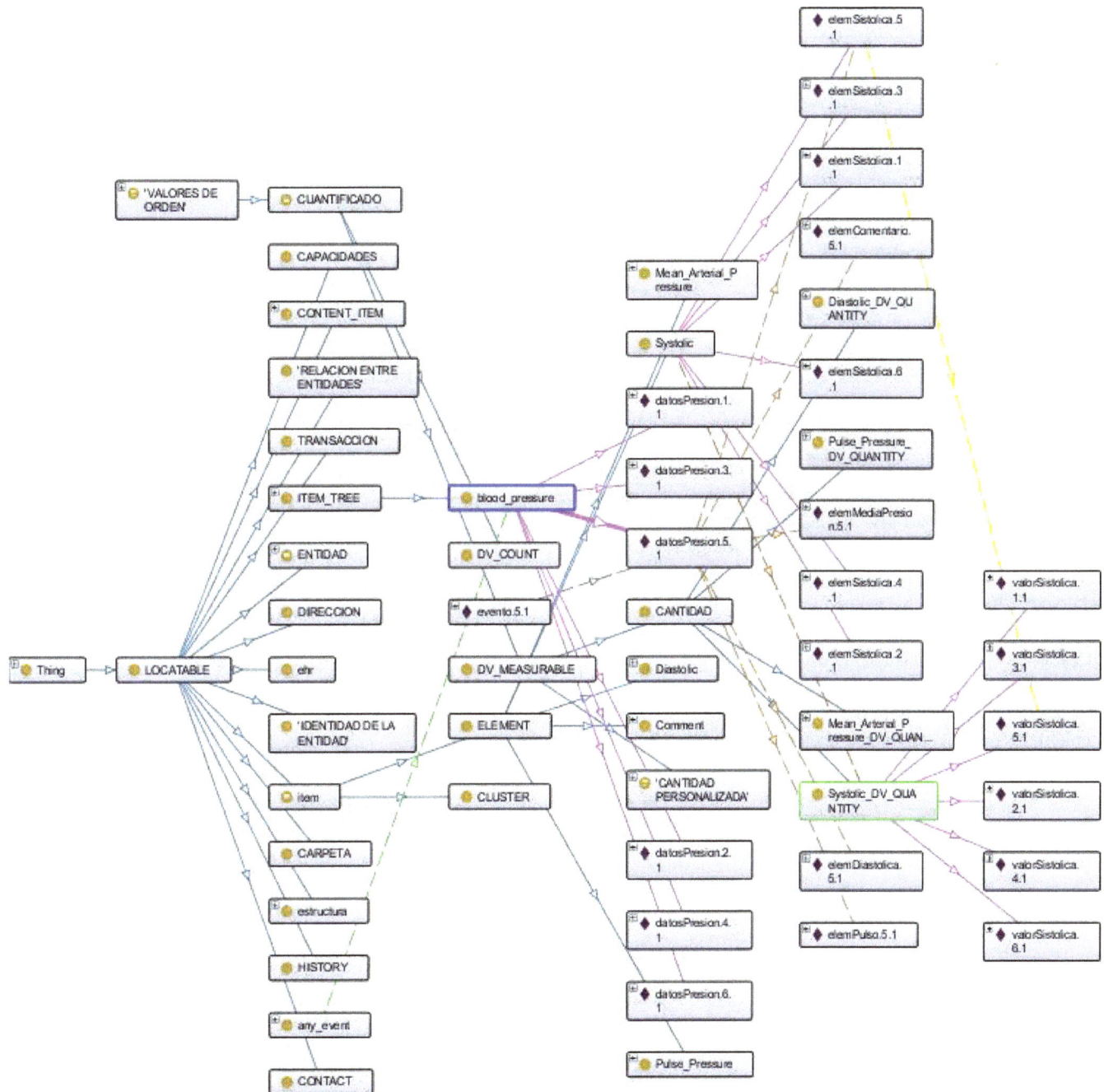

Fig. 1. Example of instances for a given data file

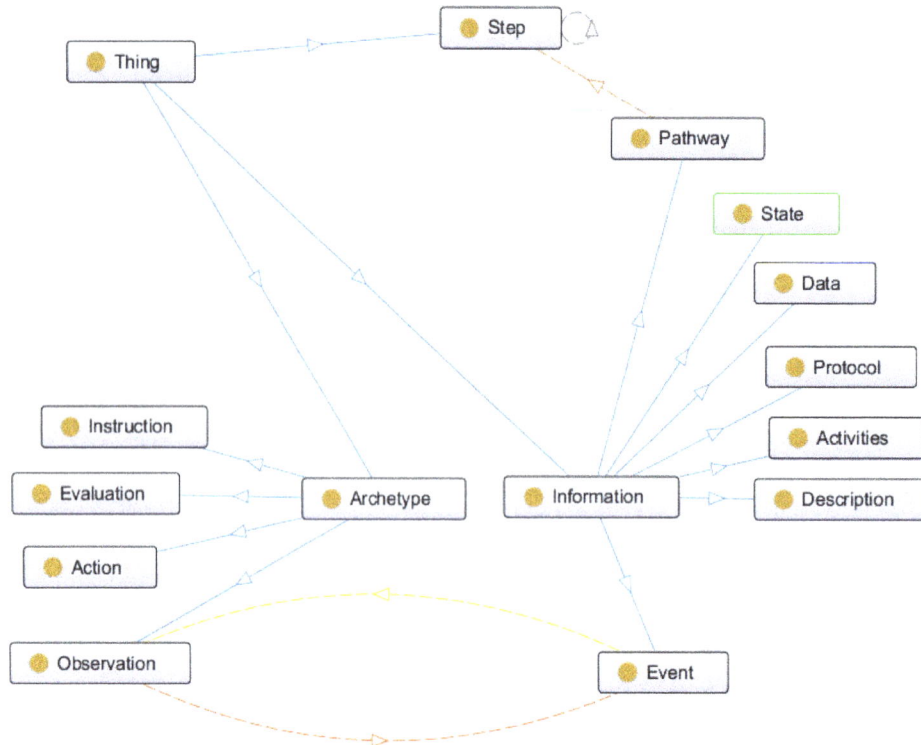

Fig. 2. Reference OWL ontology

F. 3.2   Translation results

An archetype is consistent if its set of defined constraints over both the reference model and the parent archetype are satisfiable. It is necessary to analyse the results of the translation and to check the quality of the archetypes represented in OWL. Generated instances for the current archetypes have been manually evaluated to discover translation errors. This manual process is based on the comparison of the translated archetype as an OWL ontology with the original version in ADL. Nodes are compared by their name and relationships with the other nodes. This ensures that although at first glance the archetype represented in OWL seems to have been translated correctly, there are no hidden translation failures. The quality of the translation is an important part of the translation process in order to ensure a certain level of quality of the translations offered.

IV.   REPOSITORY MANAGEMENT

The repository of archetypes is built and updated using a daily batch process connecting to a list of CKMs. This process checks all the archetypes contained in the external repositories, extracts them and compares the contents of the CKM with the local repository. If there are any differences, the process updates the archetypes in the local repository and translates the modified ones to OWL.

In order to connect to the CKM the system uses a web service that provides the CKM and returns a compressed file with all archetypes structured in directories, classified by type. The following CKMs are currently being accessed:

- NEHTA = http: //dcm.nehta.org.au/ckm/
- openEHR = http: //www.openehr.org/ckm/
- uk = http: //clinicalmodels.org.uk/ckm/
- ezdrav = http: //ukz.ezdrav.si/ckm/
- russia = http: //simickm.ru/ckm/

An archetype can pass through several states (initial, draft, review team, etc.). If an archetype is "published", it cannot be modified. In this case, modifications should be done as an archetype with the same name and higher version number. This way of managing CKM prevents the modification of published archetype contents. The contribution of updating the repository is to keep all versions of archetypes to provide users with translations to the archetype version they are using in their Health Information System, even if a new version has been published. The synchronisation process is as follows:

- If a file has been modified internally, it is replaced in the local repository by the new one and the conversion to OWL is deleted.
- If a new archetype appears, then it is copied to the local repository, this occurs when a new archetype is created in the CKM or is versioned.
- Archetypes are not deleted from the CKM rather they are labeled as rejected or obsolete. Thus, it is not necessary to check whether an archetype is missing from the local repositories.

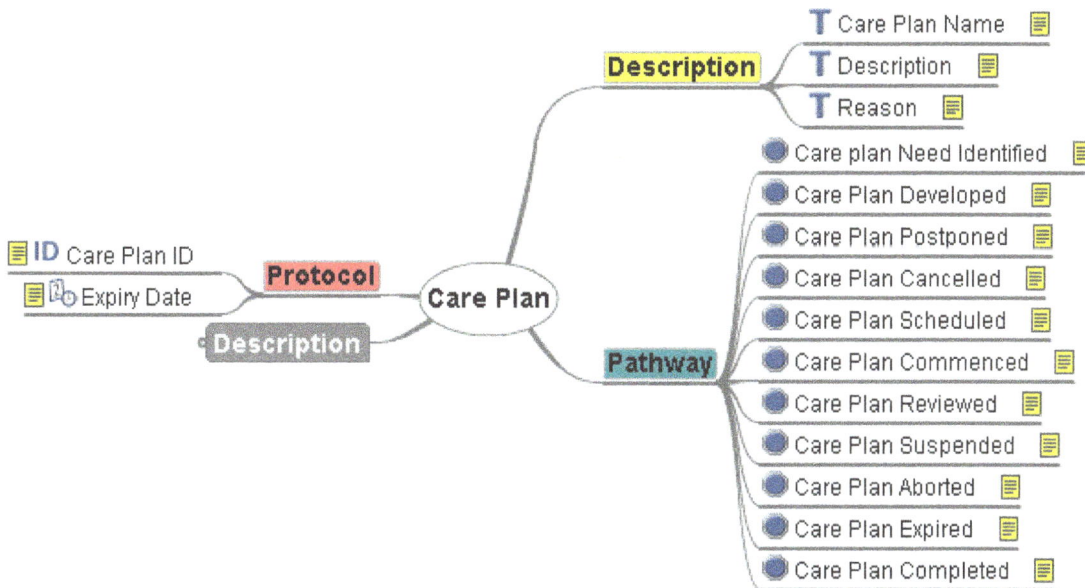

Fig. 3. Care Plan archetype

- Once the local and external repositories have been synchronised, the OWL translation process checks for each of the archetypes added or modified in the local repository.

The repository contains the automatically translated archetypes from public archetype repositories like, for example, CKM. However, ADL allows users to define their own archetypes. For this reason, the translation tool is included in the portal, so users can test its functionality. The system does not keep a copy of the archetype, or the translation, unless the user asks for them to be included in our repository.

V. DISCUSSION AND CONCLUSIONS

The use of standards such as OpenEHR will reduce the time to return on the investment of putting an EHR system to work, with the corresponding economic impact. Additionally, the use of semantics opens new ways of interoperability even with other standards making worthy this initial economic effort. The automatic translation of openEHR archetypes to OWL has been approached in the past. However, in the cur-rent climate in which the interoperability of health information systems is a priority, this topic is of strong interest. For this reason, we have started with previous work and analysed the existing problems in these types of translations. Some of the problems detected have been solved, and an improved version of the translator has been used to provide a repository of OWL ontologies representing public archetypes.

However, there is still much work to be done in this approach. The main issue we are addressing is the design of a reference OWL ontology to lead the translation process towards consistent, comprehensible and reusable ontologies. The reference mod-el (Figure 4) we have designed in the first phase simplifies the representation of archetypes. For example, for the Care Plan Archetype (Figure 5), the translation would be similar to the ontology in Figure 6.

Neither of the formal representations of ICD-10 presented in the literature has been classified nor their consistencies checked. Even more, some of then uses an OWL-Full component that prevents its use in a semantic classification system based on reasoning. Other approaches propose to model the ICD-10 exclusions using the owl:disjoint axiom, that could lead to a loss of important information and generate inconsistences in the model. There are no ontologies that combine SNOMED-CT and ICD-10-CM. SNOMED-CT and ICD-10 are broadly used in the field of medicine. In fact, SNOMED-CT is being used in most of the Health Information Systems. For this reason, our research group is working on modelling the ICD-10 (International Classification of Diseases, 10th version) [7] as an OWL ontology [8]. This medical classification standard, maintained and published by the WHO (World Health Organization) is used to classify diseases and health problems that have been recorded on death certificates and also in other records. Our ontology has also been aligned with SNOMED-CT [9]. SNOMED-CT terminology often referenced as an ontology, includes all those concepts that relate to each other logically within a specific domain [10]. As many openEHR archetypes are annotated with an ICD-10 code, this enables the possibility of aligning the OWL ontologies in our repository with our ICD-10 ontology. By means of this alignment, the reasoning capabilities of the OWL language can be exploited so as to obtain implicit information about the clinical concept described by the archetype, based on the information contained in ICD-10 and SNOMED-CT, such as its relationships with other clinical concepts, diseases and clinical procedures, to name a few.

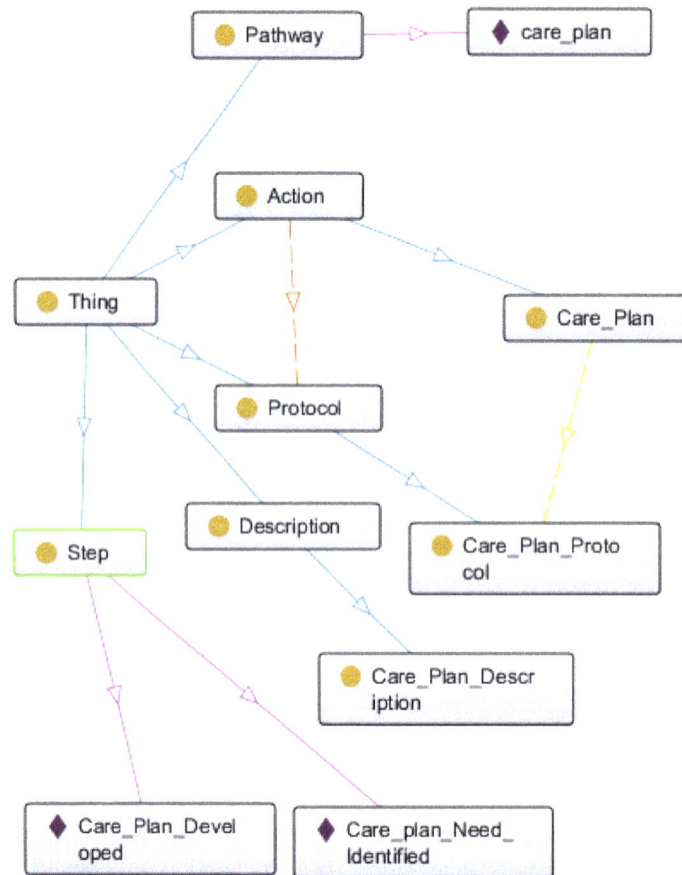

Fig. 4. Translation of Care Plan archetype

REFERENCES

[1]   The socio-economic impact of interoperable electronic health record (EHR) and ePrescribing systems in Europe and beyond. http://www.ehr-impact.eu/downloads/documents/EHRI_final_report_2009.pdf

[2]   Catalina Martinez-Costa; Marcos Menárguez-Tortosa; Jesualdo Tomás Fernández-Breis. An approach for the semantic interoperability of ISO EN 13606 and OpenEHR archetypes. Journal of Biomedical Informatics. October 2010.

[3]   María del Carmen Legaz-García; Catalina Martinez-Costa; Marcos Menárguez-Tortosa; Jesualdo Tomás Fernández-Breis. Exploitation of ontologies for the management of clinical archetypes in ArchMS. University of Murcia (España), 2012.

[4]   Isabel Román Martínez. Aportaciones metodológicas a la integración de sistemas de informa-ción sanitarios basada en gestión semántica. Supervisors: Rafael Robles Arias, Germán Madina-beitia Luque. University of Sevilla (España), 2006.

[5]   Lezcano Matías, Leonardo. Combining ontologies and rules with clinical archetypes. Universi-dad de Alcalá. Departamento de Ciencias de la Computación. Supervisor: Sicilia Urbán, Miguel Ángel. 2012.

[6]   Marcos Menárquez-Tortosa; Jesualdo Tomás Fernández-Breis. OWL-based reasoning methos for validating archetypes. Journal of Biomedical Informatics. November 2012.

[7]   International Classication of Diseases (ICD) website, 775 http://www.who.int/classifications/icd/en/.

[8]   Pastor-Rubia, D.; Rold´an-Garca, M.M.; Navas-Delgado, I. Aldana-Montes, J. F. DTEKT: semantic classification for medical diagnosis Riccardo Bellazzi and Paolo Romano (eds.). Clinical bioinformatics. 11th international workshop. NETTAB 2011. Network tools and application in biology. October 12th-14th 2011, Collegio Ghislieri, Pavia, Italy, 2011, 86-87.

[9]   SNOMED-CT website, http://www.ihtsdo.org/snomed-ct/.

[10] Jose Luis Allones; David Penas; María Taboada; Diego Martinez; Serafín Tellado. University of Santiago de Compostela (España). G. O. Young, "Synthetic structure of industrial plastics (Book style with paper title and editor)," in Plastics, 2nd ed. vol. 3, J. Peters, Ed. New York: McGraw-Hill, 1964, pp. 15–64.

5

Robust Lossless Semi Fragile Information Protection in Images

Pushkar Dixit[1], Nishant Singh[2], Jay Prakash Gupta[3]

[1]*Faculty of Engineering and Technology Agra College, Agra, India*
[2]*Poornima Institute of Engineering and Technology, Jaipur, India*
[3]*Infosys Limited, Pune, India*

Abstract – **Internet security finds it difficult to keep the information secure and to maintain the integrity of the data. Sending messages over the internet secretly is one of the major tasks as it is widely used for passing the message. In order to achieve security there must be some mechanism to protect the data against unauthorized access. A lossless data hiding scheme is proposed in this paper which has a higher embedding capacity than other schemes. Unlike other schemes that are used for embedding fixed amount of data, the proposed data hiding method is block based approach and it uses a variable data embedding in different blocks which reduces the chances of distortion and increases the hiding capacity of the image. When the data is recovered the original image can be restored without any distortion. Our experimental results indicate that the proposed solution can significantly support the data hiding problem. We achieved good Peak signal-to-noise ratio (PSNR) while hiding large amount of data into smoother regions.**

Keywords – **Data Security, Lossless Data Hiding, Semi Fragile, Encryption, Decryption, Internet Security, Steganography, Information Protection**

I. INTRODUCTION

ONE of the most important issues arising out of the wide spread use of internet is the security of information. Cryptography has long been used to maintain the secrecy of the data. There are a number of algorithms to encrypt and decrypt a message. However, sometimes it is not enough to keep a secret message. It is essential that except for the intended receiver of the message, no one should even come to know that any communication is taking place.

Nowadays there has been a rapid development in the internet and its technology, the individual prefers internet as the primary medium for communication between one parts of the world to another. There are many possible ways to transmit the data across the internet: via e-mails, chats, video streaming, video calling, etc. Internet has made the transmission of the data very simple, fast and accurate. Internet has both its merits and demerits as the intended users can view the information and one who is not intended can also do. Thus in order to make it visible only to the intended users, we must have some method such that people who are not intended may be prevented from viewing information directly. Thus we can say that one of the main problems with sending the data over the internet is 'security threat' posed, in other words personal or confidential data can be stolen or hacked in many ways. Therefore it becomes very important to take data security into consideration. It is one of the most essential factors that need attention during the process of sending or receiving of data.

Before proceeding further it is necessary to understand the conceptual difference between cryptography and steganography. Cryptography conceals information by encrypting it into cipher text and transmitting it to the user using an unknown key, whereas steganography hide the cipher text into a seemingly invisible image or other formats. The word steganography is of Greek origin which means "covered or hidden writing" [1]. Steganography is the art and science of writing hidden messages in such a way that no one, apart from the sender and intended recipient, suspects the existence of the message, a form of security through obscurity. It is therefore a book on magic. It is emerging in its peak because it does not attract anyone by itself.

Encrypting data into some form has been the most popular approach for protecting information, but this protection can be breached with enough computational power. An alternate approach to encrypting data is hiding it by making this information appear to be something else. This way only intended user can receive a true content. In particular, if the data is hidden inside an image then everyone but our intended users or the person it is meant for can view it as a picture which is transmitted. At the same time he/she could still retrieve the true information while the unintended people would view it only as an image.

Data hiding has its application in various areas due to the image being the most common digital media transmitted over the internet. Thus practically, it is very difficult for an unwanted user to masquerade the information that is transmitted over the channel by checking each and every image as it is a very time consuming job. Thus it is quite a nice option to choose images to hide the data and send them over internet and the receiver can easily extract the information from image.

Two important properties of steganographic technique are perception and data hiding capacity. Steganography generally exploits human perception because human senses are not trained to look for file that has hidden information inside it. Therefore steganography disguises information from people who try to hack it. Data hiding capacity is the amount of information that can be hidden in the cover object. The cover object means the image that we use for embedding the data and the stego object means the image obtained after embedding the data into cover object.

The different types of steganography techniques are substitution, transform domain, spread spectrum, statistical and distortion techniques and cover generation techniques. Substitution techniques replace the least significant bit of each pixel in the cover file with bits from the secret document. The transform domain techniques hide secret information in the transform space (like frequency domain) by modifying the least significant coefficient of the cover file. Most of the research work done in the area of transform domain embedding is focused on taking the advantage of redundancies in Discrete Cosine Transformation (DCT). Spread spectrum techniques spread hidden information over different bandwidths. Even if the parts of the message are removed from several bands, there would still be enough information present in other bands to make out the message. Statistical techniques change several statistics of the cover file and then split it into blocks where each block is used to hide one message bit. The most obvious limitation to these techniques is that the cover image must be very largely compared to the secret information it is going to carry fixed payload over the image. We can hide large amount of information in multiple files but it could lead to suspicion. Therefore, it is very important to use only one image file to hide the entire secret information.

Each application using information hiding technique has different requirements depending on the purpose of the application. Generally, there are four issues that we encounter while designing the algorithm: perceptibility, payload, security, robustness, and they are common to most applications. Because there are tradeoffs existing between those requirements, it is very challenging to design an algorithm that satisfies all the four requirements.

Hiding a fixed amount of data in the image may give a uniform distribution of data but it makes the image more suspicious for the attackers as the changes in the image are visible. We need to check the image by the means of threshold and make sure it is able to adapt to the amount of data that we want to embed into it. If we do not perform this step then it may result in loss of information or poor embedding of data. Also we have to select such an image which does not have large sharp details.

We present a simplified embedding algorithm based on difference expansion, which is capable of minimizing the distortion of the stego-image presents in the traditional difference expansion algorithms. The main principle underlying the proposed framework is blocks and centralized difference expansion. In the framework, the original cover image is partitioned into continuous non-overlapping blocks. The bits embedded in each block depend on its block size and the image complexity. A new approach is employed to find the image complexity of each image block, and all the blocks are categorized into three levels according to their block intensity values. Finally varying amounts of data are assigned to image blocks at different intensity levels. Although there are three types of blocks in the embedding procedure, only 1 bit is required to record these three blocks. This way, the proposed method can reach a higher hiding capacity while maintaining good visual quality of the stego-image. Our major concern is that the image should not show any visual effects and it carries as much data as possible. Some other approaches can also be utilized with to enhance the algorithm [2-9].

This paper is outlined as follows: section 2 provides some of the core concepts used in image steganography and presents a survey of efforts done by researchers in the past to address this problem; section 3 describes the framework for the scanning the image as well as embedding and extracting the data; section 4 discusses the result and compare it with classical approach; finally section 5 summarizes the novelties, achievements and limitations of the proposed solutions and indicate some future directions.

II. LITERATURE REVIEW

In this section, we focus on the previous work done by several researchers in the area of data hiding, steganography and steganalysis. Data hiding and steganography can be seen as instances of the image security. People have been resorting steganography or information hiding since Greek times. However, digital steganography is a relatively new research field. Since being undetectable is one of the essential requirements for steganographic applications, steganography and steganalysis techniques are evolving in competition with each other.

The aims of improving the original DE (data embedding) proposed by researchers are twofold: first is to make the embedding capacity as high as possible, second is to make the visible distortion as low as possible. To achieve high embedding capacity, the reviewed schemes adopted three different approaches: (i) simplifying the location map in order to increase its hiding capacity, (ii) embedding payload without location map, and (iii) expanding differences more than once which allows more data to be embedded. Meanwhile, the visual quality may be enhanced by: (i) using a predefined threshold T, (ii) selecting smooth areas to embed data, and (iii) using sophisticated classification functions. However, there is a tradeoff between distortions and embedding capacity. If distortion is minimized, lesser data can be embedded. On the other hand, if the embedding capacity is increased, it results in low visible quality.

Most of the researchers in the field of data hiding or image steganography have considered capacity and robustness as a key for their approach. Some of them have considered both and some of them have considered them individually. In most of the techniques, fragile images are used which is of no use after the extraction of data and it can't be restored to the original state. The major part of the research attention has been paid to the perception part of the topic rather than that of capacity.

Recently, Li et al. [10] proposed a reversible data hiding (RDH) scheme based on two-dimensional difference-histogram modification by using difference-pair-mapping (DPM). First, by considering each pixel-pair and its context, a sequence consisting of pairs of difference values is computed. Then, a two-dimensional difference-histogram is generated by counting the frequency of the resulting difference-pairs. Finally, reversible data embedding is implemented according to a specifically designed DPM. Where, the DPM is an injective mapping defined on difference-pairs. It is a natural extension of expansion embedding and shifting techniques used in current histogram-based RDH methods.

Faragallah [11] proposed quadruple difference expansion-based reversible data hiding method for digital images which is characterized by two aspects. First, reversible data hiding scheme is enhanced to exhibit data hiding in color palette images. Second, the embedding level is improved by using quadruple difference expansion to guarantee the embedding of 2-bit data into color images. But they have not considered the level of the details present in the image and hide 2-bit at each place (i.e. smoother and non-smoother ares) which is a drawback of this approach.

In the this section, we discuss several approaches used by researchers [12-22] with the aim of being aware to the latest research carried out our focus is on those related to the formulated problems in this paper.

A. Literature Survey

The word steganography is originally derived from Greek words which mean ''Covered Writing''. It has been used in various forms for thousands of years. In the 5th century BC, Histaiacus shaved a slave's head, tattooed a message on his skull and the slave was dispatched with the message after his hair grew back [23-25]. In Saudi Arabia at the King Abdulaziz City of science and technology, a project was initiated to translate some ancient Arabic manuscripts into English on secret writing which are believed to have been written 1200 years ago. Some of these manuscripts were found in Turkey and Germany [26].

Color palette based steganography exploits the smooth ramp transition in colors as indicated in the color palette. The LSBs here are modified based on their positions in the palette index.

Johnson and Jajodia [23] were in favour of using BMP (24 bit) instead of JPEG images. Their next-best choice was GIF files (256-color). BMP as well as GIF based steganography apply LSB techniques, while their resistance to statistical counter-attacks and compression are reported to be weak. BMP files are bigger as compared to other formats which render them improper for network transmissions. However JPEG images were avoided at the beginning because of their compression algorithm which does not support a direct LSB embedding into the spatial domain.

One of the earliest methods to discuss digital steganography is credited to Kurak and McHugh [27]. They proposed a method which resembles embedding into the 4 LSBs (least significant bits). They also examined image downgrading and contamination which is now known as image based steganography. Provos and Honeyman [24], at the University of Michigan, scrutinized three million images from popular websites looking for any trace of steganography. They have not found a single hidden message. Embedding hidden messages in video and audio files is also possible. Examples exist in [28] for hiding data in music files, and even in a simpler form such as in Hyper Text Markup Language (HTML), executable files (.EXE) and Extensible Markup Language (XML) [29].

Vleeschouwer et al. [30] solved the problem of salt-and-pepper noise artifact by using a circular interpretation of bijective transformation. The proposed algorithm guarantees the coherence of the transformation interpretation and, consequently, ensures total reversibility. To improve the performance of Fridrich et al.'s scheme in terms of message bits, Celik et al. [31] presented a high capacity, low distortion reversible data embedding algorithm by compressing quantization residues. Images can be obtained after a quantization process and then the CALIC lossless compression algorithm is used to get the compressed residues. The remainder of the compression space is used to hide the secret message. In addition, Ni et al. [32] utilizes zero or minimum point of histogram. If the peak is lower than the zero or minimum point in the histogram, it increases pixel values by one higher than the peak values to lower than the zero or minimum point in the histogram. While embedding, the whole image is searched thoroughly. Once a peak-pixel value is encountered, if the bit to be embedded is '1' the pixel is added by 1, else it is kept intact. The algorithm essentially does not follow the general principle of lossless watermarking. The advantages of this algorithm are (i) it is simple, (ii) it always offers a constant PSNR 48.0dB, (iii) distortions are quite invisible, and (iv) its capacity is high. The disadvantage is that the algorithm is time consuming because it searches the image several times.

Tian suggested multiple-layer embedding in order to achieve larger embedding capacity [33]. For example, the second layer embedding would take place in the orthogonal direction, where the difference image is obtained by performing integer Haar wavelet transform on the embedded image in column direction. If the capacity of the two-layer embedding is still insufficient for the payload, a third layer embedding is needed. One performs integer Haar wavelet transform in row direction again and repeats the embedding operation. Such a process continues until the total embedding capacity is large enough for the payload. However, multiple-layer embedding results in some unexpected problems. First, image quality (in terms of peak signal-to-noise ratio (PSNR)) drops greatly after the first layer embedding due to the use of large differences. Second, the new difference image has smaller embedding capacity than its predecessor. Each layer-embedding progressively decreases the correlation not only in the embedding directions but also of the neighborhood.

In [34], a lot of secret data bits are hidden in a vector. After the difference of that vector is expanded the difference expansion by generalized integer transform to make it work for more than two pixels per vector with k-1 bits of secret data hidden into k pixels. However when it is pixel pair difference expansion or difference expansion for more than two pixels, there is additional information to save this keeps track of the characteristics of a vector. In this case, a location map is needed because it records the characteristics of a vector.

Maniccam and Bourbakis [35] presented a lossless image compression and information hiding scheme. In their methodology, they have performed both lossless compression and encryption schemes which are based on known SCAN patterns generated by the SCAN methodology. This SCAN is a formal language based two-dimensional spatial accessing methodology which can efficiently specify and generate a wide range of scanning paths or space filling curves. This algorithm has lossless image compression and encryption abilities. The only advantage of simultaneous lossless compression and strong encryption makes the methodology very useful but the drawback of the methodology is that compression-encryption takes longer time.

Paulson [36] reported that a group of scientists at Iowa State University were focusing on the development of an innovative application which they call "Artificial Neural Network Technology for steganography (ANNTS)" aimed at detecting all present steganography techniques including DCT, DWT and DFT. The inverse discrete Fourier transform (iDFT) encompasses round-off error which renders DFT improper for steganography applications.

Abdelwahab and Hassaan [37] proposed a data hiding technique in the DWT domain. Both secret and cover images are decomposed using DWT (1st level). Each of which is divided into disjoint 4 X 4 blocks. Blocks of the secret image fit into the cover blocks to determine the best matches. Afterwards, error blocks are generated and embedded into coefficients of the best matched blocks in the HL of the cover image. But the extracted payload is not totally identical to the

embedded version as the only embedded and extracted bits belong to the secret image approximation while setting all the data in other sub images to zeros during the reconstruction process.

In [38], authors used a spatial domain technique in producing a finger print secret sharing steganography for robustness against image cropping attacks. The logic behind their proposed work is to divide the cover image into sub-images and compress and encrypt the secret data. The resulting data is then sub-divided in turn and embedded into those image portions. To recover the data, a Lagrange Interpolating Polynomial is applied along with an encryption algorithm. The computational load was high, but their algorithm parameters, namely the number of sub-images (n) and the threshold value (k) were not set to optimal values leaving the reader to guess the values. Data redundancy that they intended to eliminate does occur in their stego-image.

Lin et al. [39] created a method to restore the marked image to its pristine state after extracting the embedded data. They achieved this by applying the pick point of a histogram in the difference image to generate an inverse transformation in the spatial domain. The example shown in their hiding phase section might not be sufficient to verify the accuracy of the algorithm. Some questions remain unanswered such as what happens when we have two peak points instead of one? On which criterion will we base our selection? It is very likely that after the subtraction process we will have some values that collude with the peak value which confuses the extraction of the embedded data.

Wu and Shih [40] presented a GA-based algorithm which generates a stego-image to break the detection of the spatial domain and the frequency-domain steganalysis systems by artificially counterfeiting statistical features. This is the first paper of utilizing the evolutionary algorithms in the field of steganographic systems. Time complexity, which is usually the drawback of genetic based algorithms, is not discussed in this paper. They have only mentioned that the process is repeated until a predefined condition is satisfied or a constant number of iterations are reached.

Raja et al. [41] used wavelet transforms that map integers to integers instead of using the conventional wavelet Transforms. This overcomes the difficulty of floating point conversion that occurs after embedding. Some other approaches also can be employed to improve the performance [42-48].

III. PROPOSED METHODOLOGY

A good image steganography approach aims at concealing the highest amount of data in an image while maintaining its imperceptibility so that its visual quality is not hampered or least affected. The least significant bit scheme is one of the simplest and easily applicable data hiding methods, where bits of secret data are directly embedded in the least significant bits of each image pixel. In traditional data embedding schemes, the exact original image cannot be recovered after data embedding. Compared with loss prone embedding methods, reversible data embedding methods embed a fixed payload into a digital content in a reversible fashion. After embedding, the image changes very little or looks no different.

Another obvious feature of reversible data embedding is the reversibility, that is, when the digital content has been used for the purposed it was embedded, one can extract the embedded data and restore the original content. There are a number of challenges that must be addressed to perform data hiding in images. The issues that we must keep in mind while designing the algorithm are perceptibility, payload, security and robustness. We must maintain a trade-off among all these and find a better solution to the problems encountered in data hiding.

Steganography techniques aim at secretly hiding data in a multimedia carrier such as text, audio, image or video, without raising any suspicion of alteration to its contents. The original carrier is referred to as the cover object. In this work, we mainly focused on image steganography. Therefore, the term cover object now becomes cover image. Figure 1 illustrates a basic information hiding system in which the embedding technique takes a cover image and a secret image as inputs and produces as output a stego image. Receiver side carry out the extraction process to retrieve the secret message from the stego image sent over the communication links to the receiver.

The proposed approach is comprised of three steps as shown in Figure 1. In the first step secret message is generated and an image is being selected. If the image is too small for the data then another image is selected to make sure that the data gets embedded into the image. In the second step data embedding process is carried out. In the last step data is being extracted by the extraction algorithm and the original image and protected information are recovered.

A. Message generation and image selection
The basic step in data hiding is that first of all we should have a data or secret message to be hidden in any form i.e. it may be in the form of text or any other form. After getting the data we change its form to some digital form (i.e. binary). The algorithm to convert the message into binary array form is as follows:

1. Read and store the characters of a message in an array A.
2. Do for 1 to length of A
   a. Convert each character into its decimal value.
   b. Convert those values from decimal to binary value.
   c. Store these values in an array A'.
3. Store the length of A' in L.

After we successfully converted the secret message into binary form, we know the total number of bits of the message

that we need to embed into the image. Next we select an image and scan it to calculate the amount of data it can carry. If the message data is larger than the hiding capacity of the image then select different image otherwise continue with the same. The image that we are using here is called the cover image and the image obtained after embedding the data is known as the stego-image. Now, we have the data and the image in which the data is to be embedded.

B. Data embedding algorithm based on Difference Expansion
There are lots of redundancies in a digital image. If we change some pixel values to some extent, the appearance of the picture is similar to the original one. So, data embedding can be obtained by changing some pixel values in an image. The DE technique [33] reversibly embeds one bit data into two integers, which is explained in Figure 2. The DE technique uses the difference between two pixel values to embed one bit. Assuming there are two grayscale values x = 206, y = 201, we reversibly embed one bit b = 1. First the integer average and difference of x and y are computed,

$$l = \lfloor (x+y)/2 \rfloor \qquad (1)$$
$$h = x - y \qquad (2)$$

where the symbol $\lfloor \rfloor$ is the floor function meaning "the greatest integer less than or equal to". Next the difference value h is represented in its binary representation.

$$h = 5 = 101_2 \qquad (3)$$

Then embedding bit b is appended into the binary representation of h after the least significant bit (LSB), and the new difference value h' is obtained.

$$h' = 101b_2 = 1011_2 = 11 \qquad (4)$$

Mathematically, this is equivalent to

$$h' = 2*h + b = 2*5 + 1 = 11 \qquad (5)$$

Finally the new grayscale values are computed, based on the new difference value h' and the original integer average value l, and new two pixel values x = 209, y = 198 are obtained. After finishing this process, one bit is embedded into the two pixel values.

$$x' = l + \lfloor (h'+1)/2 \rfloor = 203 + \lfloor (11+1)/2 \rfloor = 209 \ (6)$$
$$y' = l - \lfloor h'/2 \rfloor = 203 - \lfloor 11/2 \rfloor = 198 \qquad (7)$$

This method is focused on the data rather than the image as integrity of data is more important. In this technique, we have a text message that is to be hidden in the image and an image that is available in which the data is to be embedded. We consider that the image is of size 'M x N'. We use the image in our computations as grey scale image. Further we need to consider a block of particular size let it be 'm x n'. We also

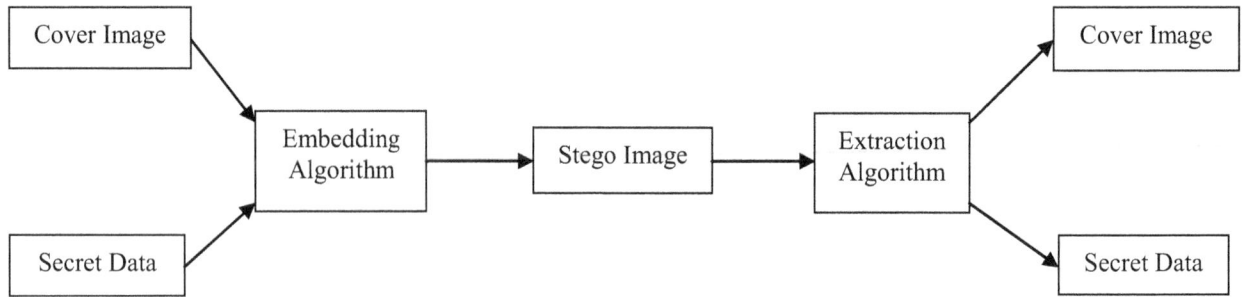

Fig. 1.   Data Hiding and Extraction Process

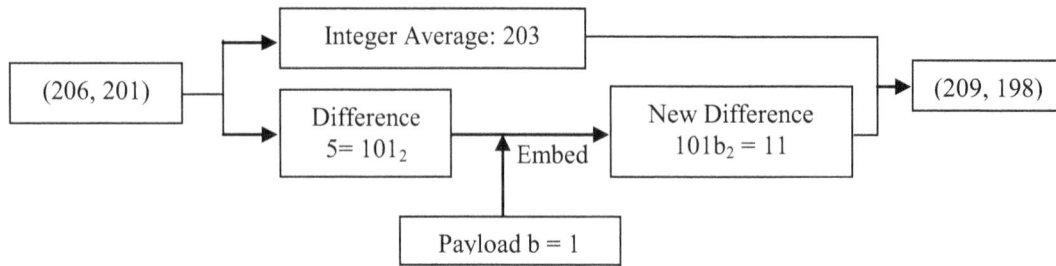

Fig. 2. An example of data embedding by Tian (2003) [33]

select a threshold value 'T' for keeping the image distortion free.

The image is divided into various non-overlapping blocks of size 'm x n'. For each block we have various components in the block. We arrange them in increasing order, select the mid value and subtract all the values from this mid value. Some values give us positive result some give negative. In the next step, we select the maximum difference of all the values and compare it with a threshold in order to decide the amount of data that we can embed into that block. The blocks are named as a, b and c. if a block belongs to type 'a', it means this block is located in a rather smooth area since the difference values are all very small. In this case, it is a very suitable block to hide more secret data bits here 3 bits of data may be embedded. Block 'b' belongs to the region of sharp detail or edges which can acquire only 1 bit of data and in the last block 'c' no secret data is embedded as to avoid distortion. We also use a record bit in order to identify the block after the embedding thus we set the record bit as 1 for all data embedded into block 'a' and record bit '0', for all data embedded into block 'b'. For block 'c', it is not needed. The input to this algorithm is an image and a secret message binary array threshold value. The output is the marked stego image. The process is carried out as shown in Figure 3.

1) Secret Data Embedding Algorithm
1. Segment the cover image into non overlapping blocks of size m x n
2. Label the components as $v_0, v_1, v_2, \ldots \ldots v_k$ where k= mn-1
3. Find v_{mid} and hence compute differences as $d_0, d_1, d_2, \ldots \ldots d_{m-1}, d_{m+1}, \ldots d_{k-1}$
4. Define threshold 'T' and 'd_{max}'.

5. Find the type of block (i.e a, b and c) using T and divide the image into smooth regions and edge regions on the basis of threshold and difference in pixel values.
6. We embed large data in smooth regions and less data in non-smooth regions. The conditions are as follows:
   a. If $d_{max} < T/8$, then it belongs to block 'a' and three bits can be embedded into each block with record bit being 1.
   b. If $T/8 \leq d_{max} < T$, then it belongs to block 'b' and one bit can be embedded to each block with record bit being 0
   c. If $d_{max} \geq T$, then it belongs to the block 'c' and it is a non-embeddable block.
7. Output: an image with data embedded (i.e. stego image).

Data embedding process is illustrated in Figure 3 and can be summarized as follows: first of all non-overlapping blocks of a particular size are extracted from the image; then for a particular block, all the pixel values are sorted and mid value are subtracted from each pixel value. Maximum difference value is selected to decide the type of the block because the magnitude of the local differences can adequately describe the edges of the local regions of the image. According to the details of the edges using maximum difference value and a threshold value we decide the type of the bloc and further the number of bit to be embedded in that block. We are not embedding any information in that block which have very high level of edge details to avoid any degradation in the image.

C. Secret Data and Cover Image Extraction
In this method, now, we have a stego image containing secret information. We consider that this image is also of size 'M x N'. In the next step, we need to consider a block of particular

size let it be 'm x n' which is same as it was for the embedding part otherwise we would not be able to extract the data from it. We have prior knowledge of threshold value 'T' and we use its same value here also. The image is divided into various blocks of size 'm x n'. For each block we have various components in the block. We arrange them in increasing order and select the mid value then subtract all the values from the mid values. Some values give us positive result while some give negative then we select the maximum difference of all the values and compare it with the threshold in order to decide the amount of data we can extract from that block. If a block belongs to type 'a', it means that this block is located in a rather smooth area since the difference values are all very small. In this case, it must contain 3 bits of concealed data in it. We check the record bit, if it is '1' we need to extract the data by subtracting the last 3 bit from the values of the component. Then with the help of the d_{max} we restore the image value to what it was earlier. On the other hand if the block is of type 'b' it contains only 1 bit of data and we extract this data in the same way as we extracted for the block 'a'.

In this situation we only need to extract the LSB (Least Significant Bit) and check the record bit being '0'. For the blocks of type 'c', there is no secret data embedded. The input to this phase is a stego-image and a threshold value and the output is original restored cover image and the secret data. During the extraction process we also notice that we are able to recover the original cover image same as it was before. Thus, this suggests that there is no noise further added during the embedding other than the secret data embedded (see Figure 4).

1) Secret Data Extraction Algorithm

1. Segment the stego image into non overlapping blocks of size m x n
2. Label components as $v_0, v_1, \ldots \ldots v_k$, where k= mn-1
3. Find v_{mid} and hence compute differences as $d_0, d_1, d_2, \ldots \ldots d_{m-1}, d_{m+1}, \ldots d_{k-1}$
4. We already know threshold 'T' and 'd_{max}'.
5. Find the type of block (i.e a, b and c) using T and divide the image into smooth regions and edge regions on the basis of threshold and difference in pixel values.
6. We extract large data from smooth regions and less from non smooth regions. The conditions are as follows
   a. If $d_{max} < T/8$, then it is block of type 'a' therefore three bits are to be extracted from each block with record bit being 1.
   b. If $T/8 \leq d_{max} < T$, then it is 'b' type block and one bit can be extracted from each block with record bit 0.
   c. If $d_{max} \geq T$, then the block belongs to type 'c' and no data can be extracted from such block.
7. Restore cover image with the extracted secret data.

The hidden information extraction approch is described in the Figure 4 and can be summrized as follows: first of all again the image is divided into non-overlapping blocks and for each block the pixel intensities are sorted and subtracted with

the mid value in order to find the type of the block using the maximum difference value (i.e. the number of bits to be extracted from a particular block), it should be noted that the maximum difference value will be same as it was previously because intensity differences are invarint to the monotonic intensity change caused by the embedding bits; three, one or none number of bits are extracted according to the type 'a', type 'b' and type 'c' of the block respectively; and finally the information and image both are recovered.

Our approach is able to embed more data in smoother images keeping the distortion not visible at all. The most important fact as well as advantage of using this approach is that we are able to restore the original image back to its initial state (i.e. the state in which it was earlier before data embedding) thus our approach makes sure that no additional noise is added into the image other than the. This is one of the features of the DE (Difference Expansion) technique that the image can be restored to its previous state.

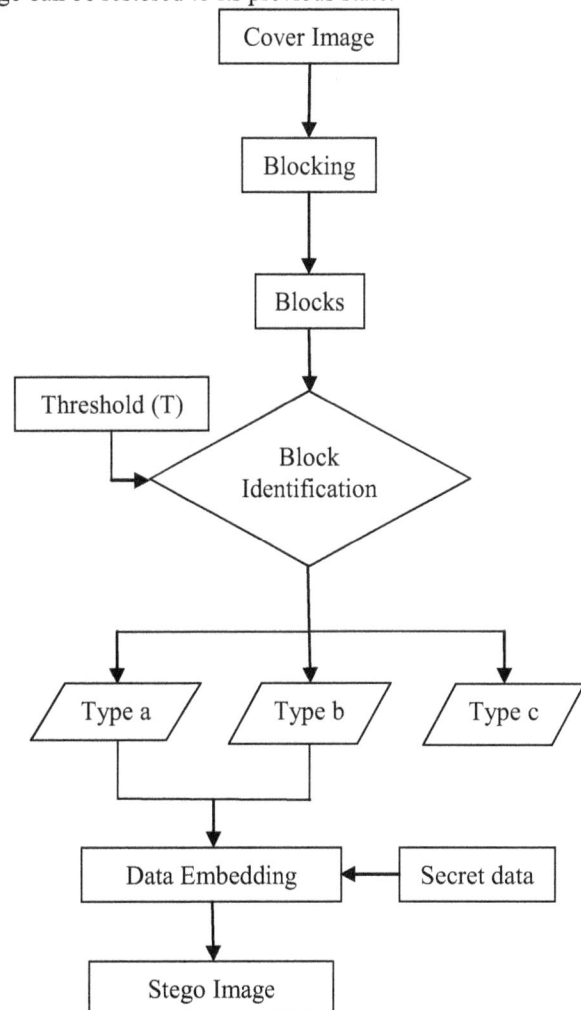

Fig. 3. Data Embedding Flowchart

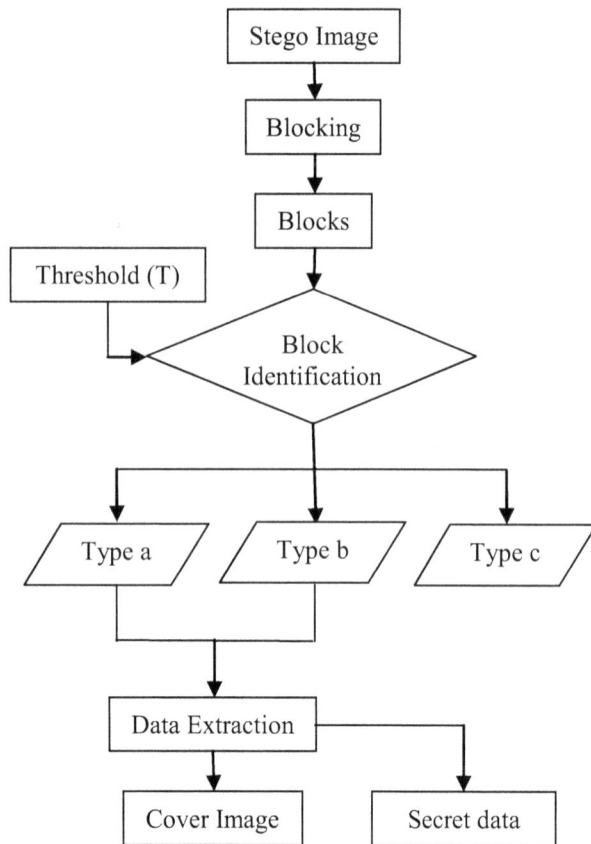

Fig. 4. Data Extraction Flowchart

IV. RESULTS AND DISCUSSIONS

This section analyses the various aspects of the proposed method. Embedding is not prone to any visual attacks as the changes are hardly visible in the image. In addition to this we use an image that has smoother regions and hence it can acquire more data as compared to the one with less smooth region. Here we test our method for the correctness and performance. The PSNR (peak signal to noise ratio) is used to measure the quality of stego image when compared with the cover image. It is calculated using equation 8, where MSE is mean square error given by equation 9.

$$PSNR = 10 \log_{10} \frac{(2^n - 1)^2}{MSE} \qquad (8)$$

Where, MSE is given by

$$MSE = \frac{\sum (\hat{x} - x)^2}{A} \qquad (9)$$

The quality of the image is higher if the PSNR value of the image is high. Since PSNR is inversely proportional to MSE value of the image, the lower MSE value yields higher PSNR value. It means, the better the stego image quality the lower the MSE value will be. Using variable data to be hidden in the image we test the images for the PSNR and the Bit per Pixel (BPP). Where BPP is given by

$$BPP = \frac{n}{MXN} \qquad (10)$$

Where n is the number of bits to be embedded in the image while M and N are the dimensions of the image.

A. Data Set Used

In order to evaluate our proposed, we have used a data set of 20 images (i.e. 20 cover images). All the images used are gray scaled images of size 512 X 512. Figure 5 shows the 20 cover images used in this paper for the experiments. The simulation for the experiment was set up and carried out on a Windows XP Professional with 1.8 GHz dual core processor and 1 GB of RAM.

The proposed approach is implemented on the publicly available MATLAB R2009b. We have further referred the images of the data set as 01.jpg, 02.jpg, 03.jpg.....20.jpg in the Figure 5.

B. Experimental Results

We have obtained the stego image after data embedding (payload size of 61376 bit with BPP of 0.2341) in all cover images of the dataset and recovered all the images after extraction of data as it is shown in Figure 6. First column shows the original input cover images, second column depicts the stego image obtained after data hiding and third column shows the recovered images obtained after data extraction.

In Figure 6, we can see that there is nearly no visual difference among input, stego and recovered images it means our embedding algorithm is robust to noise and able to recover original image accurately. Figure 7 illustrates the quality of stego image with the size of payload data embedded. It shows the graph between PSNR and BPP for all images of the dataset used. From this figure it is clear that if size of the data increases (i.e. BPP increases) quality of the stego image degrades because the value of PSNR decreases.

We are able to achieve high embedding capacity with keeping PSNR in range and also the visual quality of the image is not getting too distorted to visualize as it is depicted by Table 1. This table listed the value of the PSNR obtained for each images after data embedding of varying sizes. The value of the PSNR is highest for 13.jpg in each case of payload data because 13.jpg has largest smoother area (we can see it in the 13th image of Figure 5) and the value of the PSNR is lowest for 14.jpg in each case of payload data because 14.jpg has lowest smoother area(we can see it in the 14th image of Figure 5).

(a) 01.jpg   (b) 02.jpg   (c) 03.jpg   (d) 04.jpg   (e) 05.jpg

(f) 06.jpg   (g) 07.jpg   (h) 08.jpg   (i) 09.jpg   (j) 10.jpg

(k) 11.jpg   (l) 12.jpg   (m) 13.jpg   (n) 14.jpg   (o) 15.jpg

(p) 16.jpg   (q) 17.jpg   (r) 18.jpg   (s) 19.jpg   (t) 20.jpg

Fig. 5. 20 cover images of the data set used in this paper

(a) 01.jpg

(b) 02.jpg

(c) 03.jpg

(d) 04.jpg

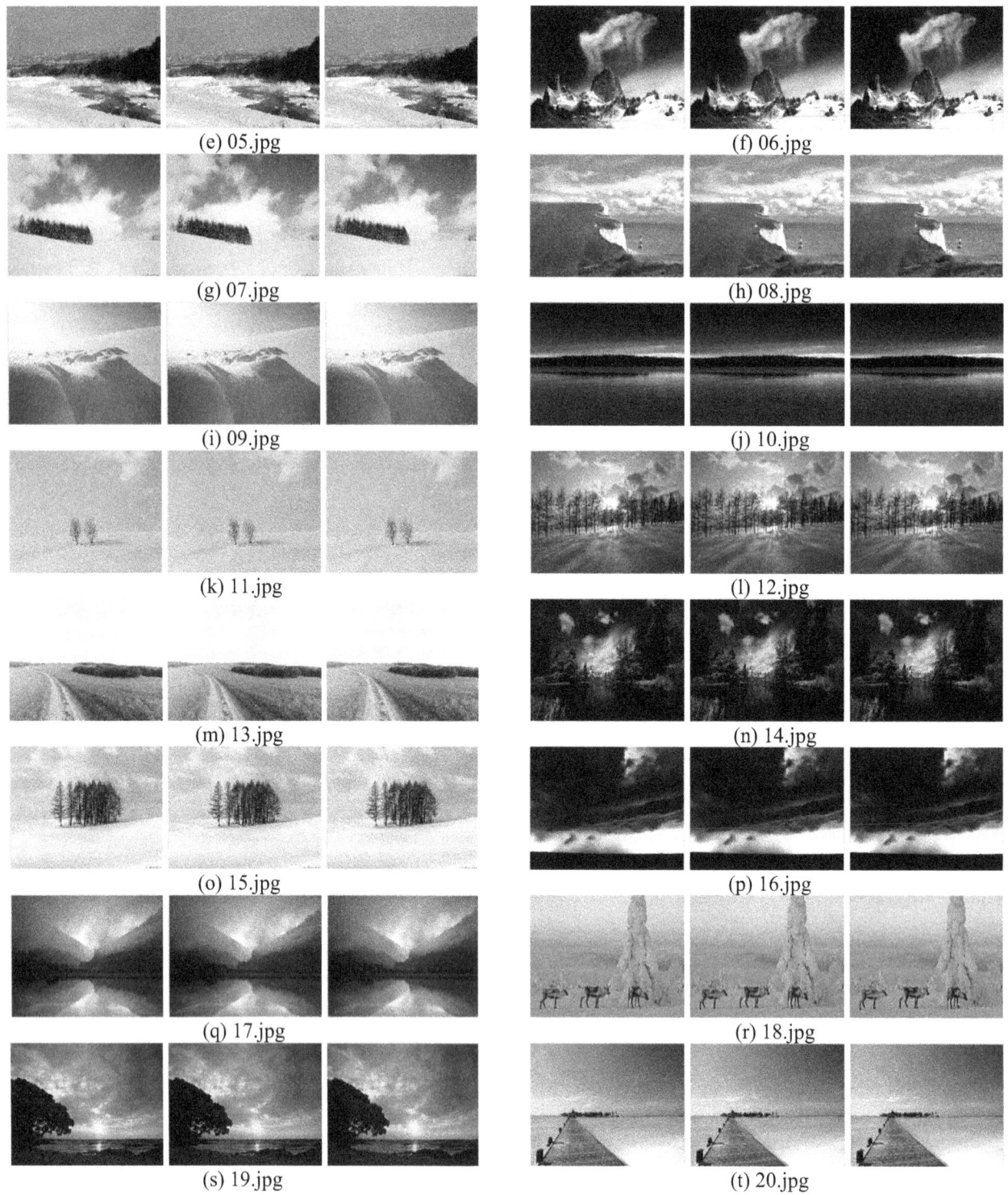

(e) 05.jpg

(f) 06.jpg

(g) 07.jpg

(h) 08.jpg

(i) 09.jpg

(j) 10.jpg

(k) 11.jpg

(l) 12.jpg

(m) 13.jpg

(n) 14.jpg

(o) 15.jpg

(p) 16.jpg

(q) 17.jpg

(r) 18.jpg

(s) 19.jpg

(t) 20.jpg

Fig. 6. Cover image (1st column), stego image (2nd column) and restored image (3rd column) for input images.

TABLE I.
RESULTS AFTER TESTING ALL THE IMAGES OF THE DATA SET

Image	Payload size in bits and their corresponding Bit Per Pixel for the images used for testing					
	14320 .0546	29896 .1140	45008 .1717	61376 .2341	75304 .2873	93568 .3569
01.jpg	43.76	40.24	38.36	36.92	35.70	34.38
02.jpg	43.26	39.76	37.36	36.35	35.59	33.53
03.jpg	43.56	40.37	39.36	37.17	34.63	33.55
04.jpg	43.91	40.57	38.55	36.78	35.53	34.15
05.jpg	42.54	39.42	37.59	36.24	35.37	34.31
06.jpg	43.08	38.66	36.69	35.21	34.26	33.42
07.jpg	42.71	39.38	37.72	36.49	35.56	34.40
08.jpg	45.87	42.37	38.99	37.04	35.88	34.57
09.jpg	44.66	41.75	39.97	38.68	37.63	36.39
10.jpg	49.67	46.23	44.44	42.95	41.10	38.14
11.jpg	42.38	39.12	37.31	35.98	35.10	34.17
12.jpg	41.46	38.07	35.78	34.01	32.94	31.85
13.jpg	49.67	46.55	44.74	43.43	42.54	41.60
14.jpg	39.92	36.20	34.18	32.89	32.08	31.18
15.jpg	41.96	38.98	37.38	36.20	35.36	34.19
16.jpg	44.98	40.35	38.24	37.18	36.79	36.41
17.jpg	48.01	44.62	42.31	40.52	39.21	37.70
18.jpg	40.21	38.53	37.23	36.22	35.41	34.53
19.jpg	45.19	41.98	39.95	38.20	36.98	35.36
20.jpg	48.88	45.64	43.83	42.40	41.32	40.23

Table 2 shows the embedding capacity of proposed approach against classical approach using difference expansion. We compared our method with the Tian's approach [33] because it is a benchmark and widely adopted for comparison purposes. Moreover this is method which applies the local differences into consideration for data embedding and we expanded this method using multiple types of embedding blocks. From this table it is clear that the embedding capacity of proposed approach is higher than the classical approach for all the images used for test purpose. The hiding capacity is highest for 13.jpg and lowest for 14.jpg using proposed approach because 13.jpg has large smoother area and 14.jpg has large fine edges. It should be noted that by embedding the data bits the stego image is not same as the original image and some degradation can be seen using the PSNR in Figure 7 and Table 1 but the amount of degradation is less such that it can't be observed visually (see the input and stego images of Figure 6). The amount of degradation is less using our approach because we have not embedded any bit in blocks having more details but we embedded more number of bits in blocks having smooth details. Moreover, if the number of blocks having less details in the image are more then the amount of data using our approach will be more without loss

of visual effect (see the number of bits embedded for each images in Table 2). Our experimental results indicate that the proposed solution significantly support the data hiding problem as well as it has higher hiding capacity than earlier approaches. Our algorithm embeds the amount of data according to the details of the image. More data can be attached with less detailed areas of the image (i.e. smoother part) and less data can be attached with the fine detailed areas.

TABLE II.
COMPARISON OF HIDING CAPACITY OF IMAGES BY PROPOSED AND CLASSICAL
APPROACH BASED ON THE DIFFERENCE EXPANSION [33]

Image	Capacity by Difference Expansion(in bits)	Capacity by Proposed Approach(in bits)
01.jpg	121421	224966
02.jpg	125815	247795
03.jpg	104685	192327
04.jpg	135254	258903
05.jpg	137055	235224
06.jpg	101839	174839
07.jpg	122550	234142
08.jpg	124331	229485
09.jpg	131545	248200
10.jpg	92291	179724
11.jpg	132201	227796
12.jpg	128108	227957
13.jpg	105920	259619
14.jpg	139650	165627
15.jpg	130403	241323
16.jpg	99237	193444
17.jpg	119110	232369
18.jpg	127461	238708
19.jpg	130935	240258
20.jpg	114030	212238

V. CONCLUSIONS AND FUTURE DIRECTIONS

An image processing based approach is proposed and experimented in this paper for information protection. The proposed approach is comprised of mainly three steps. In the first step, message generation and image selection is performed using the data that is to be embedded. In the second step, we embed the data into the corresponding image. First, image is partitioned into the different number of non-overlapping blocks and then we embedded different number of bits in different blocks of the image according to the types of the block. Three types of blocks are considered in this paper according to the intensity of details of that block i.e. smooth, average and dense and embedded three, one and none bits respectively. In the third step, we extract the data and restore the image to its initial state. An approach similar to embedding one is also employed here to know the type of block. According to the type of block the bits are extracted. We have used the difference expansion technique to embed the data into the image and to extract the data from the image. The major advantage of using difference expansion technique is that now a large amount of data can be embedded into the image and there is no visible effect on the image, moreover the image can be restored to its initial state, thus our approach shows that no

noise is added to the image except the data which is extracted in the extraction phase. Our experimental results indicate that the proposed solution can significantly support the data hiding problem. The future work includes the consideration of RGB color images and videos for information hiding.

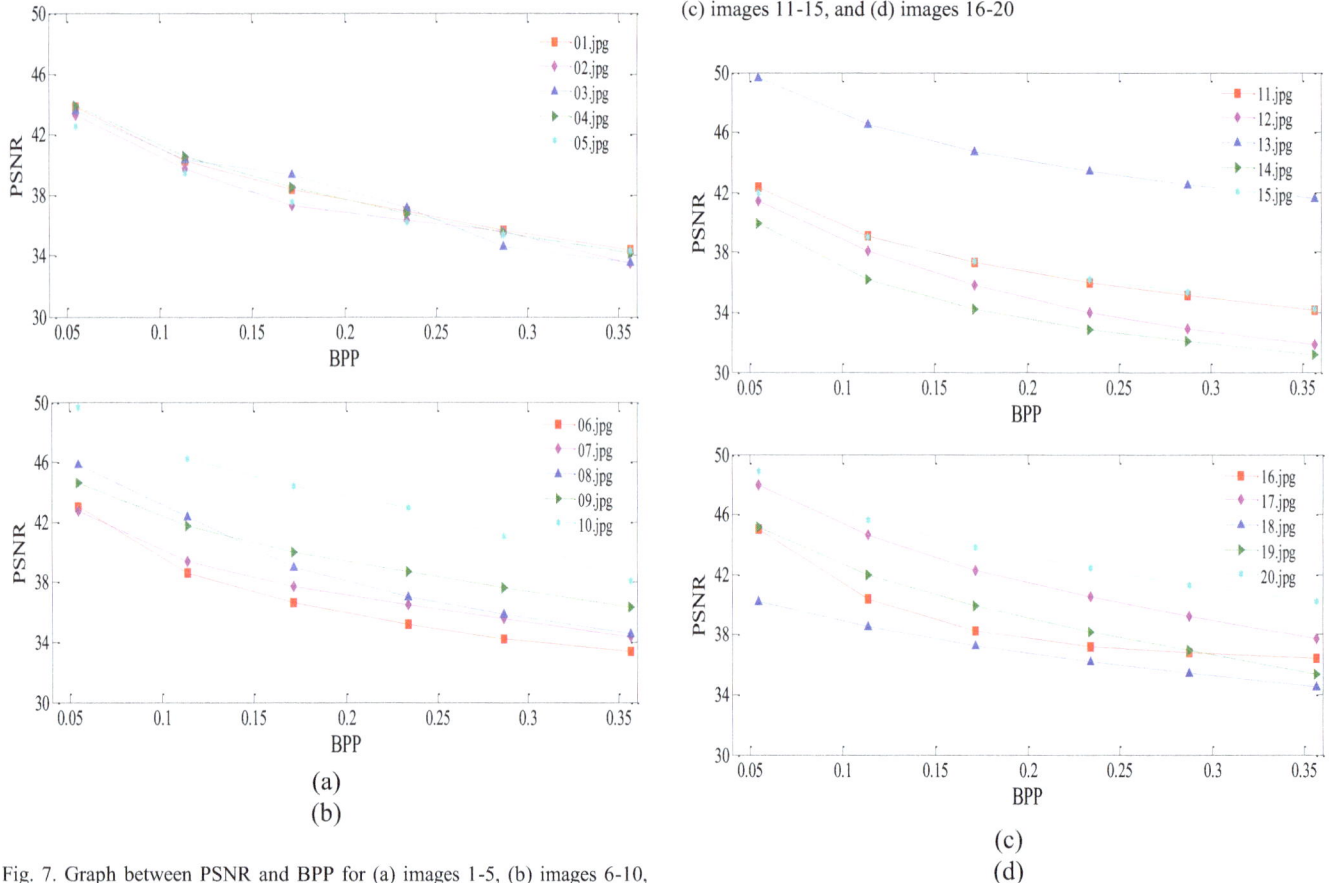

(c) images 11-15, and (d) images 16-20

(a)

(b)

(c)

(d)

Fig. 7. Graph between PSNR and BPP for (a) images 1-5, (b) images 6-10,

REFERENCES

[1] J.C. Judge, "*Steganography: past, present, future*," SANS Institute publication, 2001.

[2] S. R. Dubey and A. S. Jalal, "Adapted Approach for Fruit Disease Identification using Images," *International Journal of Computer Vision and Image Processing*, Vol. 2, No. 3, pp. 44-58, 2012.

[3] V.B. Semwal, K.S. Kumar, V.B. Semwal and M. Sati, "Accurate location estimation of moving object with energy constraint & adaptive update algorithms to save data," arXiv preprint arXiv:1108.1321, 2011.

[4] W.L. Romero, R.G. Crespo and A.C. Sanz, "A prototype for linear features generalization," *International Journal of Interactive Multimedia & Artificial Intelligence*, Vol. 1, No. 3 2010.

[5] H. Bolivar, A. Pacheco and R.G. Crespo, "Semantics of immersive web through its architectural structure and graphic primitives," *International Journal of Interactive Multimedia & Artificial Intelligence*, Vol. 1, No. 3, 2010.

[6] S.J.B. Castro, R.G. Crespo and V.H.M. García, "Patterns of Software Development Process," *International Journal of Interactive Multimedia & Artificial Intelligence*, Vol. 1, No. 4, 2011.

[7] K.K. Susheel, V.B. Semwal and R.C. Tripathi, "Real time face recognition using adaboost improved fast PCA algorithm," arXiv preprint arXiv:1108.1353, 2011.

[8] S. R. Dubey and A. S. Jalal, "Robust Approach for Fruit and Vegetable Classification," *Procedia Engineering*, Vol. 38, pp. 3449-3453, 2012.

[9] S. R. Dubey, P. Dixit, N. Singh, and J.P. Gupta, "Infected fruit part detection using K-means clustering segmentation technique," *International Journal of Artificial Intelligence and Interactive Multimedia*. Vol. 2, No. 2, pp. 65-72, 2013.

[10] X. Li, W. Zhang, X. Gui and B. Yang, "A Novel Reversible Data Hiding Scheme Based on Two-Dimensional Difference-Histogram Modification," *IEEE Transactions on Information Forensics and Security*, Vol. 8, No. 7, pp. 1091 – 1100, 2013.

[11] O.S. Faragallah, "Quadruple Difference Expansion-Based Reversible Data Hiding Method for Digital Images," *Information Security Journal: A Global Perspective*, Vol. 21, No. 5, pp. 285-295, 2012.

[12] S. R. Dubey, N. Singh, J.P. Gupta and P. Dixit, "Defect Segmentation of Fruits using K-means Clustering Technique," In *the Proceedings of the Third International Conference on Technical and Managerial Innovation in Computing and Communications in Industry and Academia*. 2012.

[13] N. Singh, S. R. Dubey, P. Dixit and J.P. Gupta, "Semantic Image Retrieval Using Multiple," In *proceeding of the 3rd International Conference on Technical and Managerial Innovation in Computing and Communications in Industry and Academia*. 2012.

[14] S. R. Dubey and A. S. Jalal, "Detection and Classification of Apple Fruit Diseases Using Complete Local Binary Patterns," In *proceeding of the Third International Conference on Computer and Communication Technology*, pp. 346-351, 2012.

[15] S. R. Dubey and A. S. Jalal, "Species and variety detection of fruits and vegetables from images," *International Journal of Applied Pattern Recognition*. Vol. 1, No. 1, pp. 108-126, 2013.

[16] N. Singh, S. R. Dubey, P. Dixit and J.P. Gupta, "Semantic Image Retrieval by Combining Color, Texture and Shape Features," In *the Proceedings of the International Conference on Computing Sciences*, pp. 116-120, 2012.

[17] J. P. Gupta, N. Singh, P. Dixit, V. B. Semwal and S. R. Dubey, "Human Activity Recognition using Gait Pattern," *International Journal of Computer Vision and Image Processing*, Vol. 3, No. 3, pp. 31 – 53, 2013.

[18] K.S. Kumar, V.B. Semwal, S. Prasad and R.C. Tripathi, "Generating 3D Model Using 2D Images of an Object," *International Journal of Engineering Science*, 2011.

[19] V.B. Semwal, V.B. Semwal, M. Sati and S. Verma, "Accurate location estimation of moving object in Wireless Sensor network," *International Journal of Interactive Multimedia and Artificial Intelligence*, Vol. 1, No. 4 , pp. 71-75, 2011.

[20] K.S. Kumar, S. Prasad, S. Banwral and V.B. Semwal, "Sports Video Summarization using Priority Curve Algorithm," *International Journal on Computer Science & Engineering*, 2010.

[21] R.G. Crespo, S.R. Aguilar, R.F. Escobar and N. Torres, "Dynamic, ecological, accessible and 3D Virtual Worlds-based Libraries using OpenSim and Sloodle along with mobile location and NFC for checking in," *International Journal of Interactive Multimedia & Artificial Intelligence*, Vol. 1, No. 7, 2012.

[22] C.J. Broncano, C. Pinilla, R.G. Crespo and A. Castillo, "Relative Radiometric Normalization of Multitemporal images," *International Journal of Interactive Multimedia and Artificial Intelligence*, Vol. 1, No. 3, 2010.

[23] N.F. Johnson and S. Jajodia, "Exploring steganography: seeing the unseen," *IEEE Computer*, Vol. 31, No. 2, pp. 26–34, 1998.

[24] N. Provos and P. Honeyman, "Hide and seek: an introduction to steganography," *IEEE Security and Privacy*, Vol. 1, No. 3, pp. 32–44, 2003.

[25] P. Moulin and R. Koetter, "Data-hiding codes," *Proceedings of the IEEE*, Vol. 93, No. 12, pp. 2083–2126, 2005.

[26] S.B. Sadkhan, "Cryptography: current status and future trends," In *the Proceedings of the IEEE International Conference on Information & Communication Technologies: From Theory to Applications*, pp. 417–418, 2004.

[27] C. Kurak and J. McHugh, "A cautionary note on image downgrading," In *the Proceedings of the 8th IEEE Conference on Computer Security Applications*, pp. 153–159, 1992.

[28] C. Hosmer, "Discovering hidden evidence," *Journal of Digital Forensic Practice*, Vol. 1, No. 1, pp. 47–56, 2006.

[29] J.C.H. Castro, I.B. Lopez, J.M.E. Tapiador and A.R. Garnacho, "Steganography in games: a general methodology and its application to the game of Go," *Computers and Security*, Vol. 25, No. 1, pp. 64–71, 2006.

[30] C.D. Vleeschouwer, J.F. Delaigle and B. Macq, "Circular interpretation of bijective transformations in lossless watermarking for media asset management," *IEEE Transactions on Multimedia*, Vol. 5, No. 1, pp. 97–105, 2003.

[31] M.U. Celik, G. Sharma, A.M. Tekalp and E. Saber, "Lossless generalized-LSB data embedding," *IEEE Transactions on Image Processing*, Vol. 14, No. 2, pp. 253–266, 2005.

[32] Z. Ni, Y.Q. Shi, N. Ansari and W. Su, "Reversible data hiding," *IEEE Transactions on Circuits and Systems for Video Technology*, Vol. 16, No. 3, pp. 354–362, 2006.

[33] J. Tian, "Reversible Data Embedding Using A Difference Expansion," *IEEE Transactions on Circuits and Systems for Video Technology*, Vol. 13, No. 8, pp. 890–896, 2003.

[34] A.M. Alattar, "Reversible watermarking using the difference expansion of a generalized integer transform," *IEEE Transactions on Image Processing*, Vol. 13, No. 8, pp. 1147–1156, 2004.

[35] S.S. Maniccam and N.G. Bourbakis, "Lossless image compression and encryption using SCAN," *Pattern Recognition*, Vol. 34, No. 6, pp. 1229–1245, 2001.

[36] L.D. Paulson, "New system fights steganography, News Briefs," *IEEE Computer Society*, Vol. 39, No.8, pp. 25–27, 2006.

[37] A.A. Abdelwahab and L.A. Hassaan, "A discrete wavelet transform based technique for image data hiding," In *the Proceedings of the 25th National Radio Science Conference*, pp. 1–9, 2008.

[38] V.M. Potdar, S. Han and E. Chang, "Fingerprinted secret sharing steganography for robustness against image cropping attacks," In *the Proceedings of the IEEE Third International Conference on Industrial Informatics*, pp. 717–724, 2005.

[39] C.C. Lin, W.L. Tai and C.C. Chang, "Multilevel reversible data hiding based on histogram modification of difference images," *Pattern Recognition*, Vol. 41, No. 12, pp. 3582–3591, 2008.

[40] Y.T. Wu and F.Y. Shih, "Genetic algorithm based methodology for breaking the steganalytic systems," *IEEE Transactions on Systems, Man, and Cybernetics—part B: cybernetics*, Vol. 36, No. 1, pp. 24–31, 2006.

[41] K.B. Raja, S. Sindhu, T.D. Mahalakshmi, S. Akshatha, B.K. Nithin, M. Sarvajith, K.R. Venugopal and L.M. Patnaik, "Robust image adaptive steganography using integer wavelets," In *the Proceedings of the Third International Conference on Communication System Software and Middleware*, Bangalore, India, pp. 614–621, 2008.

[42] S.R. Dubey, "*Automatic Recognition of Fruits and Vegetables and Detection of Fruit Diseases*", M.Tech Thesis, GLAU Mathura, India, 2012.

[43] M. Sati, V. Vikash, V. Bijalwan, P. Kumari, M. Raj, M. Balodhi, P. Gairola and V.B. Semwal, "A fault-tolerant mobile computing model based on scalable replica", *International Journal of Interactive Multimedia and Artificial Intelligence*, Vol. 2, 2014.

[44] V.B. Semwal, S.A. Katiyar, P. Chakraborty and G.C. Nandi, "Biped Model Based on Human Gait Pattern Parameters for Sagittal Plane Movement", *CARE*, 2013.

[45] V.B. Semwal, A. Bhushan and G.C. Nandi, "Study of Humanoid Push Recovery Based on Experiments", *CARE*, 2013.

[46] S.R. Dubey and A.S. Jalal, "Automatic Fruit Disease Classification using Images", *Computer Vision and Image Processing in Intelligent Systems and Multimedia Technologies*, 2014.

[47] D.D. Agrawal, S.R. Dubey and A.S. Jalal, "Emotion Recognition from Facial Expressions based on Multi-level Classification", *International Journal of Computational Vision and Robotics*, 2014.

[48] S.R. Dubey and A.S. Jalal, "Fruit Disease Recognition using Improved Sum and Difference Histogram from Images", International Journal of Applied Pattern Recognition, 2014.

6

Overlap Algorithms in Flexible Job-shop Scheduling

Celia Gutiérrez

Universidad Complutense de Madrid, Spain

Abstract — **The flexible Job-shop Scheduling Problem (fJSP) considers the execution of jobs by a set of candidate resources while satisfying time and technological constraints. This work, that follows the hierarchical architecture, is based on an algorithm where each objective (resource allocation, start-time assignment) is solved by a genetic algorithm (GA) that optimizes a particular fitness function, and enhances the results by the execution of a set of heuristics that evaluate and repair each scheduling constraint on each operation. The aim of this work is to analyze the impact of some algorithmic features of the overlap constraint heuristics, in order to achieve the objectives at a highest degree. To demonstrate the efficiency of this approach, experimentation has been performed and compared with similar cases, tuning the GA parameters correctly.**

Keywords— **Algorithm, Flexible Job-Shop Scheduling, GA parameters, Local improvement, Overlap heuristics.**

I. INTRODUCTION

A Job-shop Scheduling Problem (JSP) is based on the concept of jobs, which are composed of operations that must be processed by the resources of different type in a sequential order. Each operation has a completion time. One machine can only process one job at a time and an operation cannot be pre-empted. The objective is to minimize the total makespan (the time to complete all jobs). The simplification of this problem is enunciated like this: there are n jobs to be scheduled on m machines in a general job-shop problem, G, minimizing the total completion operation time, C_{max}, $n/m/G/C_{max}$.

Flexible Job-shop Scheduling Problem is a generalization of the JSP, where the resource is selected among a set of suitable ones, giving place to two subproblems: routing and allocation of operations. The first one produces the start-time of the operations, and the second one the assignment of operations on resources.

Both JSP and fJSP have been solved by the use of metaheuristic algorithms, like GAs. The application of a GA on the simple basis as in [1] has poor performance because no domain knowledge is inserted, leading to non-feasible results. One way to insert knowledge into the algorithm is by hybridizing the GA with heuristics that provide local search.

This paper follows the last approach, and goes beyond a deep analysis of GAs. It fact, it is an extension of [2], that explains how to achieve optimal results in the hybridization of GA with local search techniques to solve fJSP. This work provides a further analysis of the overlap constraint operators. In this way, the previous work provides a macroperspective view of the whole solution, and the present work is a microperspective view. It is structured in this way: section 2 covers the problem background; section 3 introduces the complete algorithm and the codification of information regarding the resources and fitness functions; section 4 shows the algorithms of a heuristic operator variants; section 5 shows the results of the experimentation phase; section 6 contains the comparison with similar approaches; and section 7 has the conclusions and future work.

II. PROBLEM BACKGROUND

Hybrid approaches that mix GA and heuristics are a well-known solution that has proven to be efficient, as heuristics provide domain knowledge that the simple GA cannot [3]. This focus can be applied in two ways: embedding the heuristics into the GA loop (integrated approach), or outside it (hierarchical approach), [3].

Literature shows examples of hybrid GA with intelligent genetic operators than produce optimal schedules. This is the case of [4], that describe an effective hybridization of both techniques, applying improved crossover and mutation operators when there are non-feasible schedules.[5] describes a hybrid GA solution by the use of two vector chromosome and bottleneck shifting procedure. The representation is made by two vectors: one for the machine assignment and the another one for the operation sequence. [6] solve the same problem by the use of an artificial immune algorithm. It uses several strategies for generating an initial population and selecting the best individuals. It also has operators that reorganize the operations (by a mutation). [7] adopt the hybrid GA by the use of the approach by localization to initialize the GA, and improving it by reordering jobs and machines, and by searching for a global minimum [4] have improved operators constraint and mutator operators that consider constraint violations.

The second way to include the heuristics has also been widely implemented, though the existing algorithms vary in the order of application, heuristic methods, goal of the application, and even domain. [8] follows this paradigm by means of a local search by the definition of the neighborhood.

This work follows the second approach. Having proven the efficiency of the mentioned algorithms, the objective of this research is to provide the designer with relevant issues that improve the algorithm performance when using local improvements within a hybrid GA under a hierarchical architecture. This is also considered a multi-objective fJSP, because the solution achieves three goals:

- To minimize the makespan of the operations.
- To minimize the maximal machine overload, i.e., the maximum working time spent at any machine.
- To satisfy the maximum number of constraints.

There are also recent approaches to solve the problem of JSP, like [9], where they solve the problem of scheduling independent tasks in a grid computing system. They use a new evaluation (distributed) algorithm inspired by the effect of leaders in social groups, the group leaders' optimization algorithm (GLOA). In contrast, the present work analyzes some design features of the hybrid algorithm, preferably the overlap constraint repairer.

III. HIERARCHICAL DESIGN FEATURES

This work constitutes the extended version of the previous work, providing deeper details of the heuristics design and argumentation for the parameters tuning. So, whereas [2] and [10] provide a solution to a general fJSP, the current work provides design and execution details in order to achieve the goals of the algorithm.

This research has been analysed following a hierarchical approach that decomposes the resource and the start-time assignment in two different problems solved by different and independent GA, like in [5]. Previous to both GA running, there is a module that calculates the limits for the start-time for each operation, and after both GA running the module of the heuristics solve the unfulfilled constraints. The adaptation of the algorithm to JSP claims a simpler architecture, where the resource GA module does not appear. Other variations concerning the heuristics are also discussed in the section 4.

A. Codification of the Resource GA Chromosome

The chromosome and fitness function for both GA are described in the previously cited works. There are subtle differences in the morphology of both chromosomes: while the solution for time GA is directly codified into the chromosome, the chromosome for resource GA stores as many genes as operations, which must be decoded to get the resource number. For example, for the set of 4 orders, 3 products per order (maximum), 1 product instance per product (maximum), 5 operations per instance (maximum), and 4 available resources in the job-shop, the gene value must cover 4 x 3 x 1 x 5 x 4 values, so the range is [0-239]. To decode a gene value, successive divisions must be applied using this algorithm that involves equation (1) to equation (8):

Resource number
$$= gene \ MOD \ number \ of \ resources \qquad (1)$$
cant
$$= gene \ / \ number \ of \ resources \qquad (2)$$
product instance identification
$$= gene \ MOD \ number \ of \ product \ instances \qquad (3)$$
cant
$$= cant \ / \ number \ of \ product \ instances \qquad (4)$$
operation number
$$= cant \ MOD \ number \ of \ operations \qquad (5)$$
cant
$$= cant \ / \ number \ of \ operations \qquad (6)$$
product identification
$$= cant \ MOD \ number \ of \ products \qquad (7)$$
order number
$$= cant \ / \ number \ of \ products \qquad (8)$$

For a gene value of 69, the decoding process gives the following values for the parameters:

- resource number = 1
- product instance identification = 0
- operation number = 2
- product identification= 0
- order number = 1

B. GA fitness functions

There is one fitness function for each GA. Both functions incorporate penalizations that depend on the domain they are evaluating. For both GAs, the objective is to minimize the values obtained by the fitness functions. The following subsections contain their codification:

1) Fitness function for Resource GA

This function evaluates the sums of deviations between the assignment of operations to certain resource and the ideal assignment. In other words, this fitness function penalizes non-balanced assignments of operations among the resources of the same type. The ideal assignment is the number of operations assigned to the resources of the same type, divided into the number of resources of that type, as equation (9) shows:

$$Fitness= f \times \sum_{i=0} |O_{i,t} - (O_t / R_t)| \qquad (9)$$

where:

f is a penalty factor (For simplicity, $f=1$),
i represents each resource in the job-shop,
$O_{i,t}$ is the number of operations assigned to the i resource, that belongs to the t type of resource,
O_t is the number of operations assigned to the resources of t type,
R_t is the number of resources of t type.

2) Fitness function for Time GA

This function sums up the starting times of all operations, with a penalization when an operation violates a constraint, as in equation (10):

$$Fitness = \sum_{i=0} t_i + p_i \qquad (10)$$

where:

i represents each operation in the job-shop,

t_i is the starting time of the i operation,

p_i is the sum of quantities derived from penalizations for order and overlap violated constraints, in the way equations (11) and (12) show:

-if an order constraint is violated, the fitness must be severely penalized, so that this chromosome does not to pass to the next generation:

$$p_i = p_i + 100000000 \qquad (11)$$

-if overlap constraint is violated, the fitness is penalized proportionally to the amount of the overlap. :

$$p_i = p_i + |t_{f,j} - t_i| \qquad (12)$$

where $t_{f,j}$ is the finishing time of the j overlapped operation.

Notice that range constraint is not contemplated in the penalization equation because the time GA assigns the start-times within the range limits. Therefore the solutions provided by the time GA are always valid according to this constraint.

C. Heuristic algorithm

A relevant design issue is the organization of constraints in the heuristic stage. In a Constraint Satisfaction Problem (CSP) like this, a dilemma appears on the order of repairment of the constraints, claiming a further analysis. As the repairment of a constraint can modify the degree of satisfaction of the remaining constraints, the evaluation of the constraint of each operation must be followed by each repairment, so its start-time is updated. The algorithm below shows the workflow of the heuristic stage. It ends when it reaches a maximum number of iterations (MAX_IT). This parameter is tuned depending on the size of the orders, as explained in subsection 5.2.

```
Step 1: Point to 1st operation
Step 2: Get operation data
Step 3: Point to 1st constraint
Step 4: Heuristic evaluator
Step 5: Heuristic repairer
Step 6: If no more constraints
            then go to step 8
        otherwise go to step 7
Step 7: Point to next constraint
Step 8: If more operations
            then go to step 9
        otherwise go to step 10
Step 9: Point next operation
Step 10: Termination condition.
```

```
If iterations = MAX_IT
        then exit
    otherwise go to step 1
```

IV. Variants for the Overlap Constraint

As mentioned before, each constraint has one module to evaluate, and another one to repair. Whereas Range and Order heuristics are simple and described in [2], Overlap heuristics requires a deeper design: the evaluator is more complex than the other ones, and the repairer presents different variants.

Previously to running this repairer, a conflict appears about which of the overlapped operations has the priority to get repaired, which is not necessarily the operation appointed by the main algorithm. This is solved by the designation of the *critical operation*. The overlap repairer goal is to find an interval where the operation can be shifted while respecting the range constraint, so the critical operation must have the narrowest margin for start-time assignment (i.e. it is the most restrictive), as equation (13) says:

i is critical over j if:

$$|tmax_i - tmin_i| < |tmax_j - tmin_j| \qquad (13)$$

i, j are the overlapped operations
$tmax_i$ is the start-time upper limit for i operation
$tmin_i$ is the start-time lower limit for i operation

Each overlap repairer solves one overlap of a pair of operations, so if an overlap has more than two operations like equation (14) says, it will be solved in $k+1$ iterations of the repairer. At each iteration, there will be a different designation for the critical operation.

$$k + 2, k > 0 \, | \qquad (24)$$

Apart from these variables, there are others that participate in subsequent algorithms:

- O is the current operation of the algorithm defined in section 3. It is the operation that is being evaluated/repaired at each iteration of the main program.
- J is the operation that is being compared to the O at each evaluator/repairer iteration.
- C is the critical operation in an overlap.
- t_i is the start time of i operation.
- I is the current interval of the R. An interval is considered when there is a period of time when R is not assigned to any operation, so it remains not active.
- R_i is the resource assigned to i operation.
- T_R is the type of R resource.
- S is the resource currently appointed to.
- L is the list of operations that overlap with O.
- L_I is the list of I.
- L_R is the list of resources of the same type as R_o

The structure for the evaluator and the repairer variants are described in the following subsections.

A. Overlap Evaluator

The following algorithm includes the steps to evaluate if the current operation overlaps other one(s) on the same resource:

```
Step 1: Store (O, L)
Step 2: Point J at the 1st operation
assigned to Ro
Step 3: Stop condition:
        if no more operations for Ro
        then stop
        oterwise go to step 4.
Step 4: If J not = O, and J overlaps O
        then store (J, L)
Step 5: Point J at the next operation in
Ro
Step 6: Go to step 3.
```

Operations are overlapped if an operation begins before the other one has finished. The information that results from this stage is a list of operations that overlaps the current one. This list is the input of the overlap repairer stage.

B. Overlap Repairer

The overlap repairer includes several stages (i.e. Interval Search, OperationExchange, Resource Mutation), which are successively executed if the previous one has not been successful, as [2] show.

Other design issues come out when handling constraints that interfere with others. In this case, there are two possibilities:

1. To consider a blind repairment, so that the constraint is repaired without considering the other ones. Such is the case of the order and range repairers.
2. To consider an intelligent repairment, so that the constraint is repaired taking the other ones into consideration. Overlap repairer follows this approach. There are several ways to incorporate these considerations, producing two variants for overlap repairer: the first one (pure variant) considers the range constraint for its amendments; the second one (hybrid variant) considers both the range and the order constraints. The mentioned stages can be designed in both ways:

1) Algorithms for Pure Variants.

a) Algorithm for Interval Search

```
Step 1: Find LI for Ro
Step 2: Find C among two overlapped in L
Step 3: Position I at the beginning of
LI
Step 4: Stop condition:
        if no more intervals in LI
        then go to step 8.
Step 5: If I suitable for C
        Then tc = max (tminC, tminI)
        Exit
Step 6: Position I at next interval of
LI
Step 7: Go to step 4.
Step 8: Exit.
```

An interval is suitable if it matches the assignment conditions for the critical operation, in terms of operation duration and start-time range limits.

b) Algorithm for OperationExchange.

```
Step 1: Find C among two overlapped in L
Step 2: Position J in previous operation
in Rc
Step 3: Stop condition:
if no more previous operations,
then exit.
Step 4: If J suitable for C
        then exchange (tj, tC)
        exit.
Step 5: Position J in the next previous
operation in Rc
Step 6: Go to Step 3.
```

A current operation is suitable if its start-time fulfills the range constraint of the critical one.

2) Algorithms for Hybrid Variants.

a) Algorithm for Interval Search.

It remains the same as the PureVariant, except the suitability condition is step 5. In *this case, a*n interval is suitable if it matches the assignment conditions for the critical operation, in terms of operation duration and start time range bounds, and not belonging to the same job (to assure it fulfills the order constraint).

b) Algorithm for OperationExchange .

It *remains the same as the PureVariant, except the suitability condition is step 4. In this case,* an operation is suitable if it does not belong to the same job (to assure it fulfills the order constraint), and its start-times fulfills the range constraint of the critical.

c) Algorithm for Mutation Operator.

This operator assigns the operation to another resource of the same type, while preserving the start-time. This amendment does not interfere with the other type of constraints, but it can produce overlaps in the new resource.

```
Step 1: Find C among two overlapped in L
Step 2: Position S in 1st resource in
the job-shop
Step 3: Stop condition:
        if no more resources
        then go to step 7.
Step 4: If S not = RC and Ts = TRC
        then store (S, LR)
Step 5: Position S in next resource in
the job-shop
Step 6: Go to step 3.
Step 7: Random assignment of RC among
the candidates in LR.
```

V. EXPERIMENTATION RESULTS

Tests have been performed for the complete algorithm, putting special emphasis on the variants of the overlap repairers. The machine has been a Sun Sparc workstation running Solaris operating system. There has been a preliminary stage, to configure the GA, and a main stage, to validate the complete algorithm.

A. Tuning the GA Parameters

Beside the algorithmic issues, the success of the algorithm lies on several factors, like the correct tuning of the GA parameters. Several works have inserted in the code the way to tune them dynamically like the fuzzy logic controller (FLC), which methods are described in [11]. The key of success of applying FLC to GA is a well-formed fuzzy sets and rules [12]. In this work there has been previous experimentation to analyse the best values for the GA, by testing the different GA isolatedly. The most successful configuration for the parameter set population size/number of generations/mutation rate/selection type is 50/60/0.01/tournament for the resource GA and 8/10/0.01/elite for the time GA.

B. Configuration of the Hybrid GA

Testbeds have been configured varying the number of orders from 1 to 4, number of jobs from 1 to 3, number of products from1 to 4, number of product instances from 1 to 2, number of operations from 1 to 4, and operation processing times from 24 to 100, 5 resources belonging to 4 types, with the total number of executions per testbed of 25. The number of iterations for the heuristics stage has varied with the number of orders: for one order only 100 have been needed, while for four orders more than 200. Results collect the average of the executions.

Heuristic optimization algorithms can be evaluated in two ways [13]: by measuring the solution quality and measuring the solution time. In this case we have measured the solution quality by two criteria:

Considering this problem as a CSP, the solution quality must measure the constraint satisfaction rate. In this work, we consider the mean error (ME) parameter, as the percentage of constraints not satisfied. Figure 1 shows the results for the pure and hybrid variants of interval and exchange operators, distributed horizontally by the number of orders and vertically by the ME. This figure reflects that for few operations the pure repairer is better, but when the number of operations increases, the hybrid one is better. In this case, the ME is higher than 0, due to the technological limitations, i.e. more operations for the same number of resources produces more operations with unfulfilled constraints, and therefore reduces the number of fulfilled constraints. The reason for this improvement using the hybrid repairer is that the design of that heuristics has been made in such a way that the improvement in the overlap does not worsen any other constraint, in contrast with the pure repairer. The disadvantage of that is that fewer amendments can be applied with this variant, because it is more restrictive.

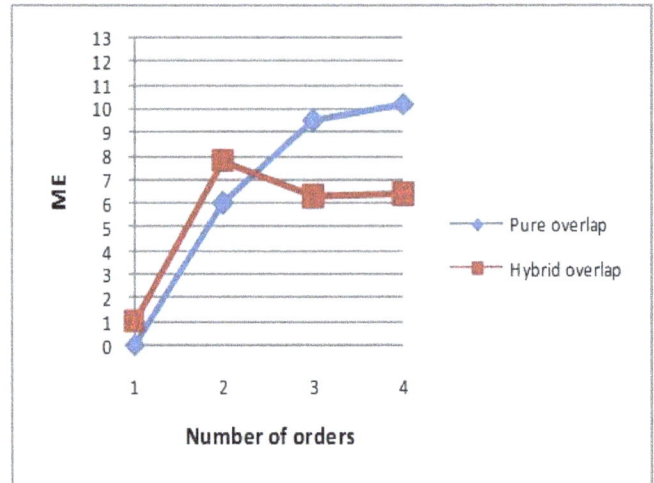

Fig. 1. ME of the two variants of overlap repairers

Considering it as a fJSP, the quality measurement is the time GA fitness. Table I shows the results for the time GA, as it is related to the constraints. PRf and HRf columns contain the Pure Repairer fitness and the Hybrid Repairer fitness respectively. Def(HRf, PRf) provides information about the percentage difference of both fitness values as equation 15 shows:

$$Def(HRf, PRf) = 100 * \frac{(HRf - PRf)}{PRf} \quad (15)$$

TABLE I
FITNESS VALUES FOR THE OVERLAP VARIANT

Number of orders	PRf	HRf	Def(HRf, PRf)
1	300	316	5.33%
2	352	379	7.67%
3	380	397	4.47%
4	411	419	1.95%

There is a relationship between the values for ME in Figure 1 and the fitness values shown in Table I. The fitness function is penalized when the range and overlap constraints are not fulfilled. The fewer the number of orders, the lower (and better) fitness results. Results are also better for the pure variant than the hybrid one. The reason is that the former reorder the overlapped operations trying to fulfill the range constraint, and the latter must also makes sure that the reorganization also fulfills the order constraint. This complexity means that the search interval does not always find the earliest interval suitable, and even does not find and interval, delaying more operations of the jobs than in the pure variant.

Besides that, the evolution of Def(HRf, PRf) is to decrease when the number of orders increases. This also shows that the fitness values in both repairers tend to be very similar for high number of orders. Therefore, it is recommendable to use the Hybrid Repairer in these cases, because they will provide similar fitness values than the Pure Repairer but with lower ME values.

VI. COMPARISON WITH ALTERNATIVE SOLUTIONS

To test the efficiency of our algorithm, Table II collects the comparison with respect the makespan using [8] benchmark. It contains the best results of a set of executions. It consists of ten problems mk1-mk10, with the number of jobs are in the range 10-20, the number of machines are in the range 6-15, number of operations are in the range 5-15. Other configuration information is: n x m, that refers to the number of jobs per number of machines; (LB, UB) with the optimum makespan if known [14]; otherwise, it reports the best lower and upper bound known; Flex. with the average number of equivalent machines per operation. This work compares the mentioned fJSP experiments of hGA from [5], AIA [6] and GA [7], and TWS for the best results achieved among the different rules in [8]. The information presented in Table 2 has been partially obtained from [2].

The proposed algorithm of GAH has achieved lower results of makespan for some fJSP instances and similar results of makespan for the remaining fJSP instances. These results combined with the ME results in section 5, demonstrate that the algorithm shows excellent quality solution as a fJSP and a CSP.

TABLE II
COMPARISON WITH BEST KNOWN MAKESPAN FOR TEN FJSP INSTANCES

Problem	n x m	Flex.	(LB, UB)	hGA	AIA	GA	TWS	GAH
Mk01	10 x 6	2.09	(36, 42)	40	40	40	42	40
Mk02	10 × 6	4.10	(24, 32)	26	26	26	32	26
Mk03	15 × 8	3.01	(204, 211)	204	204	204	211	204
Mk04	15 x 8	1.91	(48, 81)	60	60	60	81	60
Mk05	15 × 4	1.71	(168, 186)	172	173	173	186	172
Mk06	10 x 15	3.27	(33, 86)	58	63	63	86	57
Mk07	20 × 5	2.83	(133, 157)	139	140	139	157	139
Mk08	20 x 10	1.43	523	523	523	523	523	523
Mk09	20 × 10	2.53	(299, 369)	307	312	311	369	308
Mk10	20 × 15	2.98	(165, 296)	197	214	212	296	196

VII. CONCLUSIONS AND FUTURE WORK

This work has described the algorithms of a complex heuristic, like the overlap evaluator and repairers, in a hybrid GA applied to fJSP, a multi-objective problem. The most relevant issue concerns the use of two variants for the repairer: one that does not take into consideration the other constraints (pure), and the other one that incorporates them (hybrid). When adopting this approach, designers may consider what the experimentation has revealed: pure variant is better for fJSP with few operations, producing better ME results; in contrast, it is recommendable the use of the hybrid variant when the number of operations increases. It also shows that it maintains the level of quality of other algorithms, in terms of makespan. Finally, it is also recommendable an appropiate tuning of GA parameters.

The future work opens a high number of possibilities. Concerning the inclusion of intelligent operators, we are working in the design of hybrid variants for the range and precedence repairers. In the same way, we are making another variant of the ResourceMutation substage, which assures that the new resource assignment does not cause the overlap of other operations. Finally, new constraints adapted to concrete JSP and fJSP are to be incorporated and experimented. Re-design of the model is done using the FactoryMethod design patron, where a family of constraints can be chosen depending on the application that is used. The collection of classes in [2], will be transformed in the collection shown in Figure 2. The fJSP class is the superclass which the concrete application inherits from: in the described work, this application is GAH, which uses the order, range, and overlap concrete constraints. When using OtherApplication, it will use OtherConstraints (containing the measurer or evaluator), which has the corresponding OtherConstraint_unfulfilled subclass (containing the repairers for that constraint). The construction of the repairer will also contemplate the inclusion of pure and hybrid variants. The choice on which one to use will depend on the number of operations handled by the fJSP. The results of the mentioned modifications will be compared with the current version, to see how they affect to the ME and the *makespan*.

VIII. ACKNOWLEDGMENT

C. Gutiérrez thanks Jose Maria Lazaro for his continuous help on the applications of fJSP. She also wants to thank Juan Jose Merelo for his guidance in the customization of his GAGS. This research work has been funded by the Spanish Ministry for Economy and Competitiveness through the project "SOCIAL AMBIENT ASSISTING LIVING - METHODS (SociAAL)" (TIN2011-28335-C02-01).

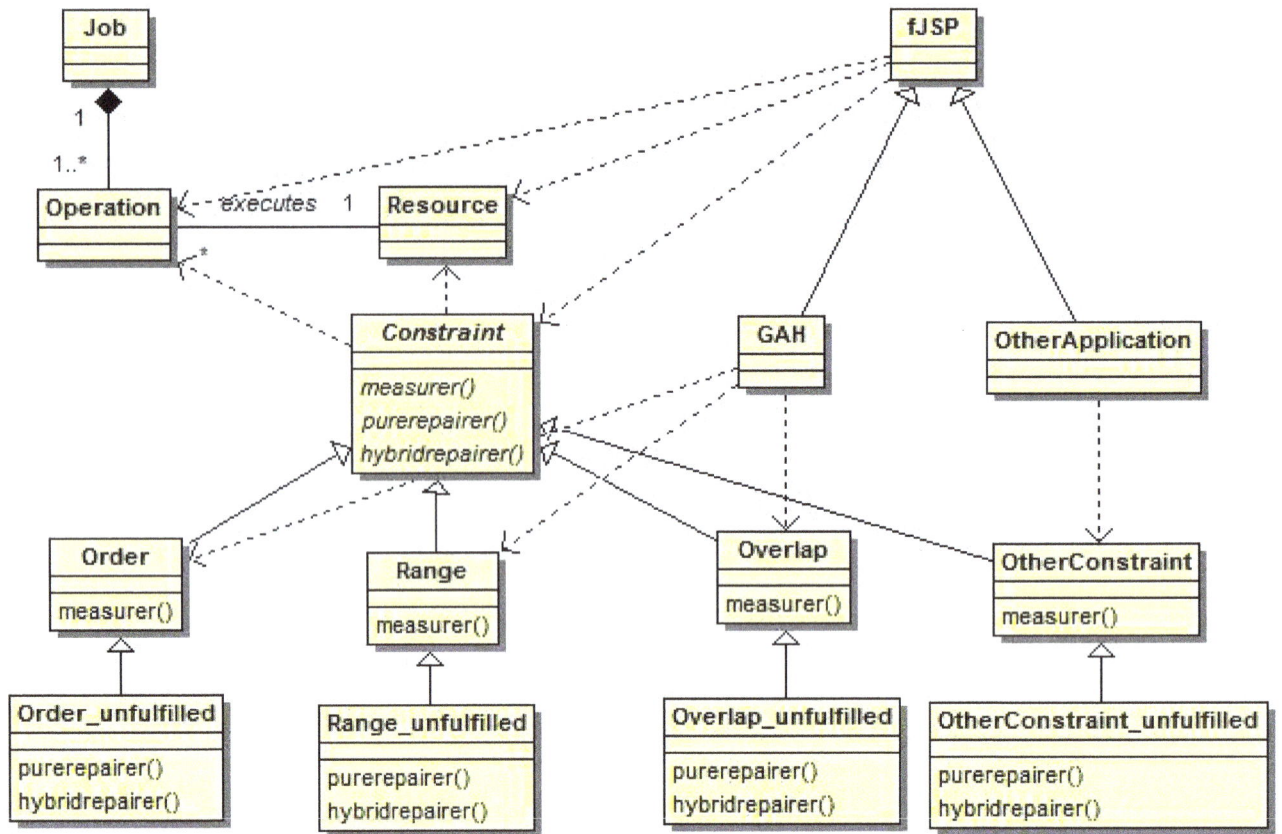

Fig. 2. Re-design of the classes for adaptation to other problems.

REFERENCES

[1] J.-J. Lin, "A GA-based Multi-Objective Decision Making for Optimal Vehicle Transportation," Journal of Information Science and Engineering, vol. 24, pp. 237-260, 2008.

[2] C. Gutierrez and I. García-Magariño, "Modular design of a hybrid genetic algorithm for a flexible job–shop," Knowledge-Based Systems, vol. 24, pp. 102–112, 2011.

[3] Y. Yun, M. Gen, M., and S. Seo, "Various hybrid methods based on genetic algorithm with fuzzy logic controller," Journal of Intelligent Manufacturing, vol. 14, pp. 401-419, 2003.

[4] J. Dorn and M. Girsch, "Genetic operators based on constraint repair," in ECAI'94 Workshop on Applied Genetic and other Evolutionary Algorithms.

[5] J. Gao, M. Gen, and L. Sun, "A hybrid of genetic algorithm and bottleneck shifting for flexible job shop scheduling problem," in 8th Annual Conference on Genetic and Evolutionary Computation, 2006, pp. 1157-1164.

[6] A. Bagheri, M. Zandieh, I. Mahdavi, and M. Yazdani, "An artificial immune algorithm for the flexible job-shop scheduling problem," Future Generation Computer Systems, vol. 26, pp. 533-541, 2010.

[7] F. Pezella, G. Morganti, and G. Ciaschetti, "A genetic algorithm for the Flexible Job-shop Scheduling Problem," Computers & Operations Research, vol. 35, pp. 3202-3212, 2008.

[8] P. Brandimarte, "Routing and scheduling in a flexible job shop by tabu search," Annals of Operations Research, vol. 41, pp. 157-183, 1993.

[9] Z. Pooranian, M. Shojafar, J. H. Abawajy, and M. Singhal, "GLOA: A New Job Scheduling Algorithm for Grid Computing," International Jorunal of Interactive Multimedia and Artificial Intelligence, vol. 2, no. 1, pp. 59-64, 2013.

[10] C. Gutierrez, "Heuristics in a General Scheduling Problem," in Proc. Conference on Evolutionary Computation, vol. 1, 2004, pp. 660-666.

[11] M. Gen and R. Cheng, Genetic Algorithms and Engineering Optimization. New Jersey: John Wiley and Sons, 2000.

[12] F. Cheong and R. Lai, "Constraining the optimization of a fuzzy logic controller using an enhanced genetic algorithm," IEEE Transactions on Systems, Man, and Cybernetics-Part B: Cybernetics, vol. 30, pp. 31-46, 2000.

[13] R. Rardin and R. Uzsoy, "Experimental Evaluation on Heuristic Optimization Algorithms: A Tutorial," Journal of Heuristics, vol. 7, pp. 261-304, 2001.

[14] M. Mastrolilli and L.-M. Gambardella, "Effective neighbourhood functions for the flexible job shop problem," Journal of Scheduling, vol. 3, pp. 3-20, 2000.

7

A Hybrid Algorithm for Recognizing the Position of Ezafe Constructions in Persian Texts

Samira Noferesti, Mehrnoush Shamsfard

Faculty of Electrical and Computer Engineering,

Shahid Beheshti University, Iran.

Abtract — In the Persian language, an Ezafe construction is a linking element which joins the head of a phrase to its modifiers. The Ezafe in its simplest form is pronounced as –e, but generally not indicated in writing. Determining the position of an Ezafe is advantageous for disambiguating the boundary of the syntactic phrases which is a fundamental task in most natural language processing applications. This paper introduces a framework for combining genetic algorithms with rule-based models that brings the advantages of both approaches and overcomes their problems. This framework was used for recognizing the position of Ezafe constructions in Persian written texts. At the first stage, the rule-based model was applied to tag some tokens of an input sentence. Then, in the second stage, the search capabilities of the genetic algorithm were used to assign the Ezafe tag to untagged tokens using the previously captured training information. The proposed framework was evaluated on Peykareh corpus and it achieved 95.26 percent accuracy. Test results show that this proposed approach outperformed other approaches for recognizing the position of Ezafe constructions.

Keywords — Ezafe construction, genetic algorithm, genitive construction, rule-based model.

I. INTRODUCTION

THE "Ezafe"[1] is a Persian language grammatical construct which links two words together. Ezafe means "addition" and is an unstressed vowel –e– which marks genitive cases. The constructs linked by the Ezafe particle are known as "Ezafe constructions". Some common uses of the Persian Ezafe are [1]:

— a noun before an adjective:

e.g. توپ قرمز[2] (tu:p-Ezafe Germez) "red ball"

— a noun before a possessor:

e.g. کتاب علی (ketä:b-Ezafe Ali:) "Ali's book"

— some prepositions before nouns:

e.g. زیر میز (zi:r-Ezafe mi:z) "under the table"

The Ezafe in its simplest form is pronounced as –e, but generally not indicated in writing. In some cases the Ezafe has an explicit sign in writing. For example, with nouns ending in ا (ä:) or و (ou), the Ezafe appears as an ی (j) at the end; with nouns ending in a silent ه (h) (short e followed by a mute h), the Ezafe may appear as a superscript ء (hamze) or a ی (j).

The Ezafe is also found in Urdo [2], Kurdish [3] and Turkish [4]. The Persian Ezafe has been discussed extensively [5]-[8]. This construction raises several issues in syntax and morphology. There are three issues on the function of the Ezafe in the literature: (1) the Ezafe is a case marker [9], (2) the Ezafe is inserted at PF to identify constituenthood [10], and (3) the Ezafe is a phrasal affix [11].

Determining the position of an Ezafe construct may facilitate text processing activities in natural language processing (NLP) applications, such as segmenting a phrase or detecting the head word of a phrase [12]. Moreover, recognizing words which need an Ezafe is advantageous for tokenization [13], morphological analysis, and syntax parsing [14], and it is essential for speech synthesis [15].

Some NLP tasks in Persian, such as machine translation [16], construction of morphological lexicons [14], and grammar construction [17], have benefited from the availability of an Ezafe construction. However, they have determined the position of the Ezafe manually, exploited cases in which the Ezafe is visually represented, or extracted some insertion rules which are not general; therefore, they could not determine the Ezafe tags for all tokens in a text.

The Persian Ezafe has been discussed extensively in theory [5], [18], but there are few works on the automatic detection of this construction in Persian texts. The most completely reported works on this subject are one work based on probabilistic context free grammar (PCFG) [19] and another based on classification and regression tree (CART) [20]. The former uses a bank including trees of noun groups in Persian for training PCFG. Then, a bottom-up parser extracts the most probable noun groups of the input. Finally, using lexical analysis, the system determines which words need an Ezafe. The disadvantage of this method is that writing a PCFG requires a large amount of linguistic knowledge. In addition, it is not sensitive to lexical information. The latter uses morpho-syntactic features of words to train and construct binary classification trees to predict the presence or absence of an

---

[1] It is also known as Kasreh.
[2] For each Persian word or phrase we wrote its transliteration within parenthesis and its English meaning within double quotes. International Phonetic Alphabet (IPA) was used to represent Persian language pronunciations.

Ezafe between two adjacent words. In fact, there are two kinds of rules: rules which predict Ezafe words, and rules which predict non-Ezafe words. Although this method can predict the absence of an Ezafe with high accuracy, it is not sufficient in detecting words which need an Ezafe. In other words, the rules which predict the non-Ezafe words act more precisely.

The main contributions of this paper were (1) introducing a framework for combining genetic algorithms with rule-based models, and (2) using the proposed framework to develop an Ezafe tagger.

Combining genetic algorithms with rule-based models brings the advantages of both approaches and overcomes their problems. Genetic algorithms can detect general patterns in text, but sometimes they cannot handle exceptions and special cases. In such cases, rule-based models can provide significant improvements by defining rules for handling special cases and exceptions. In our proposed framework, for the rule-based model, linguistic rules were extended by analyzing errors of the genetic algorithm and defining new rules for handling these errors (named as correction rules). In contrast, a rule-based model needs a great deal of knowledge external to the corpus that only linguistic experts can generate. In fact, acquiring rules through interviews with experts is cumbersome and time-consuming. Furthermore, certain application domains are very complicated and may require a large number of rules. Therefore, the acquired rules may be incomplete or even partially correct. In order to overcome these problems, we can handle general patterns by genetic algorithms and only define correction rules to handle special cases, which means less time handling by expert humans. We can also define a set of general rules besides correction rules in order to reduce the run time of the genetic algorithm.

There is a remarkable amount of ongoing research on applying machine learning approaches to different tasks of NLP in the English language. Most machine learning approaches such as those methods based on hidden Markov models, use information extracted from a tagged corpus to assign a suitable tag to each word according to preceding tags. Since these approaches are purely statistical, as such they are most suitable for cases that have a corpus large enough to contain all possible combinations of n-grams. In contrast, evolutionary algorithms offer a more generalized method that can be applied to any statistical model. For example, they can be applied to perform tagging operations according to the Markov model (tag prediction for a current word based on preceding tags) or improve the Markov results by using more contextual information (for example, using tags of preceding words or those of following words). In other words, HMM or other models can be used as part of the fitness function in a genetic algorithm. Therefore, a genetic algorithm provides more flexibility than any of the other classical approaches such as HMM based methods. On the other hand, the effectiveness of using hybrid approaches has been demonstrated in different NLP tasks [21]. Thus a hybrid approach for determining the position of Ezafe construction was chosen for this study.

Results of the tests in this study show that our proposed algorithm outperformed other algorithms for Persian Ezafe tagging as well as the classical MM based method.

The rest of the paper is organized as follows. Section 2 introduces the annotated corpus of Persian texts. Section 3 explains our proposed model. Experiment results are discussed in section 4. Finally, section 5 concludes the paper.

II. THE CORPUS

An annotated corpus of Persian text is needed in order to train and evaluate the Ezafe tagger. This corpus must be annotated with POS and Ezafe tags. For the current work, a subset of Persian POS tagged corpora known as Peykareh[3] [21] was used. This collection was gathered from daily news and common texts and contained about 2.6 million, manually tagged tokens. The main corpus was tagged with a rich set of POS tags consisting of 550 different tags from which 20 tags were selected for the system. Those that could be detected by the present Persian POS taggers were selected for use in the system applied to this study.

The tagged corpus was divided into three sets: (1) a training set including 423,721 tokens, (2) a held-out data set containing 1,010,375 tokens, and (3) a test set containing 39,850 tokens.

A big portion of the Peykareh corpus was set aside as a held-out dataset. The held-out dataset was used to find exceptions to general rules in the rule-based model. Since the exceptions occur only rarely, much more data was needed to determine the exceptions. Furthermore, to determine the classes of conjunctions and prepositions which never take an Ezafe or always require an Ezafe, the held-out data set was searched. Thus, a sufficiently large data set was needed to find as many words as possible.

III.   THE PROPOSED ALGORITHM

This paper proposes a hybrid approach to determine the position of Ezafe constructions in Persian texts. The Ezafe tagger contained two phases. The first phase used the rule-based model to tag as many words as possible. Then the second phase ran the genetic algorithm to assign tags to the tokens which had not been assigned an Ezafe tag in the previous phase. Therefore, a faster genetic algorithm was achieved by producing more tagged tokens that had been generated from the rule-based model.

The Ezafe tagger assigned each word of an input sentence with one of two tags: *true* or *false*. Tag *true* for a word meant that it requires an Ezafe, and the tag *false* meant it does not require one.

A. The rule-based model

Initially, some general rules such as "verbs do not take an Ezafe" were defined using linguistic knowledge. Although the genetic algorithm could detect these tags correctly by training on annotated examples, we preferred to define such rules in

---

[3] This corpus also is known by its author's name, Bijankhan.

order to reduce the run time of the algorithm. The more tokens detected by the rule-based model there were, the less chromosome length and lower number of generations were needed.

Next, the exceptions of each rule were explored on the held-out data, and some new rules were defined to handle exceptions. This process was repeated. In other words, if the generated rules had exceptions, new rules were defined. In some cases we could not find suitable rules to fix errors. In these cases, probabilistic rules were used.

The genetic algorithm tagged tokens according to the context; however, experiments showed that words which appeared in an infrequent context usually took an incorrect tag from the genetic algorithm. We tried to handle these cases through hand-crafted rules. Thus, the initial rule-based model and the genetic algorithm were run on the held-out data, and errors were analyzed to introduce rules that would fix them.

In this way, a set of 53 hand-crafted rules was developed. Then, the most suitable sequence of rules was determined in terms of avoiding bleeding and creeping; in fact, a tree was constructed. The first level of the tree contained some general rules. Level 2, consisted of some rules for handling exceptions of the first level and so it continued. However, each node in level *i* handled an exception of the rule of its parent node in level *i-1* (if that rule had exceptions).

At the first stage of the proposed algorithm, each rule was taken individually from the rule-set one at a time and the function was performed only if the rule was applicable to the input word.

Rules were categorized according to various dimensions:

— Deterministic vs. probabilistic rules: Deterministic rules are those which are always valid and correct; probabilistic rules may have exceptions. In other words, probabilistic rules are valid most of the time, but as they may have exceptions we apply them with a probability of less than 100%. This probability is extracted from the corpus.

— Negative vs. positive rules: Negative rules find and tag negative examples which are the structures which never take an Ezafe, while the positive rules determine structures which need an Ezafe.

— Syntactic, morphological and lexical rules: Syntactic rules use part-of-speech to tag words (either as the target word or a neighboring word) in a sequence to determine the Ezafe tag, while morphological rules consider internal and morphological structures of a word to do this task, and lexical rules consider real words.

In the rest of this section these categories are discussed in more detail some examples are given from each category.

1) Syntactic Rules
Some POS categories enforced a special tag on words or on neighboring words. The accusative case marker ‌را (rã:) and verbs were among this set.

— Verbs

o In the Persian language the Ezafe is not used with verbs. The following rule dictated that verbs, which were shown in the corpus with the POS tag V, never take an Ezafe.

If POS(X) =V Then EZ-Tag(X) =false

o If a verb appears as a stand-alone, the word before it does not take an Ezafe. We presented this by the following rule:

If POS(X) =V Then EZ-Tag(X-1) =false

o If the verb is not a stand-alone and appears as an attachment (enclitic) to another word, then the previous word (before the combination) may take an Ezafe. This is the case for some of the enclitics representing the copula verb 'to be' such as ی (i:) "to be- single second person" and م٘ (æm) "to be- single first person". These enclitics are ambiguous and, in addition to copula verbs, can be interpreted as an indefinite marker or as a single first-person possessive pronoun, respectively. For example, the word شاعری (ʃã:?eri:) may mean شاعر هستی (ʃã:?er hæsti:) "you are poet" or یک شاعر (jek ʃã:?er) "a poet". In the first case, even though the whole word was tagged as a verb in the corpus, it is actually a combination of a noun and a verb. Even though its verb part and its previous word do not take an Ezafe, the word before the noun part of it may take one. As another example, in the following sentence the word دولتم (dolætæm) "I'm government" is an abbreviation of دولت هستم (dolæt hæstæm) "I am government". In Peykareh corpus, this word was tagged as verb with POS tag V,AJCC. However, the previous word takes an Ezafe.

من در استخدام دولتم.

I am a government employee.

Thus, we used POS tag V, ACJJ for this kind of verbs to prevent applying the previous rule for them.

— The accusative case marker ‌را (rã:)

o The Persian language has an accusative case marker ‌را (rã:) that follows the direct object, adverb or prepositional object. The following rule dictated that the accusative case marker, which was shown in the corpus with the POS tag POSTP, never takes an Ezafe.

If POS(X) =POSTP Then EZ-Tag(X)=false

o The word before ‌را (rã:) does not take an Ezafe too.

If POS(X) =POSTP Then EZ-Tag(X-1)=false

o In some cases the accusative case marker is attached to the previous noun or pronoun. For example the word مرا[4] (mærã:) is an abbreviation of من را (mæn rã:), in which ‌را (rã:) is an object marker and من (mæn) "me" is a pronoun. This rule was written as follows:

If postfix(X) = accusative-case-marker EZ_Tag(X) =false

2) Morphological rules
The following rules are examples of morphological rules that determine structures that take an Ezafe.

—When a word ending in the plural suffix ها (hã:) needs the

---

[4] Sometimes it means 'my' and other times it means 'me'

Ezafe, the letter ی (j) must be attached to the end of the word in writing. Thus, if a plural word ends in ها (hã:), this word should not be followed by an Ezafe unless it is followed by a ی (j) clitic.

If postfix(X)= ها Then EZ-Tag(X)=false

Consider the following example.

نامه های علی را خواندم.

I read Ali's letters.

The word نامه ها (nã:me hã:) "letters" is the plural form of نامه (nã:me) "letter". When this word requires Ezafe, we add ی (j) at the end of it.

—If the last character of a word is اً (Tanvin)[5], then it does not take an Ezafe.

If LastChar(X) = اً Then EZ-Tag(X)=false

3) Lexical rules

Lexical rules consider real form of words as shown in the following examples:

—Prepositions

Reference [23] showed that the class of prepositions in the Persian language is not uniform with respect to the Ezafe. Some prepositions reject the Ezafe (These prepositions were called Class P1.), while others either permit or require it. We divided the latter group into two classes. The first class which always requires an Ezafe was called Class P2. The other class which permits an Ezafe but does not necessarily require one was called Class P3. Table I shows some examples of each class. We applied the following rules to handle prepositions:

If POS(X) =P and WORD(X)∈ClassP1 Then EZ-Tag(X)=false
If POS(X) =P and WORD(X)∈ClassP2 Then EZ-Tag(X)=true

TABLE I
EXAMPLES OF PREPOSITION CLASSES

Class name	Examples
Class P1	به (be) "to" از (æz) "from" با (bã:) "with" در (dær) " in, on"
Class P2	وسط (væsæt) "in the middle" دور (du:r) "around" بیرون (bi:ru:n) "outside" داخل (dã:xel) "inside"
Class P3	زیر (zi:r) "under" رو (ru:) "on" بالا (bã:lã:) "up" جلو (dӡoulou) "in front of"

— Conjunctions

Same as prepositions, we divided conjunctions into two classes. Some conjunctions never take an Ezafe (These conjunctions were called Class C1), while others always take an Ezafe (These conjunctions were called Class C2). In order to determine these classes we searched 300 files of Peykareh corpus which were selected as held-out data. Some examples of Class C1 and Class C2 are presented in Table II. The

following rules applied to conjunctions:

If POS(X) =CONJ and WORD(X)∈ClassC1 Then EZ-Tag(X)=false
If POS(X) =CONJ and WORD(X)∈ClassC2 Then EZ-Tag(X)=true

TABLE II
EXAMPLES OF CONJUNCTION CLASSES

Class name	Examples
Class C1	و (væ) "and" زیرا (zi:rã:) "because" یا (jã:) "or" که (ke) "that" یعنی (jæni:) "means"
Class C2	علی رغم (?ælã:ræγ_me) "in spite of" باستثناء (beestesnã:?e) "except" سوای (sævã:je) "except" برخلاف (bærxælã:fe) "in spite of"

—Adverbs

We also divided Persian adverbs into three classes. Class A1 contained adverbs which never take an Ezafe; class A2 included adverbs with an obligatory Ezafe; class A3 contained adverbs with an optional Ezafe. Examples of these classes are shown in Table III. The following rules applied to adverbs:

If POS(X) =ADV and WORD(X)∈ClassA1 Then EZ-Tag(X)=false
If POS(X) =ADV and WORD(X) ∈ClassA2 Then EZ-Tag(X)=true

TABLE III
EXAMPLES OF ADVERB CLASSES

Class name	Examples
Class A1	بویژه (bevi:ӡe) "specially" هیچگاه (hi:tʃgã:h) "never" شاید (ʃã:jæd) "maybe"
Class A2	مثل (mesle) "like" مانند (mã:nænde) "like" از قبیل (æz Gæbi:le) "such as"
Class A3	گذشته (gozæʃte) "past" سالیانه (sã:li:jã:neh) "annual"

4) Probabilistic rules

We also defined 5 probabilistic rules which were correct and valid in most cases but had some exceptions in a few cases. Defining each rule, the probability of that rule was calculated according to the corpus. The lowest probability among these rules was 0.95. Here, we discuss some of the probabilistic rules.

—Long vowels

There are three long vowels in Persian: ا (ã:), ی (i:) and و (u:). Generally, when a word ending in ا (ã:) or و (u:) needs an Ezafe, the letter ی (j) is added to the end of it. However, this rule has some exceptions.

In the case of ا (ã:), these exceptions happen when we replace أ (Alef Hamze) by ا (ã:) (single alef). Alef Hamze is a single Arabic character that represents the two-character

---

[5] This sign was taken from Arabic alphabet

combination of Alef plus Hamze and in Persian writing is sometimes replaced by the letter ‍ا (ã:). Consider the following example:

آقای احمدی منشا فساد را فقر می داند.

"Mr. Ahamdi believes that the source of evil is poverty."

The word آقا (ã:Gã:) "Mr." takes a ی (j) at the end because it requires the Ezafe; however, the word منشا (mænʃæ) "source" also ends in ا (ã:) and needs the Ezafe, but it does not get ی (j). In fact, the last character of this word is أ (Alef Hamze) which is written the same as ا (ã:).

To compute the probability of the rule, the algorithm searched the held-out data set and computed the percentage of words ending with ا (ã:) and the Ezafe which had the letter ی (j) added. In other words, this probability was computed by the following formula:

$$p = \frac{count(\text{words ending with } \text{اى} \ (\text{ã:je}) \text{ and } True \text{ Tag})}{count(\text{words ending with } \text{اى} \ (\text{ã:je}))} \tag{1}$$

Thus, the following rule was defined with a 95% probability:

If LastChars(X) = اى Then with a 0.95 probability EZ-Tag(X) =true

In the same way, the following rules were defined:

If LastChars(X) = ا Then with a 0.9978 probability EZ-Tag(X) =false
If LastChars(X) = وی Then with a 0.96 probability EZ-Tag(X) =true

—Tanvin

Tanvin is a sign which is derived from Arabic. The following rule says that with a probability of 96.24% the word preceding a word that has أ (Tanvin) as the final character does not take an Ezafe:

If LastChars(X) = أ Then with a 0.9624 probability EZ-Tag(X-1) =false

Frequency counts for the rule-categories are shown in Table IV.

TABLE IV
NUMBER OF EXTRACTED RULES IN EACH CATEGORY

Syntactic	Morphological	Lexical	Probabilistic
25	4	19	5

After running the rule-based model, some of the tokens remained untagged. Thus, a genetic based algorithm was used to tag the remaining words.

B. Genetic tagging algorithm

The proposed genetic algorithm receives a natural language sentence and assigns a corresponding tag according to previously computed training information from the annotated corpus. Formally, given a sequence of n words and corresponding POS tags, the aim is to find the most probable Ezafe tag sequence.

In our implementation, each gene can take values: *true* or *false*. Individuals of the first generation were produced randomly. After producing an individual, all tokens of a given sentence were assigned Ezafe tags (some of tokens get Ezafe

tag by the rule-based model and others get Ezafe tag by the genetic algorithm).

An initial population was created randomly by assigning a random value to each untagged gene (some genes were assigned Ezafe tags from the rule-based model). These individuals were sorted according to fitness value of individuals from high to low.

Three genetic operations were used for producing the next generation.

—Selection: All individuals in the population are sorted according to fitness, so the first individual was the best fit in the generation. To perform crossover, the ith and $(i+1)$th individuals of the current generation were selected, where $i=1,2,...,[(p+1)/2]$ and p was the population size. The aim of selection was to choose the fitter individuals.

— Crossover: Selected two chromosomes, crossover exchanges portioned of a pair of chromosomes at a randomly chosen point called the crossover point.

— Mutation: Selected an untagged gene randomly and toggled its value, for example if its value was *true,* it was reset to *false* and vice versa.

1) Fitness Functions

To evaluate the quality of Ezafe tags generated for an individual, four functions were used; F1, F2, F3 and F4. These functions considered the context in which a word appeared. Context consisted of a current word, one tag to the left and another to the right and the previous and next word.

F1 considered the sequence of POS tags of a sentence. The probability of the sequence of POS tags of a sequence of n words was as follows:

$$F1 = \sum_{i=1}^{n} P(EZ - POS_i | POS_{i+1}) \tag{2}$$

Where, $P(EZ - POS_i | POS_{i+1})$ represents the probability that the current word with POS$_i$ tag gets the Ezafe when the next word has the POS$_{i+1}$ tag. The probability of assigning the Ezafe to a word given the next POS tag was computed as:

$$P(EZ - POS_i | POS_{i+1}) = \frac{count(EZ - POS_i, POS_{i+1})}{count(POS_i, POS_{i+1})} \tag{3}$$

Where, count(POS_i, POS_{i+1}) was the number of occurrences of the (POS_i, POS_{i+1}) sequence within the training corpus, and $count(EZ - POS_i, POS_{i+1})$ was the number of (POS_i, POS_{i+1}) occurrences when the first token has an Ezafe within the same corpus. In order to compute F1 function, the HMM model can be used with the Viterbi algorithm [24].

For computing F2 function, a data driven approach was applied to calculate the probability that a specific word has the Ezafe.

$$F2 = \sum_{i=1}^{n} P(EZ_i | word_i) \tag{4}$$

F2 was defined because some words in Persian are mostly assigned a special tag. For example, the word أتــقــریــب (tæGri:bæn) "approximately" never take an Ezafe.

F3 function was the probability that a token gets an Ezafe when it occurs before a specific word in the training corpus.

$$F3 = \sum_{i=1}^{n} P(EZ_i | word_i , word_{i+1}) \qquad (5)$$

This function was defined to handle compound words such as اخــتلاف نـظر (extelã:f-Ezafe næzær) "difference in opinion".

In Persian, some words such as ســایــر (sã:jer) "other" get the Ezafe most of the time. Therefore, we defined the F4 function to consider these words. The F4 function was the probability that a specific word occurs after a word with the Ezafe in the training corpus.

$$F4 = \sum_{i=1}^{n} P(word_{i+1} | EZ_i) \qquad (6)$$

The following fitness function was used to evaluate the genetic algorithm:

$$fitness - function = \sum_{i=1}^{4} w_i . F_i \qquad (7)$$

Where $w_i s$ are constant parameters chosen from [0,1) and show relative importance of syntactic and lexical information. It was assumed that 0 is a legal value to show the effect of removing one or more functions from the formula. To adjust w_i parameters in the fitness function formula, variable structure learning automata were applied on chunked held-out data. For more information you can see [25]. Finally, the values of w_1, w_2, w_3 and w_4 were set to 0.8, 0.5, 0.1 and 0.1 respectively.

IV. EXPERIMENTS

The proposed algorithm was implemented using java language and was run on a Pentium IV processor. First, the rule-based model was run followed by the genetic algorithm, and the best solution was selected. Approximately 78% of the tokens were tagged with the rule-based model, because about 80% of the tokens selected as test data did not require an Ezafe, and most of them were tagged by the rule-based system.

To evaluate the performance of our proposed algorithm, three measures were taken: accuracy (the percentage of correctly tagged tokens), precision (the percentage of predicted tags that were correct) and recall (the percentage of predictable tags that were found).

Since performance was related to both precision and recall, the F-measure was given as the final evaluation.

$$F - measure = \frac{2.precision.recall}{precision+recall} \qquad (8)$$

A. Tuning Parameters of the Genetic Algorithm

The efficiency of a genetic algorithm greatly depends on how its parameters are tuned. To adjust the genetic parameters, a subset of 34,832 tokens from held-out data set was selected. Then, the proposed algorithm was run on this set.

Beginning with a baseline configuration, such as Dejong's setting [26] with 1000 generations, 50 chromosomes in each generation and 0.6 for crossover probability, the algorithm was run for different mutation probabilities (P_m) from 0.01 to 0.3. Fig. 1 shows that the best results were obtained using the mutation probability 0.05.

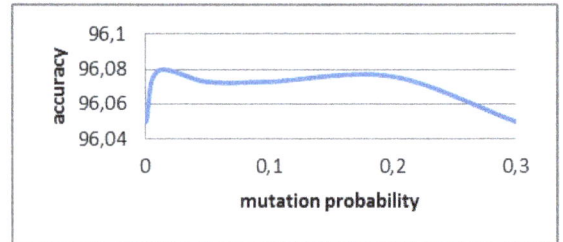

Fig. 1. Average fitness values of executing of the GA using different mutation probabilities

In the same way, crossover probability was set to 0.6. In Fig. 2 the results of running the genetic algorithm using mutation probability 0.05, crossover probability 0.6, population size 50 and different number of generations are shown.

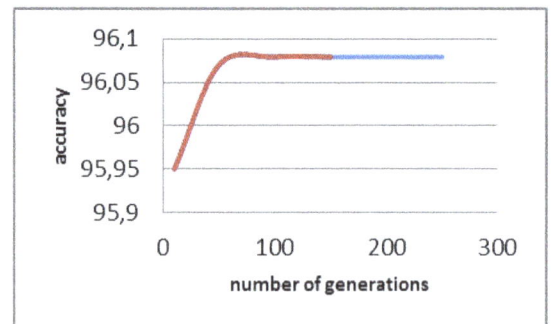

Fig. 2. Average fitness values of executions of the GA using different number of generations

Table V shows the optimum values of genetic algorithm parameters.

TABLE V
SELECTED VALUES FOR GENETIC ALGORITHM PARAMETERS

P_m	P_c	population size	generations
0.05	0.6	50	150

B. Effectiveness of the Proposed Algorithm

The experiment applied 423,721 annotated tokens as the training set and 39,850 tokens as the test set. Parameter settings shown in Table V were used for the genetic algorithm.

Table VI compares our approach with a baseline method and other available methods based on PCFG [19] and CART [20]. We also implemented the binary Markov model with Viterbi decoding (a typical algorithm widely used for stochastic tagging). As can be seen, our proposed algorithm outperformed these algorithms in terms of F-measure. The baseline assumed all words have an Ezafe, resulting in 100% recall but very low precision (15.79%). We could define another baseline where no word has the Ezafe tag. In this case, we would achieve 84.21% precision, but the recall would be 0.

TABLE VI
EVALUATING OUR PROPOSED ALGORITHM IN TERMS OF F-MEASURE

	recall	precision	F-measure
Our proposed algorithm	88.81	87.85	88.33
Baseline	100	15.79	27.27
PCFG method [19]	86.74	87.54	87.14
CART method [20]	88.85	84.13	86.43
Viterbi method	95.51	78.63	86.25

Since we had no access to the corpus that was used for training in the CART method [20] or a description of the exact features used, we could not regenerate the exact results. For this reason, we used two approaches for comparison. In the first approach, we compared results of our proposed method with the best results reported in [20], and in the second one, we implemented the CART method using the same features as our proposed method and tested it on our test corpus. Since the performance of the second approach was much lower than what was reported in [20], we only presented the results of the first approach in Table VI.

In the above-mentioned experiments, correct POS tags were used, because results from the proposed algorithm were compared to those from other available Ezafe taggers. Since these taggers had used correct tags, we also used the correct tags to enable the comparison. By using a Tnt tagger, the proposed algorithm achieved a 95.08% accuracy, while with correct tags it achieved an accuracy of 95.49%. This indicates that the tagging error decreased the Ezafe detection accuracy by only about 0.41%. The reason for this is that both the rule-based model and the genetic algorithm consider other features besides POS tags, and these features can, to some extent, cover the errors of the POS tagger.

Considering accuracy as the percentage of correctly assigned tags, we evaluated the performance of the proposed algorithm from two different aspects: (1) the overall accuracy by taking all tokens in the test corpus into account, and (2) the accuracy for words with an Ezafe and without an Ezafe, respectively. Table VII shows that the overall accuracy of the proposed algorithm was around 95.26%. Additionally, the accuracy for detecting words without an Ezafe was significantly higher than that for words with an Ezafe (96.89% versus 88.81%).

TABLE VII
EVALUATING OUR PROPOSED ALGORITHM IN TERMS OF ACCURACY

	Number of correctly tagged tokens		
	with Ezafe	without Ezafe	Total
Corpus	8054	31796	39850
Our proposed Algorithm	7153	30807	37960
Accuracy	88.81	96.89	95.26

Table VIII compares overall accuracy from the combination

of the rule-based model and the genetic algorithm. Approximately 78% of tokens were tagged by the rule-based model with 99.21% accuracy. In fact, from tokens in the test set, 30,972 tokens were tagged by the rule-based model and among them 30,728 tokens were assigned correct tags. In contrast, the genetic algorithm assigned correct tags to 7,232 tokens from 8,878 tokens and achieved 81.46% accuracy.

TABLE VIII
COMPARING THE ACCURACY OF THE RULE-BASED MODEL VERSUS GENETIC ALGORITHM

	Number of tagged tokens	Number of correctly tagged tokens	Accuracy
Rule-based model	30972	30728	99.21
Genetic algorithm	8878	7232	81.46

Table IX compares the accuracy of the rule-based model versus the genetic algorithm. In RBM1, we ran the rule-based model and assigned the false tag to tokens which did not get the Ezafe tag after applying the rules. In contrast, the untagged tokens got true tags in RBM2. We also ran the genetic algorithm alone (without the rule-based model). Results show that the combination of the rule-based model and the genetic algorithm outperformed both individual algorithms. As might be expected, the main problem of the RBM models was missing rules, which caused some tokens remained untagged, and the main problem of the genetic algorithm was special cases that could not be handled by general patterns.

TABLE IX
COMPARING THE ACCURACY OF THE RULE-BASED MODEL VERSUS GENETIC ALGORITHM

	Accuracy
RBM1	85.29
RBM2	91.21
GA	89.21
Combination of rule-based and GA	95.26

Since the ratio of words with an Ezafe to words without an Ezafe was low, the Kappa coefficient was used to evaluate the proposed algorithm. This measure was first suggested for linguistic classification tasks [27] and has since been used to avoid dependency of the score on the proportion of non-breaks in the text. The Kappa coefficient (K) was calculated as:

$$K = \frac{Pr(A) - Pr(E)}{1 - Pr(E)} \qquad (9)$$

Where, Pr(A) was accuracy, and Pr(E) was the ratio of words without an Ezafe to total words. Table X shows how to evaluate an algorithm in terms of Kappa value. Using (9) the Kappa coefficient became 0.77. According to Table X, our proposed algorithm is assessed as good.

TABLE X
DECISION MAKING BY USING KAPPA [19]

Kappa values	Strength of agreement
K<0.2	bad
0.2<K≤0.4	average
0.4<K≤0.6	relatively good

| 0.6<K≤0.8 | good |
| 0.8<K≤1 | very good |

Table XI shows that our proposed algorithm outperformed previously reported algorithms in terms of Kappa value.

TABLE XI
COMPARING THE PERFORMANCE OF THE PROPOSED ALGORITHM WITH OTHER METHODS

Kappa value		
Our proposed algorithm	PCFG method	CART method
0.77	0.74	0.72

In the final experiment, we assessed the impact of training corpus size on the performance of the proposed algorithm. The corpus size was reduced slightly until it reached 32% of the initial training corpus size. The results are presented in Fig. 3. As can be seen, the proposed algorithm's accuracy did not show a significant drop when reducing the training corpus size from 100% to 60%.

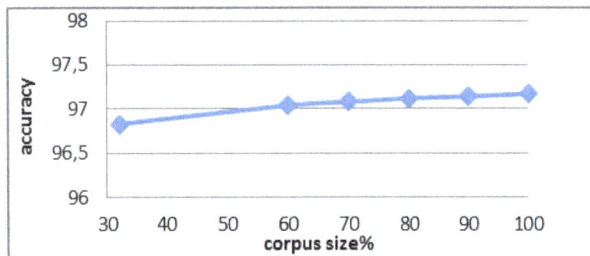

Fig. 3. The impact of training corpus size on performance

V. CONCLUSION AND FUTURE WORK

This paper proposes a framework for recognizing the position of Ezafe constructions in Persian written texts that combines genetic algorithms with rule-based models. Genetic algorithms provide a search strategy to learn general Ezafe patterns in text optimizing a measure of probability that is effective globally. However, the rule-based model handles special cases and exceptions to general patterns. Results of the tests reported in this study show that the proposed algorithm outperformed other algorithms for Persian text Ezafe tagging and classical HMM based methods.

Although this paper presents an algorithm for Persian Ezafe tagging, the principles can be applied to other NLP tasks such as POS tagging or chunking in any language. A genetic algorithm can be used for any language to find common statistical patterns for tagging. Obviously, there may be exceptions to these patterns, so some rules are defined to handle exceptions in the rule-based model that serve to improve performance of the genetic algorithm. In fact, combining a genetic algorithm with the rule-based model improves performance of the tagging process.

In addition, we showed that the accuracy of the proposed algorithm does not depend highly on the training corpus size. This feature is advantageous for practical applications, because annotating training corpora for text analysis purposes

is an extremely demanding task.

In future work, linguistic rules may be extended by analyzing errors of test data. It is also observed that input of the Ezafe-tag set has a major influence on accuracy. Errors in the training data have caused some problems, and these can be reduced by correcting the training data.

In addition, it is intended that new attributes be added to the fitness function of the genetic algorithm. One advantage of the genetic algorithm compared to other classical approaches such as HMM based methods is that new attributes can be added to the system and this facilitates examination of the effect of different attributes on tagging without altering the system's basic structure. Thus, tests will be done on new attributes applied to the fitness function of the genetic algorithm and to evaluate effects on tagging accuracy.

It was also observed that high accuracy is extremely influenced by input tag set. A richer tag set with POS information produces more accurate results. For example, we can consider additional information with a POS noun, such as time, location, and so on. In addition, in [5] there is a class of lexical words called eventive adjectives, and they cannot co-occur with an Ezafe in contrast with other lexical words. Consider the following examples. Predicative adjectives may only appear in Light Verb Constructions (a) and not in Ezafe Constructions (b).

(a) علی کتاب را فراموش کرد "Ali forgot the book."

(b) * فراموش کتاب توسط علی "forgetting the book by Ali"

We are going to enrich the tagset with more POS tags such as eventive adjectives to define more accurate rules.

REFERENCES

[1] A. Kahnemuyipour, "Persian Ezafe construction: case, agreement or something else," in *Proceedings of the 2nd Workshop on Persian Language and Computer, Tehran University, Tehran, Iran*, 2006.
[2] T. Bögel, M. Butt, and S. Sulger, "Urdu ezafe and the morphology-syntax interface," *Proceedings of LFG08*, 2008.
[3] A. Holmberg and D. Odden, "The Izafe and NP structure in Hawrami," Durham Working Papers in Linguistics, 2004.
[4] G. Van Schaaik, *The noun in Turkish: Its argument structure and the compounding straitjacket*: Otto Harrassowitz Verlag, 2002.
[5] G. Karimi-Doostan, "Lexical categories in Persian," *Lingua*, vol. 121, pp. 207-220, 2011.
[6] A. K. Ahranjani, "The head noun in possessive construction in English and Persian languages," *International Journal of Academic Research*, vol. 2, 2010.
[7] A. K. Ahranjani and R. Tohidian, "Ezafe construction in complex noun phrases in Persian medieval poems," *International Journal of Academic Research*, vol. 3, 2011.
[8] A. Moinzadeh, "The Ezafe phrase in Persian: How complements are added to Ns and As," *Journal of Social Sciences & Humanities of Shiraz University*, vol. 23, pp. 45-57, 2006.
[9] R. Larson and H. Yamakido, "Ezafe and the deep position of nominal modifiers," in *Barcelona Workshop on Adjectives and Adverbs, Barcelona*, 2005.
[10] J. Ghomeshi, "Non-projecting nouns and the ezafe: construction in Persian," *Natural Language & Linguistic Theory*, vol. 15, pp. 729-788, 1997.

[11] P. Samvelian, *A (phrasal) affix analysis of the Persian Ezafe*: Cambridge Univ Press, 2007.

[12] P. Samvelian, "The Ezafe as a head-marking inflectional affix: Evidence from Persian and Kurmanji Kurdish," *Aspects of Iranian Linguistics: Papers in Honor of Mohammad Reza Bateni,* pp. 339-361, 2007.

[13] M. Ghayoomi and S. Momtazi, "Challenges in developing Persian corpora from online resources," in *Asian Language Processing, 2009. IALP'09. International Conference on*, 2009, pp. 108-113.

[14] B. Sagot and G. Walther, "A morphological lexicon for the Persian language," in *Proceedings of the 7th Language Resources and Evaluation Conference (LREC'10)*, 2010.

[15] M. Sheikhan, M. Tebyani, and M. Lotfizad, "Continuous speech recognition and syntactic processing in Iranian Farsi language," *International Journal of Speech Technology,* vol. 1, pp. 135-141, 1997.

[16] J. W. Amtrup, H. M. Rad, K. Megerdoomian, and R. Zajac, *Persian-English machine translation: An overview of the Shiraz project*: Citeseer, 2000.

[17] S. Muller and M. Ghayoomi, "PerGram: A TRALE implementation of an HPSG fragment of Persian," in *Computer Science and Information Technology (IMCSIT), Proceedings of the 2010 International Multiconference on*, 2010, pp. 461-467.

[18] M. Bijankhan, J. Sheykhzadegan, M. Bahrani, and M. Ghayoomi, "Lessons from building a Persian written corpus: Peykare," *Language resources and evaluation,* vol. 45, pp. 143-164, 2011.

[19] S. Isapour, M. Homayounpour, and M. Bijabkhan, "The Prediction of Ezafe Construction in Persian by Using Probabilistic Context Free grammar," in *In Proceedings of 13th Annual Conference of Computer Society of Iran*, Kish Island, 2008.

[20] A. Koochari, B. Qasemzade, M. Kasaeiyan, and M. Namnabat, "Ezafe Prediction in Phrases of Farsi Using CART," in *Proceedings of the I International Conference on Multidisciplinary Information Sciences and Technologies*, 2006, pp. 329-332.

[21] R. Dehkharghani and M. Shamsfard, "Mapping Persian Words to WordNet Synsets," *International Journal of Interactive Multimedia and Artificial Intelligence*, vol. 1, 2009.

[22] M. Bijankhan, "The role of the corpus in writing a grammar: An introduction to a software," *Iranian Journal of Linguistics,* vol. 19, 2004.

[23] V. Samiian, "The Ezafe construction: some implications for the theory of X-bar syntax," *Persian Studies in North America,* pp. 17-41, 1994.

[24] G. D. Forney Jr, "The viterbi algorithm," *Proceedings of the IEEE*, vol. 61, pp. 268-278, 1973.

[25] S. Noferesti and M. Rajaei, "A Hybrid Algorithm Based on Ant Colony System and Learning Automata for Solving Steiner Tree Problem," *International Journal of Applied Mathematics and Statistics*, vol. 22, pp. 79-88, 2011.

[26] K. A. De Jong and W. M. Spears, "An analysis of the interacting roles of population size and crossover in genetic algorithms," in *Parallel problem solving from nature*, ed: Springer, 1991, pp. 38-47.

[27] J. Carletta, "Assessing agreement on classification tasks: the kappa statistic," *Computational linguistics,* vol. 22, pp. 249-254, 1996.

8

An Example in Remote Computing Over the Internet applied to Geometry

Ferreira, M[1] and Casquilho, M.[2]

[1] *Department of Computer Science and Engineering,*
Instituto Superior Técnico, Technical University of Lisbon, Lisboa, Portugal
[2] *Centre for Chemical Processes, Department of Chemical Engineering,*
Instituto Superior Técnico, Technical University of Lisbon, Lisboa, Portugal

Abstract — **Scientific computing over the Internet can suit many activities that have not, in the authors' opinion, been explored enough in general. Resources such as executables, languages, packages, can be used from a remote computing system. In this study, largely based on academic practice, a simple illustrative example in Geometry is implemented on a distributed system that outsources the computing-intensive tasks to remote servers that may be located in other universities or companies, linked to grids and clusters and so on. The software stack and software developed to support the communication is explained in detail. The architecture developed stresses the interoperability of the software, and a suitable high degree of decoupling between components hosted in various locations. The results of this study motivate further work and serve a practical purpose that may be useful to everyone doing scientific computing.**

Keywords — **Internet, remote executables, Scientific computing, university-industry links.**

I. INTRODUCTION

MANY areas of scientific computing can be addressed over the Internet, but this approach has not, in general — in these authors' opinion — been appropriately explored, all the more if compared with most uses of that ubiquitous communication network. One of the authors has, since more than a decade, intensively used this mode of computing in research and teaching at his university work, in domains related to Mathematics, namely Operational Research, Statistics or Chemical Engineering. The computing has been mainly done in a server of the university's information technology centre, intended typically to host faculty and students' webpages. The present study, largely based on that previous academic practice, focuses on the establishment of a link between two universities, one supposedly wishing to execute software made available by the other. This would also apply to any two entities, such as a set of two companies or a university-company linkage (a particular application of [5]). In the Internet context, resources adequate to the particular technical purpose, such as executables, languages, or packages, can be used, if accessible at this level with due permissions, from a remote computing system.

The Internet affords nowadays an unprecedented ease of communication at a very low cost, so that a step can be taken to reap benefits from using remote resources. There are, of course, many resources for computing on the Web, dealing with small tasks, ranging from conversions of units to more complex mathematical problems. Regarding scientific computing over the Web, an extensive example of this activity in the academic environment is the original work by Ponce ([7]), containing a large number of (Fortran) programs to solve problems dealing with Hydraulics and related areas in Civil Engineering. These applications are presumably (as all of our previous work) deployed wholly on single nodes, which also host the web interface and logic. Building on such projects as the excellent one referred above and our own previous projects, the present work intends to take this topology into a next stage, allowing further decoupling of components, by introducing an intermediate communication layer between distributed nodes, which together form the web computing system.

Internet-based computing as an everyday activity has been deemed by one of the authors indispensable to his activities as a tool in the academic practice, and a gateway to the university-industry linkage — widely praised but often scanty — in an era of cheap information technology gear.

The present study is based on a simple, yet surprising, illustrative example in Geometry — an example that might be used in a lecture — chosen both to be clear to a wide readership and to avoid beclouding the underlying software structure. Thus: the problem is started in a webpage of one entity; and the computation is done, without the user's perception, at another machine (suggesting the extension to more), allowing a certain software to be accessed.

In the following sections: the illustrative example is briefly described in its mathematical aspects; the developed resolution based on network computing is presented, with the implemented software architecture to support it; and finally some conclusions are drawn about the proposed solution and system developed.

II. ILLUSTRATIVE EXAMPLE

A problem in Geometry, otherwise conceived as a simple template for more applied cases, was chosen as an illustrative example for the technique. Let the minimum distance be sought between points A, source, and B, destination, as seen in Fig. 1, both on the X-axis, passing by point P, to be determined, on the half line s making an angle γ with the horizontal axis. The problem is treated in [1] and solved by simple differential calculus. The analytical solution for $P = (X, Y)$ is given in (1).

$$\frac{1}{X} = \frac{1}{2}\left(\frac{1}{x_1} + \frac{1}{x_2}\right)\sec^2\gamma \qquad (1)$$

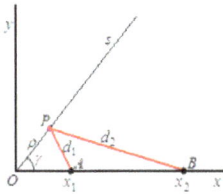

Fig. 1. Route from A to B, passing by P on s, for minimum distance.

With $Y = X\tan\gamma$, (1) leads to (2):

$$X = 2\frac{x_1 x_2}{x_1 + x_2}\cos^2\gamma$$

$$\qquad (2)$$

$$Y = 2\frac{x_1 x_2}{x_1 + x_2}\sin(2\gamma)$$

More concisely, in polar coordinates, (ρ, θ), with $\theta \equiv \gamma$, the radial coordinate is

$$\rho = X/\cos\gamma = \left(2\frac{x_1 x_2}{x_1 + x_2}\cos^2\gamma\right)\Big/\cos\gamma =$$

$$= 2\frac{x_1 x_2}{x_1 + x_2}\cos\gamma \qquad (3)$$

The interest of this problem — the reason it was chosen — lies in the unexpected result as γ decreases towards 0. In Fig. 2, the optimum routes are shown, to which correspond the optimum positions of P, for various descending values of γ, always with $x_1 = 1$ and $x_2 = 3$. The results come from the authors' website ([2]). Now, intuition would possibly lead ρ to the *arithmetic mean* of x_1 and x_2.

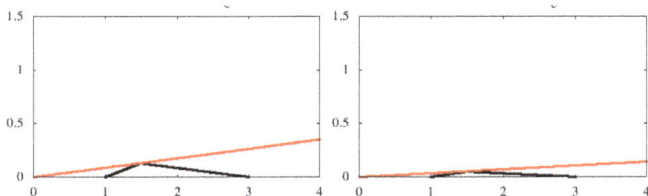

Fig. 2. Optimum routes for $\square$ = 60, 40, 20, 10° (from left to right).

Observation of the sequence in Fig. 2, however, disputes intuition, and confirms (3): ρ tends to the *harmonic* mean of x_1 and x_2. Images for small angles, 5 and 2°, in Fig. 3, show the limiting ρ to be not 2, the arithmetic mean of $x_1 = 1$ and $x_2 = 3$, but 1.5, their harmonic mean.

The adequacy of the arithmetic mean in its own right should be noted (for $\gamma = 0$), notwithstanding, by verifying that, just by letting x_1 and x_2 grow indefinitely, with $x_2 - x_1 = \delta$ (δ constant), the harmonic mean tends to the arithmetic mean, as seen in (4).

$$\lim_{x_1 \to \infty} \frac{1}{\rho} = \frac{1}{2}\left(\frac{1}{x_1} + \frac{1}{x_1 + \delta}\right) =$$

$$= \frac{1}{2}\frac{x_1 + \delta + x_1}{x_1(x_1 + \delta)} = \frac{1 + \dfrac{\delta/x_1}{2}}{x_1 + \delta} = \qquad (4)$$

$$= \frac{1}{\dfrac{x_1 + \delta}{1 + \dfrac{\delta/x_1}{2}}}$$

Considering the infinitesimal δ/x_1, (4) becomes (5), where the arithmetic mean is now visible ($\rho = x_1 + \delta/2$).

$$\lim_{x_1 \to \infty}\frac{1}{\rho} = \frac{1}{(x_1 + \delta)\left(1 - \dfrac{\delta}{2x_1}\right)} =$$

$$= \frac{1}{x_1 + \delta - \dfrac{x_1\delta}{2x_1} - \dfrac{\delta^2}{2x_1}} = \frac{1}{x_1 + \dfrac{\delta}{2}} \qquad (5)$$

Another interesting property of the optimum routes is that, for varying γ (with fixed x_1, x_2), the locus of the optimum points P is a circle with radius $R = x_1 x_2 / (x_1 + x_2)$ (same physical units of the x's, of course) centred at $(0, R)$, here $R = 3/4$. These facts, out of the scope of this study, corroborate the adequacy of the Internet also to openly reveal noteworthy features.

III. SOFTWARE ARCHITECTURE

This study is based on previous applications for many types of scientific problems and expands their capacity using the Internet, following past and current academic practice. In this work, we developed a decentralized computing architecture, distributed on a network, using the HTTP protocol to communicate between the servers, in what is usually known as a web4 service. The architecture is composed by servers playing two separate roles:

a) a front end role, providing the computing services to the clients, with a simple, practical web interface that can be easily accessed through any browser; and

[4] In *web* (as attributive) or *Web*, the Chicago Manual of Style Online ([4]) was roughly followed.

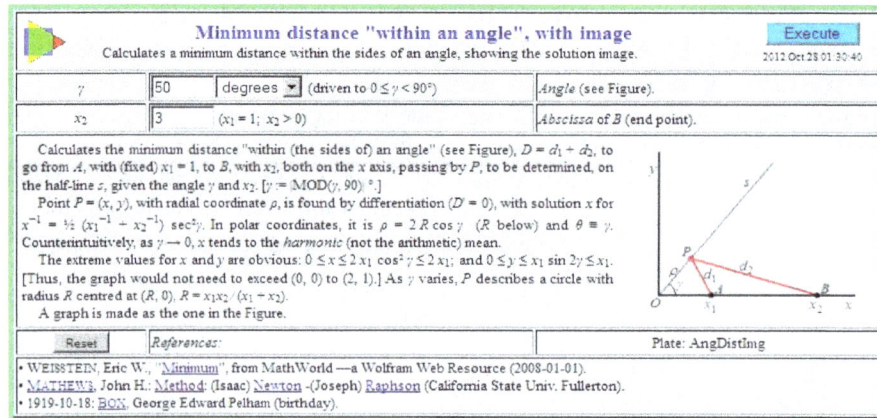

Fig. 3. Optimum routes for □ = 5, 2° (from left to right).

b)   a back end role, receiving the computing tasks from the front end and having the required software to execute them.

The remote call may incur a substantial delay, depending only on the network latency, and mainly the complexity of the problem and computing power of the remote server machine.

The back end addresses must be known to the front end servers, so that they can be located on the public network. Likewise, the front end must be publicly accessible to the users/clients, and have a well-known address.

In the architectural layout described, both the front and back end servers are highly decoupled between them and from the other servers, having no structural dependencies on any single network point [no SPOF5 (e.g., [4])]. Therefore, they can be easily brought up and down, and change location, without disturbing the overall functioning of the system, which grants a very valuable comparative advantage. The only requirement for the system to work is just one front end and one back end servers online at any given time.

The decoupling is highly beneficial for two reasons: i) load balancing of requests between the front ends, and of computing tasks between the back ends; and ii) fault tolerance against possible node crashes.

The front end and the back end support parallel task/requests that require a separation and isolation of execution contexts. This is guaranteed by the HTTP server and the script engine used, which is PHP, with additional safeguards required in the code to carefully avoid any conflict in the resources used (filenames, etc.).

The system is illustrated in this study with the geometric example above, implemented on an Internet link between two semi-closed local networks, the Sigma cluster of IST, and the web servers of FCUL, following the steps described in the next two subsections.

The IST server is deployed on a cluster of AMD64 Opteron processors (2.4 GHz) running Debian Linux, Apache 2.2.16, and PHP 5.3.3-7. The FCUL server runs on a cluster of i386 Intel Xeon processors with Red Hat Linux, Apache 2.2.3 and PHP 5.1.6.

Local execution

The starting point of the study, based on previous work done, was a system deployed in a single local server. This system combines the front end and back end functionalities locally. This is a simple case scenario that served to develop and test the basic computing service.

The system uses the following five files in turn:

a)   Webpage, such as [2], in a well-known address of a front end server — It is a PHP file containing an HTML 'form' to receive the user's data, which is then sent via an HTTP POST method to a processing PHP script (following item);

b)   PHP script 'interface.php', which

  1. Extracts the user's arguments from the HTTP request;

  2. Launches the required program in a new process (via PHP's 'proc_open') with redirected streams to new process pipes, open to the calling PHP process;

  3. Feeds it with the given arguments through the child process read pipe;

  4. Waits to read the output of the called program from the other, write pipe; and

  5. Closes the pipes and terminates the child process.

c)   Binary program ('angDist.exe', compiled from a Fortran 90 source), which also writes to the output stream the data required for a graphic to be created afterwards.

Now, the 'interface.php' script [in b)] constructs a dynamic webpage from:

a)   'interfacetop.php' (constant), the top of the webpage;

b)   body (main) section, in HTML 'pre' format, with the results of the program call, and (typically) a graphic with plotted results, closing HTML bottom.

The screenshots are shown in Fig. 4 for the user data and in Fig. 5 for the results of the computation.

---

[5] "single point of failure"

```
                          Results

2012-10-28 UTC+0000  1:36:55.952
*** Distance within an angle ***                    (Oct-2012, MC)

gamma (deg., rad),  50.0     0.8727   |
x_1, x_2,           1.       3.000    |
------------------------------------- +------------------------------------
COS, TAN(gamma),  0.64279   1.1918    | COS^2,                    0.41318
                                      |
If x_2 = x_1:                         | (absolute minimum, d0)
   x_P, y_P,      0.41318   0.49240   | d0 = x_1 cos^2 = x_P =    0.41318
If x2 = infinity:                     |
   x_P, y_P,      0.82635   0.98481   | x_P = 2 d0
                                      |
MINIMUM DISTANCE for given x_2         |
x_P, y_P,         0.61976   0.73861   | Cartesian coordinates
rho_P,                      0.96418   | (Polar) radial coordinate
d_1, d_2,         0.8307    2.492     | A--P, P--B
Min. distance,              3.3229    | A--P--B

2012-10-28 UTC+0000  1:36:55.952
2012-10-28 UTC+0000  1:36:56.386     CPU:     0.4 sec.            End
```

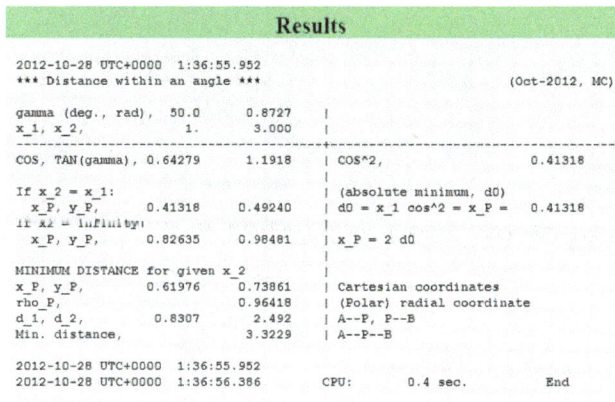

Fig 4.. Webpage for the user data..

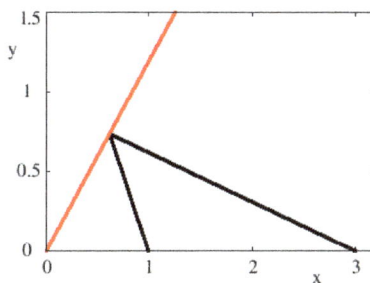

Fig. 5. Webpage for the results.

A. Remote execution

The remote execution mode is the focus of the present work. In this mode, the computing component was distributed to a remote network allowing for a scalable expansion of the system by adding more computing nodes. The decoupling adopted thus requires the development of a middle-ware communication layer between the web-interface (front-end) and the computing nodes (back-end). The decoupled architecture also provides scalability for the front-end, allowing the deployment of multiple interface nodes, scaling up according to the number of incoming requests. The accessible web front-end is available in [3]. The system can be easily deployed throughout many nodes, which can be switched on and off depending on the desired system throughput and efficiency.

Starting from the local execution system described in the previous section, the interface between the computing program and the web front end was greatly modified to support the distribution of both parts, mainly the computing intensive tasks. A muddle-ware was developed to implement the network communication, with the required transfers and conversions of data. The process of service lookup by the remote servers is done by a semi-static approach, i.e., a list of hostnames of the known service providers, contacted in sequence until a live one replies.

To the desired end, the following changes were made:

a) Refactoring the PHP complete service module, into two separate modules: a local front end component, and a back end web service interface for the remote program;

b) The front end interface loads the list of known back end servers' addresses, and polls them to find one available;

c) The front end makes an HTTP request to the available server, by invoking the PHP script on the back end. The front end forwards the input data using the HTTP POST method, specifying in the request which service is required (i.e., 'angDist' in the example);

d) The back end interface calls the binary program in a manner similar to the local execution mode, executing the requested task in isolation in that node;

e) The back end sends the results back to the front end, i.e., both the main results and the parameters of the to-be-created graphic, formatted following a well-defined template, and packaged in the same HTTP response body;

f) The front end process receives the output of the task, and unpacks the two blocks of data (results and graphic's parameters), which have been pre-formatted accordingly; and

g) The front end retains the responsibility of generating the graphic with the parameters received from the remote request, using the GNU tool gnuplot.

The choice was made not to send the graphic itself over the Web, for it could lead to problems of text data encoding (one of the tenets of web services being the use of textual ASCII data), and it would considerably increase the messages' payload size.

The results for the user are, of course, the same as previously. A different HTML background image was chosen to differentiate between a service running in local execution mode (the front end at IST) and another in remote mode (the one at FCUL [3]). The remote execution network is schematically shown in Fig. 6.

The system performed as expected, namely, the communication latency introduced by the network was negligible when compared to the typical computing time for scientific problems, and it is a constant delay depending only on the size of input and output data, and the underlying network infrastructure.

The remote execution mode is the focus of the present work. In this mode, the computing component was distributed to a remote network allowing for a scalable expansion of the system by adding more computing nodes. The decoupling adopted thus requires the development of a middle-ware communication layer between the web-interface (front-end) and the computing nodes (back-end). The decoupled architecture also provides scalability for the front-end, allowing the deployment of multiple interface nodes, scaling up according to the number of incoming requests. The accessible web front-end is available in [3]. The system can be easily deployed throughout many nodes, which can be switched on and off depending on the desired system throughput and efficiency.

Starting from the local execution system described in the previous section, the interface between the computing program and the web front end was greatly modified to support the distribution of both parts, mainly the computing intensive tasks. A muddle-ware was developed to implement the network communication, with the required transfers and conversions of data. The process of service lookup by the remote servers is done by a semi-static approach, i.e., a list of hostnames of the known service providers, contacted in sequence until a live one replies.

To the desired end, the following changes were made:

a) Refactoring the PHP complete service module, into two separate modules: a local front end component, and a back end web service interface for the remote program;

b) The front end interface loads the list of known back end servers' addresses, and polls them to find one available;

c) The front end makes an HTTP request to the available server, by invoking the PHP script on the back end. The front end forwards the input data using the HTTP POST method, specifying in the request which service is required (i.e., 'angDist' in the example);

d) The back end interface calls the binary program in a manner similar to the local execution mode, executing the requested task in isolation in that node;

e) The back end sends the results back to the front end, i.e., both the main results and the parameters of the to-be-created graphic, formatted following a well-defined template, and packaged in the same HTTP response body;

f) The front end process receives the output of the task, and unpacks the two blocks of data (results and graphic's parameters), which have been pre-formatted accordingly; and

g) The front end retains the responsibility of generating the graphic with the parameters received from the remote request, using the GNU tool gnuplot.

The choice was made not to send the graphic itself over the Web, for it could lead to problems of text data encoding (one of the tenets of web services being the use of textual ASCII data), and it would considerably increase the messages' payload size.

The results for the user are, of course, the same as previously. A different HTML background image was chosen to differentiate between a service running in local execution mode (the front end at IST) and another in remote mode (the one at FCUL [3]). The remote execution network is schematically shown in Fig. 6.

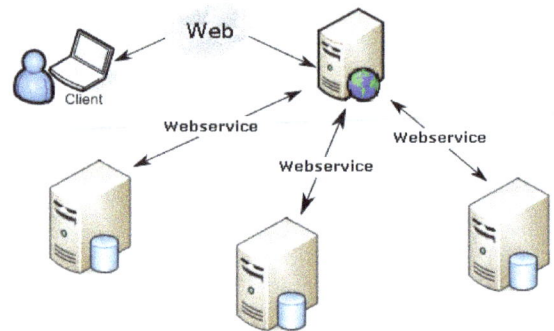

Fig. 6. Remote execution network.

The system performed as expected, namely, the communication latency introduced by the network was negligible when compared to the typical computing time for scientific problems, and it is a constant delay depending only on the size of input and output data, and the underlying network infrastructure.

IV. CONCLUSIONS

The present study inherits former extensive work in scientific computing over the Internet by one of the authors, akin to the work by [7]. Our work has been done in one server of IST, where the webpages and their respective executables are located. The study extrapolates that approach to a two-server solution permitting a webpage on a new server, at FCUL, to access an executable placed on the other server, at IST, without the user's perception. The access is governed by two PHP scripts, each placed in one of the servers.

This shows the ease of use of an executable in a remote locus possessing required resources (executables, languages, packages), thus avoiding the breach of the source webpages' style. With the current ease of communication, this points to the use of remote software among collaborating entities, such as companies or universities or in the university-industry linkages. Thus, some software components topologically isolated from a web gateway or from unsecure locations outside its LAN may be accessed by a trusted web server and provided to the worldwide web users.

ACKNOWLEDGMENT

The research was done at (1.st author) the Department of Computer Science and Engineering and (2.nd author) Centro de Processos Químicos (Centre for Chemical Processes), Department of Chemical Engineering, both of the Technical University of Lisbon, Lisbon, Portugal. Thanks are due to the CIIST (Centro de Informática, Informatics Centre) of Instituto Superior Técnico and, for their special effort, CI of Faculdade de Ciências (Faculty of Sciences) of the Lisbon University.

REFERENCES

[1] Casquilho, M., Buescu, J.: A minimum distance: arithmetic and harmonic means in a geometric dispute, International J. of Mathematical Education in Science and Technology, 147, 399–405 (2011).
[2] Ferreira, M., http://web.ist.utl.pt/ist11038/compute/com/Fx-angdistImg.php

[3] Ferreura, M., http://webpages.fc.ul.pt/~maxxxxxxxxx/compute/Fx-angdistRemote.php

[4] Chicago Manual of Style Online (The), http://www.chicagomanualofstyle.org/

[5] Dikaiakos, M. D., Katsaros, D., Mehra, P., Pallis, G., Vakali, A.: Cloud computing: distributed internet computing for IT and scientific research, Internet Computing, IEEE, 13(5), 10–13 (2009).

[6] Dooley, K.: Designing large-scale LAN's. O'Reilly Media, Inc., Sebastopol, Ca. (USA) (2002).

[7] Ponce, V. M.: Vlab, http://onlinecalc.sdsu.edu/, San Diego State University.

9

Analysis of Gait Pattern to Recognize the Human Activities

Jay Prakash Gupta[1], Pushkar Dixit[2], Nishant Singh[3], Vijay Bhaskar Aemwal[4]

[1]*Infosys Limited, Pune, India*
[2]*Faculty of Engineering and Technology Agra College, Agra, India*
[3]*Poornima Institute of Engineering and Technology, Jaipur, India*
[4]*SiemensInformation System, India*

Abstract – **Human activity recognition based on the computer vision is the process of labelling image sequences with action labels. Accurate systems for this problem are applied in areas such as visual surveillance, human computer interaction and video retrieval. The challenges are due to variations in motion, recording settings and gait differences. Here we propose an approach to recognize the human activities through gait. Activity recognition through Gait is the process of identifying an activity by the manner in which they walk. The identification of human activities in a video, such as a person is walking, running, jumping, jogging etc are important activities in video surveillance. We contribute the use of Model based approach for activity recognition with the help of movement of legs only. Experimental results suggest that our method are able to recognize the human activities with a good accuracy rate and robust to shadows present in the videos.**

Keywords – **Feature Extraction, Gait Pattern, Human Computer Interaction, Activity Recognition, Video Surveillance**

I. INTRODUCTION

THE goal of automatic video analysis is to use computer algorithms to automatically extract information from unstructured data such as video frames and generate structured description of objects and events that are present in the scene. Among many objects under consideration, humans are of special significance because they play a major role in most activities of interest in daily life. Therefore, being able to recognize basic human actions in an indispensable component towards this goal and has many important applications. For example, detection of unusual actions such as jumping, running can provide timely alarm for enhanced security (e.g. in a video surveillance environment) and safety (e.g. in a life-critical environment such as a patient monitoring system). In this paper, we use the concept of Gait for human activity recognition. The definition of Gait is defined as: "A particular way or manner of moving on foot". Using gait as a biometric is a relatively new area of study, within the realms of computer vision. It has been receiving growing interest within the computer vision community and a number of gait metrics have been developed. We use the term Gait recognition to signify the identification of an individual from a video sequence of the subject walking. This does not mean that Gait is limited to walking, it can also be applied to running or any means of movement on foot. Gait as a biometric can be seen as advantageous over other forms of biometric identification techniques for the following reasons: unobtrusive, distance recognition, reduced detail, and difficult to conceal. This paper focuses on the design, implementation, and evaluation of activity recognition system through gait in video sequences. It introduces a novel method of identifying activities only on the basis of leg components and waist component. The use of waist below components for recognizing the activities makes it to achieve fast activity recognition over the large databases of videos and hence improves the efficiency and decreases the complexity of the system. To recognize the actions, we establish the features of each action from the parameters of human model. Our aim is to develop a human activity recognition system that must work automatically without human intervention. We recognized four actions in this paper namely walking, jumping, jogging and running. The walking activity is identified by the velocities of all components superior to zero but lesser than a predefined threshold. In case of jumping activity, every part of human moves only vertically and in the same direction either up or down. Therefore, jumping action can be identified by the velocities of all the three components to be near or equal to zero in horizontal direction but greater than zero in vertical direction. The only differences between jogging and running activities are that travelling speed of running is greater than jogging and other difference is of distance ratio between the leg components to the axis of ground. In case of running activity, speed of travelling is greater than jogging and the other difference is of distance ratio between leg components to the axis of ground.

The rest of the paper is structured as follows: Section 2 discusses the trend of activity recognition research area in the past decade which introduces the fundamentals of gait recognition systems and human activity recognition models; Section 3 presents the proposed work of human activity recognition using Gait; Section 4 analyzes and evaluates the empirical results of experiments to validate the proposed framework. Before evaluating the proposed system, some hypotheses are established and the evaluations are conducted against these hypotheses; finally section 5 summarizes the novelties, achievements, and limitations of the framework, and proposes some future directions of this research.

II. LITERATURE REVIEW

In recent years, various approaches have been proposed for human motion understanding. These approaches generally fall under two major categories: model-based approaches and model-free approaches. Poppe has made a survey on vision based human action recognition [1]. When people observe human walking patterns, they not only observe the global motion properties, but also interpret the structure of the human body and detect the motion patterns of local body parts. The structure of the human body is generally interpreted based on their prior knowledge. Model-based gait recognition approaches focus on recovering a structural model of human motion, and the gait patterns are then generated from the model parameters for recognition. Model-free approaches make no attempt to recover a structural model of human motion. The features used for gait representation includes: moments of shape, height and stride/width, and other image/shape templates.

Leung & Yang reported progress on the general problem of segmenting, tracking, and labeling of body parts from a silhouette of the human [2]. Their basic body model consists of five U-shaped ribbons and a body trunk, various joint and mid points, plus a number of structural constraints, such as support. In addition to the basic 2-D model, view-based knowledge is defined for a number of generic human postures (e.g., "side view kneeling model," "side horse motion"), to aid the interpretation process. The segmentation of the human silhouette is done by detecting moving edges. Yoo et al. estimate hip and knee angles from the body contour by linear regression analysis [3]. Then trigonometric-polynomial interpolant functions are fitted to the angle sequences and the parameters so-obtained are used for recognition.

In [4], human silhouette is divided into local regions corresponding to different human body parts, and ellipses are fitted to each region to represent the human structure. Spatial and spectral features are extracted from these local regions for recognition and classification. In model-based approaches, the accuracy of human model reconstruction strongly depends on the quality of the extracted human silhouette. In the presence of noise, the estimated parameters may not be reliable.

To obtain more reliable estimates, Tanawongsuwan and Bobick reconstruct the human structure by tracking 3D sensors attached on fixed joint positions [5]. However, their approach needs lots of human interaction because they have considered and identified only walking type of activity whereas our method has considered four type of activities and the performance is reasonable for each type of activity. Wang et al. build a 2D human cone model, track the walker under the Condensation framework, and extract static and dynamic features from different body part for gait recognition [6]. Their approach has fused static and dynamic features to improve the gait recognition accuracy but extraction of both static and

dynamic features required more computation which lacks its applicability in real time scenario.

Zhang et al. used a simplified five-link biped locomotion human model for gait recognition [7]. Gait features are first extracted from image sequences, and are then used to train hidden Markov models for recognition. In [8], an approach for automatic human action recognition is introduced by using the parametric model of human from image sequences using motion/texture based human detection and tracking. They used the motion/texture of full body part whereas proposed approach used only the gait pattern of the lower body part which is more time efficient. Bobick & Davis interpret human motion in an image sequence by using *motion-energy* images (MEI) and *motion-history* images (MHI) [9]. The motion images in a sequence are calculated via differencing between successive frames and then thresholded into binary values. These motion images are accumulated in time and form MEI, which are binary images containing motion blobs. The MEI is later enhanced into MHI, where each pixel value is proportional to the duration of motion at that position. Moment-based features are extracted from MEIs and MHIs and employed for recognition using template matching. Because this method is based on the whole template matching instead of the only gait pattern of the legs, it does not take the advantage of recent development whereas we incorporated the matching only based on the gait analysis. Recent Gait studies for activity recognition suggest that gait is a unique personal characteristic, with cadence and cyclic in nature [10]. Rajagopalan & Chellappa [11] described a higher-order spectral analysis-based approach for detecting people by recognizing human motion such as walking or running. In their proposed method, the stride length was determined in every frame as the image sequence evolves.

TABLE 1. COMPARISON OF EXISTING APPROACHES

Ref.	Method	Advantage	Disadvantage	Uses
[8]	Gait recognition	Locomotion human model	Insensitive to noise	Indoor scenario
[9]	Model-based Action Recognition	Inclusion of motion texture	Poor performance in walking case	Indoor environment
[12]	Spectral analysis of human motion	Higher-order Spectral	Periodic detection	Differentiate between people and vehicular objects
[13]	View based motion analysis	Object models are not required	Need to reduce the distribution combinatory	Outdoor scenario
[27]	Activity recognition using smartphones	Real time application	More than one classifier reduces the accuracy	Indoor/Outdoor both

Vega and Sarkar [12] offered a novel representation scheme for view-based motion analysis using just the change in the relational statistics among the detected image features, without the need for object models, perfect segmentation, or part-level tracking. They modeled the relational statistics using the probability that a random group of features in an image would exhibit a particular relation. To reduce the representational combinatorics of these relational distributions, they represented them in a Space of Probability Functions (SoPF). Different motion types sweep out different traces in this space. They also demonstrated and evaluated the effectiveness of that representation in the context of recognizing persons from gait. But, there method requires multiple cameras from different viewpoints to model multi-view recognition system which requires extra setup and also computation, whereas the proposed approach is able to achieve high recognition performance from only a single viewpoint. Several other approaches and features used in [13-25] may be tied with gait analysis to predict the human actions. Human activity recognition using smartphones is also studied [26] but its recognition rate can be improved using gait analysis with more time efficiently. Table 1 compares the existing approaches.

III. PROPOSED METHODOLOGY

The proposed technique of human activity recognition is based on the foreground extraction, human tracking, feature extraction and recognition. Figure 1 shows the framework of the introduced human activity recognition system using Gait to identify four basic human activities (i.e. walking, running, jogging and jumping). The proposed method has following main steps: Foreground Extraction, Human Tracking, Feature Extraction and Activity Recognition. In this framework, the video is given as an input to the system from the activity database and frames are extracted from that video. The parametric model of human is extracted from image sequences using motion/texture based human detection and tracking. After that the results are displayed as the recognized activities like walking, running, jogging and jumping; and finally the performance of the method is tested experimentally using the datasets under indoor and outdoor environments.

A. Foreground Extraction

The first step is to provide a video sequence of an activity as an input in the proposed system from the dataset. That video contains a number of continuous frames. After that background subtraction technique is used to separate moving object present inside those frames. But these frames contain some noises which may lead to incurrent foreground subtraction. So first of all, we remove these noises. Some of the small noises are removed by using morphological image processing tools such as Erosion, Dilation, or Gaussian Filters. Generally, an object might be detected in several fragmented image regions. In that case, a region-fusion operation is needed. Two regions are considered to be the same object if they are overlapped or their distance less than a specific threshold value. With these constraints, the method is again very sensible to light condition, such as shadow, contrast changing and sudden changes of brightness.

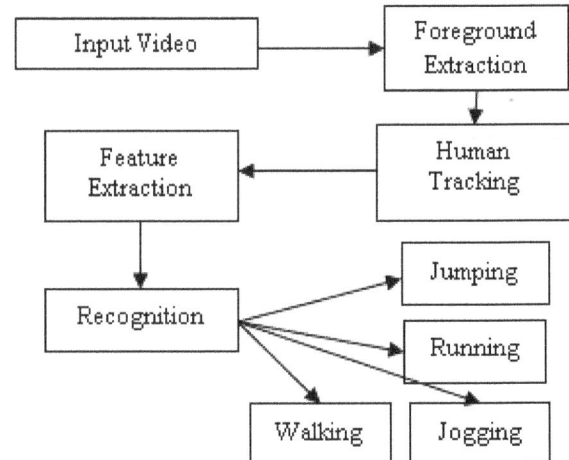

Fig. 1. Framework of Proposed System of Human Activity recognition

Intuitively, introducing some special characteristics of object, for instance texture properties, will probably improve the better results. Therefore, in the fusion process the color probability density of object's texture is additionally applied for computing the similarity between regions using Mean-shift algorithm [27]. This mixture of motion and texture of object for detection and tracking can reduce significantly noises and increases consequently the effectiveness of our tracking algorithm. However, there are always additive noises superposed with detected objects that will be eliminated later by human model constraints. The mean shift algorithm is a nonparametric clustering technique which does not require prior knowledge of the number of clusters, and does not constrain the shape of the clusters. Hence, mean shift represents a general non-parametric mode finding/clustering procedure.

B. Human Tracking and Activity Recognition

In this phase, we apply Hu-moments [28] for shape analysis in which Zero- to third-order moments are used for shape recognition and orientation as well as for the location tracking of the shape. Hu-moments are invariant to translation, rotation and scaling. Hu derived expressions from algebraic invariants applied to the moment generating function under a rotation transformation. They consist of groups of nonlinear centralized moment expressions. The result is a set of absolute orthogonal (i.e. rotation) moment invariants, which can be used for scale, position, and rotation invariant pattern identification. The advantage of using Hu invariant moment is that it can be used for disjoint shapes. In particular, Hu invariant moment set consists of seven values computed by normalizing central moments through order three. In terms of central moment the seven moments are given as below:

$$M_1 = \eta_{20} + \eta_{02}$$

$$M_2 = (\eta_{20} - \eta_{02})^2 + 4\eta_{11}^2$$

$$M_3 = (\eta_{30} - 3\eta_{12})^2 + (3\eta_{21} - \eta_{03})^2$$

$$M_4 = (h_{30} + h_{12})^2 + (h_{21} + h_{03})^2$$

$$M_5 = (h_{30} - 3h_{12})(h_{30} + h_{12})[(h_{30} + h_{12})^2 - 3(h_{21} + h_{03})^2]$$
$$+ (3h_{21} - h_{03})(h_{21} + h_{03})[3(h_{30} + h_{12})^2 - (h_{21} + h_{03})^2]$$
$$M_6 = (h_{20} - h_{02})(h_{30} + h_{12})^2 - (h_{21} + h_{03})^2$$
$$+ [4h_{11}(h_{30} + h_{12})(h_{21} + h_{03})]$$
$$M_7 = (3h_{21} - h_{03})(h_{30} + h_{12})[(h_{30} + h_{12})^2 - 3(h_{21} + h_{03})^2]$$
$$+ (3h_{21} - h_{30})^2(h_{21} + h_{03})[3(h_{30} + h_{12})^2 - (h_{21} + h_{03})^2]$$

These seven values given by Hu are used as a feature vector for centroid in the human model.

C. Feature Extraction

We employed a model based approach to extract the features. The extracted foreground that supposed to be a human is segmented into centroid and two leg components. We use Mean-shift algorithm again for computing the similar regions below the centroid of the human body for each leg components that will serve for tracking legs. We assume that with only these three components of human model the four basic actions could be identified correctly. The human model constraints are used for noise suppression. The three components namely centroid, left leg and right leg (i.e. vm1, vm2, vm3 respectively), are used in order to model parametric approach. The threshold concept is also used along with the defined method. Threshold calculation is applied as follows: Video sequences from the KTH and Weizmann datasets are normalized on the basis of number of frames and the time of a particular sequence for an activity. The threshold is calculated on the basis of a case study given in [29]. To recognize the actions, we establish the features of each action from the parameters of human model as follows: **Walking feature**: In case of walking action, every part of human move generally and approximately in the same direction and speed. Therefore, the walking activity can then be identified by the velocities of all components superior to zero but lesser than a predefined threshold for walking. Note that the significant difference between running and walking strides is that at least one of the feet will be in contact with the principal axis (ground) at any given time as shown in Figure 2 (a). **Jumping feature**: In case of jumping activity, every part of human moves only vertically and in the same direction either up or down [30-39]. Therefore, jumping action can be identified by the velocities of all the three components to be near or equal to zero in horizontal direction but greater than zero in vertical direction as shown in Figure 2(b). **Jogging feature**: The only differences between jogging and running activities were that travelling speed of running is greater than jogging and other difference is of distance ratio between the leg components to the axis of ground as shown in Figure 2(c). **Running feature**: Similarly in case of running activity, speed of travelling is greater than jogging and the other difference is of distance ratio between leg components to the axis of ground as shown in Figure 2 (d).

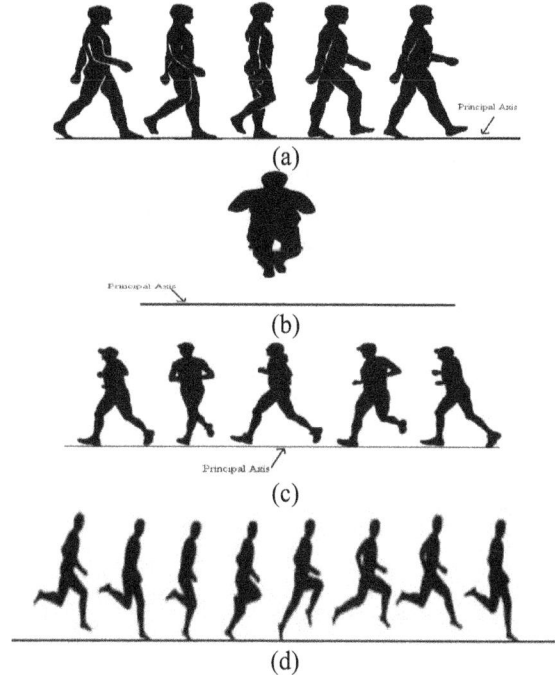

Fig. 2. Silhouette pattern for (a) Walking, (b) Jumping, (c) Jogging and (d) Running

Algorithm for Human Activity Recognition

1) Input is fed to the system as a single video sequence.
2) Frames are extracted from the input video, which are used for further processing.
3) Background subtraction technique is implemented to subtract background from the frames in order to obtain the foreground moving object.
4) Morphological operators are used to remove additional noises in the frames.
5) Mean-shift algorithm is used to track the human; based on the texture similarities in the frames.
6) Hu-moments are calculated to recognize the centroid of the tracked human. Again the Mean-shift algorithm is used to recognize each leg components of the model.
7) For feature extraction, model based approach is employed. The extracted foreground that supposed to be human is then segmented into centroid and the two leg components i.e., total three components.
8) The features of each action from the parameters of human model acts as the features for classifying all four activities (walking, jumping, jogging and running).
9) The features depend on the following criteria: Walking, Jumping, Jogging and Running.

(a)   (b)

(c)   (d)

Fig. 3. Templates of (a) jogging, (b) running, (c) walking, and (d) jumping for human activities.

IV. RESULTS AND DISCUSSIONS

This section analyses the various aspects of the proposed method. In activity recognition through gait, feature requirement is the main issue to model the human according to the parameters to fulfill the criteria.

A. Data Set Used

In order to evaluate our proposed approach of human activity recognition, we have used two datasets: (1) KTH Human Actions dataset (http://www.nada.kth.se/cvap/actions) and (2) Weizmann Actions dataset (http://www.wisdom.weizmann.ac.il/~vision/SpaceTimeActions.html).

KTH Human Actions dataset: KTH video dataset uses six types of human actions such as "walking", "jogging", "running", "boxing", "hand waving" and "hand clapping", which were performed by 25 subjects in different scenarios with different clothing conditions as well.

The video sequences are down sampled to 160*120 pixels and an average length varying from 4 to 41 seconds. This dataset contains 2391 activity sequences. All videos are having static background with 25 fps. We use walking, jogging and running sequences of KTH actions data set for evaluation.

Weizmann Actions dataset: Weizmann Actions dataset uses ten types of natural human actions such as "run," "walk," "skip," "jumping-jack", "jump-forward-on-two-legs", "jump-in-place-on-two-legs", "gallop sideways", "wave-two-hands", "wave-one-hand", or "bend" which are performed by 9 different people in different scenarios with different clothing conditions as well. The video sequences are down sampled to 184*144 pixels and an average length varying from 2 to 4 seconds. This dataset contains 90 low resolution activity sequences. All the videos are having static background and running with 50 fps. We use walking, jogging and jumping sequences of Weizmann Actions dataset in this paper.

We have used templates of Mean Shift Clustering and Hu-Moments for jogging, running, walking and jumping activities as shaown in Figure 3. It is assumed that using centroid and two legs only these four activities can be identified.

B. Experimental Results

We have performed the human activity recognition experiments, with the proposed technique, on several videos, captured in outdoor and indoor environment. We have used two standard dataset namely KTH action dataset and Weizmann action dataset. In this paper, we have performed the experiments considering both indoor and outdoor scenario using KTH action dataset. But we have performed on only outdoor images of Weizmann action dataset.

1) Results on KTH dataset

Figure 4, 5 and 6 show the different frames of experimental results at different time instances on a standard KTH actions dataset. In Figure 4, first image of frame 5 shows that a human is walking. Second image of frame 5 shows the corresponding recognition result as walking with good accuracy. In Figure 5, first image of frame 10 shows that a human is jogging. Second image of frame 10 shows the corresponding recognition result as jogging. In Figure 6, first image of frame 3 shows that a human is running. Second image of frame 3 shows the corresponding recognition result as running with good accuracy.

2) Results on Weizmann dataset

To validate the robustness of our proposed method, we experimented on a standard Weizmann dataset. Figure 7, 8 and 9 shows the frame by frame result analysis of different human activity on this dataset at different time instances. In Figure 7, first image of frame 5 shows that a human is walking in outdoor environment. Second image of frame 5 shows the corresponding recognition result as walking with good accuracy.

In Figure 8, first image of frame 10 shows that a human is running in outdoor environment. Second image of frame 1 shows the corresponding recognition result as running with good accuracy. In Figure 9, first image of frame 1 shows that a human is jumping in outdoor environment. Second image of frame 1 shows the corresponding recognition result as jumping with good accuracy.

C. Result Analysis

Accuracy of proposed method is measured based on the number of frames recognized and number of frames not recognized by the following formulae:

$$Accuracy\ (\%) = \frac{No.\ of\ frames\ currectly\ recognized}{Total\ no.\ of\ video\ frames\ in\ a\ sequence} \times 100$$

Table 2 shows the accuracy of introduced approach over two large datasets with encouraging results; up to 95.01% of activities are recognized correctly in KTH dataset and 91.36% of activities are recognized correctly in Weizmann dataset. We have calculated the accuracy in both indoor and outdoor scenarios in the case of KTH dataset. Table 3 shows that the proposed method outperforms other existing methods.

Zhang et al. achieved 61% gait recognition accuracy over USF dataset of 4-7 activities using a simplified five-link biped locomotion human model [8]. Over indoor dataset of 5 activities, 93% accuracy is gained using the parametric model of human from image sequences using motion/texture based human detection and tracking [9]. Vega and Sarkar reported 90% accuracy using 3 actions over 71 subjects using the change in the relational statistics among the detected image features, without the need for object models, perfect segmentation, or part-level tracking [13]. Whereas, we are able to gain upto 95% and 91% accuracy using just gait analysis over KTH and Wiezmann datasets respectively. From the experimental results it is deduced that the introduced approach is more robust and able to achieve high accuracy over large datasets by considering more activities.

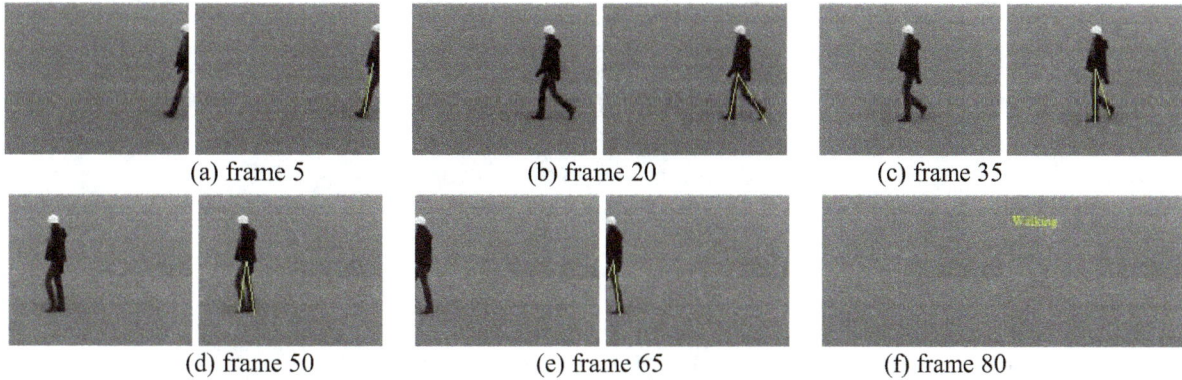

| (a) frame 5 | (b) frame 20 | (c) frame 35 |
| (d) frame 50 | (e) frame 65 | (f) frame 80 |

Fig. 4. Result on standard KTH dataset from of walking; first image shows input frame, second image shows corresponding output image; at the end, it recognize human activity as "Walking".

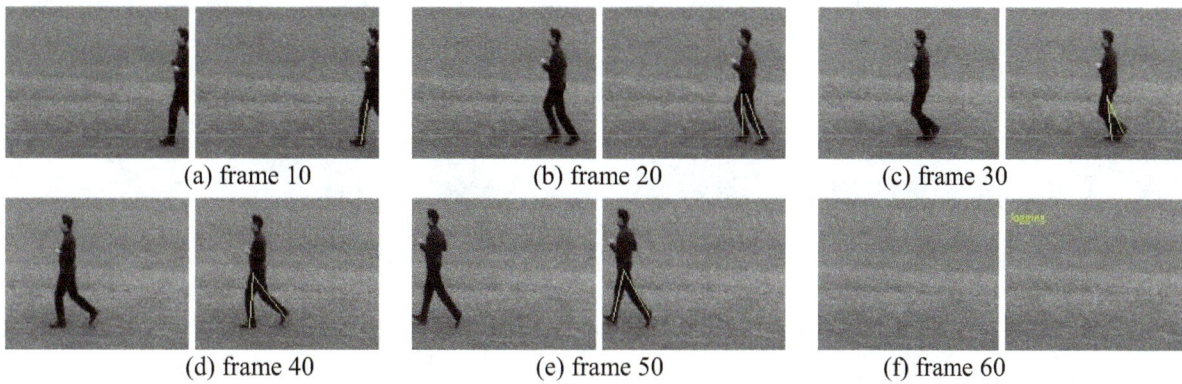

| (a) frame 10 | (b) frame 20 | (c) frame 30 |
| (d) frame 40 | (e) frame 50 | (f) frame 60 |

Fig. 5. Experimental result on standard KTH dataset of jogging; first image shows input frame, second image shows corresponding output image; at the end, it recognize human activity as "Jogging".

TABLE 2. TABLE SHOWS THE RESULT ANALYSIS OF PROPOSED METHOD ON KTH HUMAN ACTIONS DATASET AND WEIZMANN ACTIONS DATASET ON THE BASIS OF FRAMES

Name of Dataset	Environment condition	Human Activities	Number of Frames	Number of Frames recognized	Recognition rate
KTH Dataset	Outdoor	Walking	1443	1434	99.3%
	Indoor	Walking	1415	1383	97.7%
	Outdoor	Jogging	1525	1425	93.4%
	Indoor	Jogging	1218	1157	94.9%
	Outdoor	Running	1089	980	89.9%
	Indoor	Running	1137	1080	94.9%
				Avg. %	**95.01%**
Weizmann Dataset	Outdoor	Walking	678	650	95.8%
	Outdoor	Running	588	552	93.8%
	Outdoor	Jumping	756	642	84.5%
				Avg.%	**91.36%**

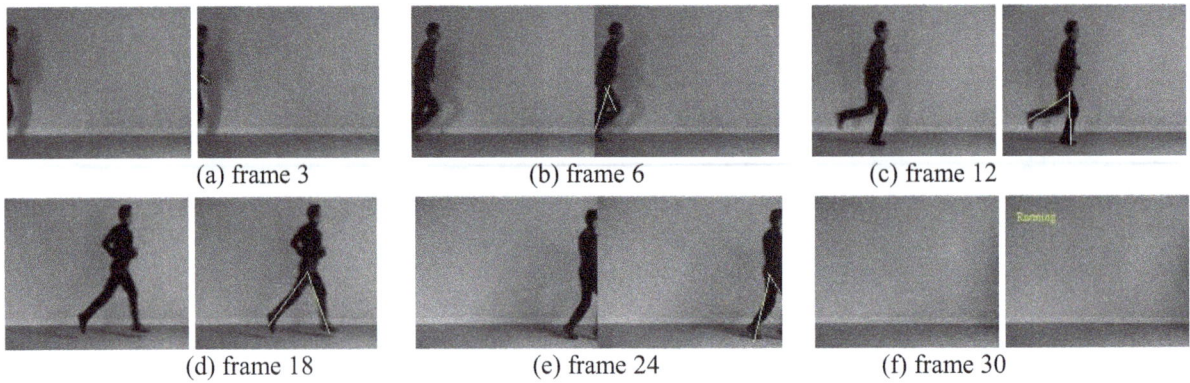

(a) frame 3   (b) frame 6   (c) frame 12

(d) frame 18   (e) frame 24   (f) frame 30

Fig. 6. Result on standard KTH dataset of running; first image shows input frame, second image shows corresponding output image; at the end, it recognize human activity as "Running".

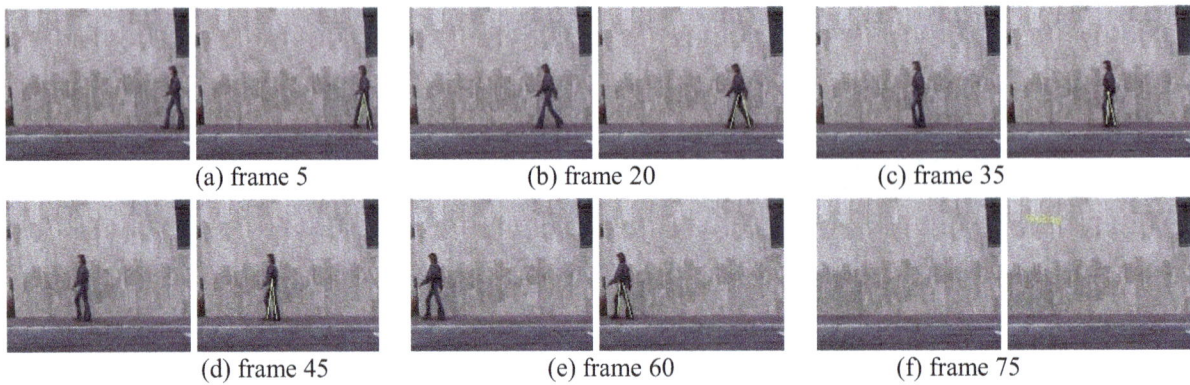

(a) frame 5   (b) frame 20   (c) frame 35

(d) frame 45   (e) frame 60   (f) frame 75

Fig. 7. Experimental result on standard Weizmann dataset of walking; first image shows input frame, second image shows corresponding output image; at the end, it recognize human activity as "Walking".

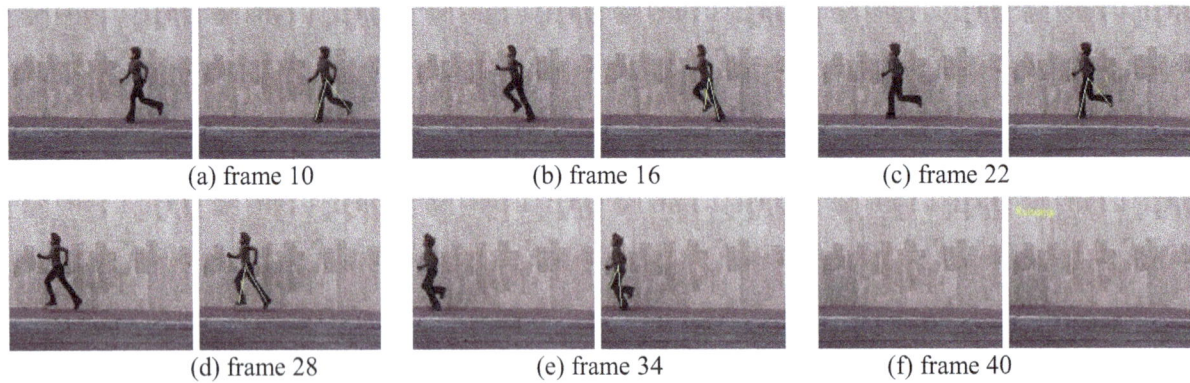

(a) frame 10   (b) frame 16   (c) frame 22

(d) frame 28   (e) frame 34   (f) frame 40

Fig. 8. Experimental result on standard Weizmann dataset of running; first image shows input frame, second image shows corresponding output image; at the end, it recognize human activity as "Running".

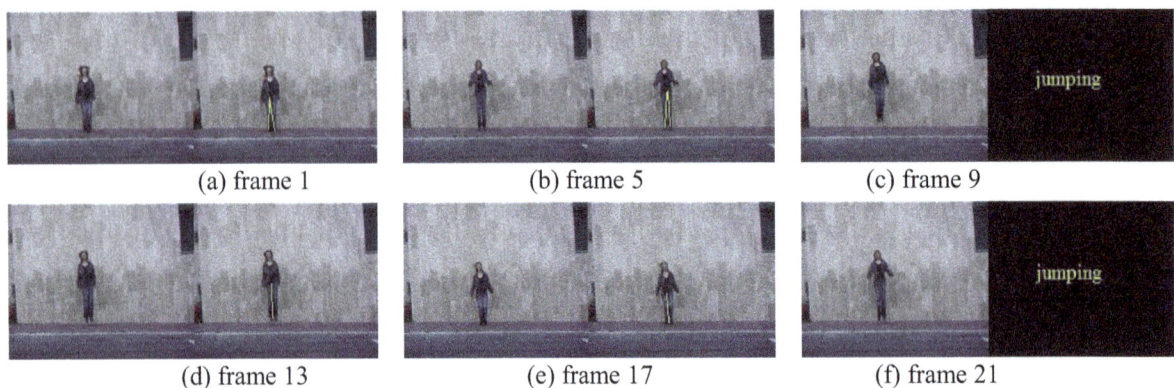

(a) frame 1   (b) frame 5   (c) frame 9

(d) frame 13   (e) frame 17   (f) frame 21

Fig. 9. Experimental result on standard Weizmann dataset of jumping; first image shows input frame, second image shows corresponding output image; at *the* end of each sub-sequence it recognize human activity as "Jumping".

TABLE 3. COMPARISON OF RESULTS WITH EXISTING METHODS

Method	Dataset	No. of Subjects	Number of Frames	Human Activities	Recognition rate
Proposed	KTH Dataset	25	7827	Walking Jogging Running	**95.01%**
	Weizmann Dataset	9	2022	Walking Jumping Running	**91.36%**
[8]	USF Dataset	75	2045	4-7 activities	**61%**
[9]	Indoor Dataset	-	9933	Standing Sitting Bending Walking Laying	**93%**
[13]	-	71	-	Walking Jogging Running	**90%**

In Figure 8, first image of frame 10 shows that a human is running in outdoor environment. Second image of frame 1 shows the corresponding recognition result as running with good accuracy. In Figure 9, first image of frame 1 shows that a human is jumping in outdoor environment. Second image of frame 1 shows the corresponding recognition result as jumping with good accuracy.

D. Result Analysis

Accuracy of proposed method is measured based on the number of frames recognized and number of frames not recognized by the following formulae:

$$Accuracy\ (\%) = \frac{No.\ of\ frames\ currectly\ recognized}{Total\ no.\ of\ video\ frames\ in\ a\ sequence} \times 100$$

Table 2 shows the accuracy of introduced approach over two large datasets with encouraging results; up to 95.01% of activities are recognized correctly in KTH dataset and 91.36% of activities are recognized correctly in Weizmann dataset. We have calculated the accuracy in both indoor and outdoor scenarios in the case of KTH dataset. Table 3 shows that the proposed method outperforms other existing methods.

Zhang et al. achieved 61% gait recognition accuracy over USF dataset of 4-7 activities using a simplified five-link biped locomotion human model [8]. Over indoor dataset of 5 activities, 93% accuracy is gained using the parametric model of human from image sequences using motion/texture based human detection and tracking [9]. Vega and Sarkar reported 90% accuracy using 3 actions over 71 subjects using the change in the relational statistics among the detected image

features, without the need for object models, perfect segmentation, or part-level tracking [13]. Whereas, we are able to gain upto 95% and 91% accuracy using just gait analysis over KTH and Wiezmann datasets respectively. From the experimental results it is deduced that the introduced approach is more robust and able to achieve high accuracy over large datasets by considering more activities.

V. CONCLUSIONS

An efficient human activity recognition using gait technique based on model based approach is introduced in this paper which uses Mean shift clustering algorithm and Hu-Moments to construct the activity templates. This method has a promising execution speed of 25 frames per second and good activity recognition accuracy. The experimental results demonstrate that the proposed method accurately recognizes different activities in various video frames considering both indoor and outdoor scenarios while maintaining a high recognition accuracy rate. Currently our method determines key poses of each activity independently using parametric model only. Different activity classes may give similar key poses which may cause confusion and redundancy in recognition. More discriminative key poses can be applied jointly using some more refined and sophisticated algorithms such as Support Vector Machine (SVM). We found promising recognition performance more than 95% over 3-4 activities. Experimental results suggest that the proposed method outperforms other existing methods.

REFERENCES

[1] R. Poppe, "A survey on vision-based human action recognition," Image and Vision Computing, Vol. 28, No. 6, pp. 976-990, 2010.

[2] M.K. Leung and Y.H. Yang, "First Sight: A human body outline labeling system," IEEE Transaction on Pattern Analysis and Machine Intelligence, Vol. 17, No. 4, pp. 359–377, 1995.

[3] J.H. Yoo, M.S. Nixon and C.J. Harris, "Model-driven statistical analysis of human gait motion," In the Proceedings of the IEEE International Conference on Image Processing, pp. 285–288, 2002.

[4] L. Lee and W.E.L. Grimson, "Gait analysis for recognition and classification," In the Proceedings of the Fifth IEEE International Conference on Automatic Face and Gesture Recognition, pp. 148–155, 2002.

[5] R. Tanawongsuwan and A. Bobick, "Gait recognition from time-normalized joint-angle trajectories in the walking plane," In the Proceedings of the IEEE Conference on Computer Vision and Pattern Recognition, pp. 726–731, 2001.

[6] L. Wang, H. Ning, T. Tan and W. Hu, "Fusion of static and dynamic body biometrics for gait recognition," IEEE Transactions on Circuits and Systems for Video Technology, Vol. 14, No. 2, pp. 149–158, 2004.

[7] R. Zhang, C. Vogler and D. Metaxas, "Human gait recognition at sagittal plane," Image and Vision Computing, Vol. 25, No. 3, pp. 321–330, 2007.

[8] N. Nattapon, S. Nikom and K. Montri, "Model-based Human Action Recognition," In the Proceedings of SPIE, the International Society for Optical Engineering, pp. 111–118, 2008.

[9] A.F. Bobick and J.W. Davis, "The recognition of human movement using temporal templates," IEEE Transactions on Pattern Analysis and Machine Intelligence, Vol. 23, No. 3, pp. 257–267, 2001.

[10] J. P. Gupta, N. Singh, P. Dixit, V. B. Semwal and S. R. Dubey, "Human Activity Recognition using Gait Pattern," International Journal of Computer Vision and Image Processing, Vol. 3, No. 3, pp. 31 – 53, 2013.

[11] A.N. Rajagopalan and R. Chellappa, "Higher-order spectral analysis of human motion," In the Proceedings of the International Conference on Image Processing, pp. 230–233, 2000.

[12] I.R. Vega and S. Sarkar, "Statistical motion model based on the change of feature relationships: human gait-based recognition," IEEE Transactions on Pattern Analysis and Machine Intelligence, Vol. 25, No. 10, pp. 1323–1328, 2003.

[13] S. R. Dubey and A. S. Jalal, "Detection and Classification of Apple Fruit Diseases Using Complete Local Binary Patterns," In proceeding of the Third International Conference on Computer and Communication Technology, pp. 346-351, 2012.

[14] S. R. Dubey and A. S. Jalal, "Species and variety detection of fruits and vegetables from images," International Journal of Applied Pattern Recognition. Vol. 1, No. 1, pp. 108-126, 2013.

[15] W.L. Romero, R.G. Crespo and A.C. Sanz, "A prototype for linear features generalization," International Journal of Interactive Multimedia & Artificial Intelligence, Vol. 1, No. 3 2010.

[16] K.S. Kumar, V.B. Semwal and R.C. Tripathi, "Real time face recognition using adaboost improved fast PCA algorithm," arXiv preprint arXiv:1108.1353, 2011.

[17] S. R. Dubey and A. S. Jalal, "Adapted Approach for Fruit Disease Identification using Images," International Journal of Computer Vision and Image Processing, Vol. 2, No. 3, pp. 44-58, 2012.

[18] H. Bolivar, A. Pacheco and R.G. Crespo, "Semantics of immersive web through its architectural structure and graphic primitives," International Journal of Interactive Multimedia & Artificial Intelligence, Vol. 1, No. 3, 2010.

[19] S. R. Dubey, P. Dixit, N. Singh, and J.P. Gupta, "Infected fruit part detection using K-means clustering segmentation technique," International Journal of Artificial Intelligence and Interactive Multimedia. Vol. 2, No. 2, pp. 65-72, 2013.

[20] N. Singh, S. R. Dubey, P. Dixit and J.P. Gupta, "Semantic Image Retrieval by Combining Color, Texture and Shape Features," In the Proceedings of the International Conference on Computing Sciences, pp. 116-120, 2012.

[21] S.J.B. Castro, R.G. Crespo and V.H.M. García, "Patterns of Software Development Process," International Journal of Interactive Multimedia & Artificial Intelligence, Vol. 1, No. 4, 2011.

[22] S. R. Dubey and A. S. Jalal, "Robust Approach for Fruit and Vegetable Classification," Procedia Engineering, Vol. 38, pp. 3449-3453, 2012.

[23] R.G. Crespo, S.R. Aguilar, R.F. Escobar and N. Torres, "Dynamic, ecological, accessible and 3D Virtual Worlds-based Libraries using OpenSim and Sloodle along with mobile location and NFC for checking in," International Journal of Interactive Multimedia & Artificial Intelligence, Vol. 1, No. 7, 2012.

[24] V.B. Semwal, V.B. Semwal, M. Sati and S. Verma, "Accurate location estimation of moving object in Wireless Sensor network," International Journal of Interactive Multimedia and Artificial Intelligence, Vol. 1, No. 4 , pp. 71-75, 2011.

[25] C.J. Broncano, C. Pinilla, R.G. Crespo and A. Castillo, "Relative Radiometric Normalization of Multitemporal images," International Journal of Interactive Multimedia and Artificial Intelligence, Vol. 1, No. 3, 2010.

[26] P. Siirtola and J. Röning, "Recognizing Human Activities User-independently on Smartphones Based on Accelerometer Data," International Journal of Interactive Multimedia & Artificial Intelligence, Vol. 1, No. 5, pp. 38-45, 2012.

[27] Y. Cheng, "Mean shift, mode seeking, and clustering," IEEE Transactions on Pattern Analysis and Machine Intelligence, Vol. 17, No. 8, pp. 790-799, 1995.

[28] M.K. Hu, "Visual pattern recognition by moment invariants," IRE Transactions on Information Theory, Vol. 8, No. 2, pp. 179-187, 1962.

[29] M.G. Gazendam and A.L. Hof, "Averaged EMG profiles in jogging and running at different speeds," Gait Postures, Vol. 25, No. 4, pp. 604–614, 2007.

[30] S. R. Dubey, N. Singh, J.P. Gupta and P. Dixit, "Defect Segmentation of Fruits using K-means Clustering Technique," In the Proceedings of the Third International Conference on Technical and Managerial Innovation in Computing and Communications in Industry and Academia. 2012.

[31] N. Singh, S. R. Dubey, P. Dixit and J.P. Gupta, "Semantic Image Retrieval Using Multiple," In proceeding of the 3rd International Conference on Technical and Managerial Innovation in Computing and Communications in Industry and Academia. 2012.

[32] K.S. Kumar, V.B. Semwal, S. Prasad and R.C. Tripathi, "Generating 3D Model Using 2D Images of an Object," International Journal of Engineering Science, 2011.

[33] V.B. Semwal, V.B. Semwal, M. Sati and S. Verma, "Accurate location estimation of moving object in Wireless Sensor network," International Journal of Interactive Multimedia and Artificial Intelligence, Vol. 1, No. 4 , pp. 71-75, 2011.

[34] K.S. Kumar, S. Prasad, S. Banwral and V.B. Semwal, "Sports Video Summarization using Priority Curve Algorithm," International Journal on Computer Science & Engineering, 2010.

[35] S.R. Dubey, "Automatic Recognition of Fruits and Vegetables and Detection of Fruit Diseases", M.Tech Thesis, GLAU Mathura, India, 2012.

[36] M. Sati, V. Vikash, V. Bijalwan, P. Kumari, M. Raj, M. Balodhi, P. Gairola and V.B. Semwal, "A fault-tolerant mobile computing model based on scalable replica", International Journal of Interactive Multimedia and Artificial Intelligence, Vol. 2, 2014.

[37] S.R. Dubey and A.S. Jalal, "Automatic Fruit Disease Classification using Images", Computer Vision and Image Processing in Intelligent Systems and Multimedia Technologies, 2014.

[38] D.D. Agrawal, S.R. Dubey and A.S. Jalal, "Emotion Recognition from Facial Expressions based on Multi-level Classification", International Journal of Computational Vision and Robotics, 2014.

[39] S.R. Dubey and A.S. Jalal, "Fruit Disease Recognition using Improved Sum and Difference Histogram from Images", International Journal of Applied Pattern Recognition, 2014.

10

Infected Fruit Part Detection using K-Means Clustering Segmentation Technique

Shiv Ram Dubey[1], Pushkar Dixit[2], Nishant Singh[3], Jay Prakash Gupta[4]

[1]*GLAU, Mathura, India*
[2]*Dept. of Inform. Tech., Dr. M.P.S Group of Institutions College of Business Studies, Agra, India*
[3]*Dept. of Comp. Engg. & Applications, Poornima Group of Colleges, Jaipur, India*
[4]*Systems Engineer in Infosys Limited, Bangalore, India*

Abstract — **Nowadays, overseas commerce has increased drastically in many countries. Plenty fruits are imported from the other nations such as oranges, apples etc. Manual identification of defected fruit is very time consuming. This work presents a novel defect segmentation of fruits based on color features with K-means clustering unsupervised algorithm. We used color images of fruits for defect segmentation. Defect segmentation is carried out into two stages. At first, the pixels are clustered based on their color and spatial features, where the clustering process is accomplished. Then the clustered blocks are merged to a specific number of regions. Using this two step procedure, it is possible to increase the computational efficiency avoiding feature extraction for every pixel in the image of fruits. Although the color is not commonly used for defect segmentation, it produces a high discriminative power for different regions of image. This approach thus provides a feasible robust solution for defect segmentation of fruits. We have taken apple as a case study and evaluated the proposed approach using defected apples. The experimental results clarify the effectiveness of proposed approach to improve the defect segmentation quality in aspects of precision and computational time. The simulation results reveal that the proposed approach is promising.**

Keywords — **K-Means, Defect Segmentation, Fruit Images, Image Processing.**

I. INTRODUCTION

Digital images are one of the most key medium of conveying information. Extracting the information from images and understanding them such that the extracted information can be used for several tasks is an important characteristic of Machine learning. Using images for the navigation of robots is an example of the same. Other applications such as extracting malign tissues from the body scans etc form an integral part of Medical diagnosis. Image segmentation is one of the initial steps in direction of understanding images and then finds the different objects in them.

Modern agricultural science and technology is extreme advance. The value of fruit depends on the quality of fruit. It is an important issue how to assay quality of fruit in agricultural science and technology. The classical approach of fruits quality assessment is done by the experts and it is very time consuming. Defect segmentation of fruits can be seen as an instance of the image segmentation in which we are interested only to the defected portion of the image.

Image segmentation entails the separation or division of the image into areas of similar attributes. In another way, segmentation of the image is nothing but pixel classification. The difficulty to which the image segmentation process is to be carried out mostly depends on the particular problem that is being solved. It is treated as an important operation for meaningful interpretation and analysis of the acquired images. It is one of the most crucial components of image analysis and pattern recognition and still is considered as most challenging tasks for the image processing and image analysis. It has application in several areas like Analysis of Remotely Sensed Image, Medical Science, Traffic System Monitoring, and Fingerprint Recognition and so on.

Image segmentation methods are generally based on one of two fundamental properties of the intensity values of image pixels: similarity and discontinuity. In the first category, the concept is to partition the image into several different regions such that the image pixels belonging to a region are similar according to a set of predefined criteria's. Whereas, in the second category, the concept of partition an image on the basis of abrupt changes in the intensity values is used. Edge detection technique is an example of this category which is similar to the boundary extraction. Researchers have been working on these two approaches for years and have given various methods considering those region based properties in mind. But, still, there is no fixed approach for the image segmentation. Based on the discontinuity or similarity criteria, many segmentation methods have been introduced which can be broadly classified into six categories: (1) Histogram based method, (2) Edge Detection, (3) Neural Network based segmentation methods, (4) Physical Model based approach, (5) Region based methods (Region splitting, Region growing & merging), (6) Clustering (Fuzzy C-means clustering and K-Means clustering).

Histogram based image segmentation techniques are computationally very efficient when compared to other image segmentation techniques because they usually require only a single pass through the image pixels. In this technique, a histogram is calculated from all of the image pixels, and the peaks and valleys are detected in the histogram. Now the image pixels between two consecutive peaks can be considered to a single cluster. A disadvantage of this method is that it is not able to categorize when the image has no clear gray level histogram peak. Another disadvantage of this method is that the continuity of the segmented image regions cannot be ensured. We should focus on global peaks that are likely to correspond to the dominant image regions for the histogram based segmentation method to be efficient.

The edge detection method is very widely used approaches to the image segmentation problems. It works on the basis of the detection of points considering abrupt changes at gray levels. A disadvantage of the edge detection method is that it does not work well when there are many edges in the image because in that case the segmentation technique produces an over segmented output, and it cannot easily identify a boundary or closed curve. For an edge based segmentation method to be efficient, it should identify the global edges and these edges have to be continuous.

Neural Network based image segmentation relies on processing small regions of an image using a neural network or a set of different artificial neural networks. After this, the decision-making method marks the regions of an image on the basis of the category recognized by the artificial neural network. Kohonen self organizing map is a type of network designed especially for such type of problems.

The physical model based image segmentation technique assumes that for an image, individual regions follow a recurring form of geometrical structure. This type of segmentation methods uses texture feature.

The region based image segmentation method uses the similarity of pixels within a region in an image. Sometimes a hybrid method incorporating the region based and edge based methods have been proved to be very useful for some applications. The seeded region growing method was the first region growing method.

Clustering based image segmentation methods are also used by many researchers [1] [2]. The segmentation method incorporating clustering approaches encounters great difficulties when computing the number of clusters that are present in the feature space or extracting the appropriate feature. This type of image segmentation is widely used due to the simplicity of understanding and more accurate result.

This paper presents an efficient image segmentation approach using K-means clustering technique based on color features from the images. Defect segmentation is carried out into two stages. At first, the pixels are clustered based on their color and spatial features, where the clustering process is accomplished. Then the clustered blocks are merged to a specific number of regions. Using this two step procedure, it is possible to increase the computational efficiency avoiding feature extraction for every pixel in the image of fruits. Although the color is not commonly used for defect segmentation, it produces a high discriminative power for

different regions of the image.

The rest of the paper is organized as follows: Section 2 presents a brief overview of the related work. Section 3 describes the K-means clustering method. In section 4 the proposed method for the defect segmentation of fruits based on color using K-means clustering technique is presented and discussed. Section 5 demonstrates the experimental results obtained with apple as a case study. Finally, section 6 concludes with some final remarks.

II. A Brief Overview of Related Work

Color image segmentation has been a difficult task for the researchers over the past two decades. It is an essential operation in image processing and in many computer vision, pattern recognition, and image interpretation system, with applications in industrial and scientific field(s) such as Remote Sensing, Microscopy, Medicine, content-based image and video retrieval, industrial automation, document analysis and quality control [3]. The efficiency of color image segmentation may significantly influence the quality of an image understanding system [4]. A detail review on various image segmentation techniques are provided by Pal & Pal [5].

Among myriads of existing segmentation techniques, many have used unsupervised clustering methods. For example, image segmentation on the basis of region merging is analogue of agglomerative clustering [6]. Graph cut methods such as normalized cut and minimal cut characterize the problem of clustering in a graph theoretic way [7]. A major problem for this kind of methods known as the problem of validity is how to decide the number of clusters in any image. Since the problem is basically unresolved, most techniques need that the user should provide a terminating criterion.

Soft computing techniques have been used for segmenting color image by Sowmya and Sheelarani [8]. The soft computing techniques they used were competitive neural network and Possibilistic C means algorithm (PCM). Researchers also used Fuzzy set and Fussy logic techniques for solving segmentation problem. Borji et al. presented CLPSO-based Fuzzy color image segmentation [9]. Cheng et al. used Fuzzy homogeneity approach for the segmentation of color image [10]. Besides this, Genetic algorithm (GA) and artificial neural network (ANN) techniques also have been used for the image segmentation [11].

There are various segmentation techniques in medical imaging problems depending on the region of interest in the image. There are region growing segmentation methods and atlas-guided techniques. Some of them use a semi-automatic method and still need some operator relations. Other techniques use fully automatic methods and the operator has just a verification role.

Automatic image segmentation by integrating seeded region growing and color edge detection was proposed by Fan et al. [12]. They have used fast Entropy thresholding for the extraction of edges. After they have obtained color edges that provided the foremost geometric structures in an image, then they have determined the centroids between these adjacent regions and considered it as the initial seeds. These seeds were then replaced by centroids of the generated homogeneous edge regions by incorporating the additional pixels step by step.

Another method using seeded region growing was proposed by Adams and Bischof [13]. Shih and Cheng proposed another image segmentation method using regions in the image [14] where based on the standard deviation in a neighbor, initial seeds are selected. This method assigns each pixel in that region as seeds after checking whether the value is under a threshold. They have applied region growing and region merging techniques after the selection of seeds. As discussed above color image segmentation has been widely used by the researchers.

Authors in [17], [18] have used the concept of k-means clustering for background subtraction. They segmented the region of interest (i.e. foreground) with the background by making two clusters one for foreground and one for background. In the case of fruit diseases more than one disease may be present at a time so we have to use more than two clusters to segment the infected part with fruit and background.

III. K-MEANS CLUSTERING ALGORITM

The food image processing using clustering is an efficient method. Clustering technique classifies the objects into different groups, or more specifically, partitioning of a data set into clusters (subsets), so that the data in each cluster (ideally) shares some common trait - often according to some defined distance measurement. Data partitioning is a usual technique for the analysis of statistical data, which is used in many areas, including machine learning, image analysis, pattern recognition, bioinformatics and data mining. The computational task of partitioning the data set into k subsets is often referred to unsupervised learning.

There are many approaches of clustering designed for a wide variety of purposes. K-means is a typical clustering algorithm (MacQueen, 1967) [15]. K-means is generally used to determine the natural groupings of pixels present in an image. It is attractive in practice, because it is straightforward and it is generally very fast. It partitions the input dataset into k clusters. Each cluster is represented by an adaptively changing center (also called cluster center), starting from some initial values named seed-points. K-means clustering computes the distances between the inputs (also called input data points) and centers, and assigns inputs to the nearest center.

K-means method is an unsupervised clustering method that classifies the input data objects into multiple classes on the basis of their inherent distance from each other [16]. Clustering algorithm assumes that a vector space is formed from the data features and tries to identify natural clustering in

them. The objects are clustered around the centroids $\mu i \forall i = 1 . . . k$ which are computed by minimizing the following objective

$$V = \sum_{i=1}^{k} \sum_{x_j \in S_i} (x_j - \mu_i)^2 \quad (1)$$

Where k is the number of clusters i.e. S_i, $i = 1, 2, \ldots, k$ and μ_i is the mean point or centroid of all the points $x_j \in S_i$.

As a part of this work, we implemented an iterative version of K-means algorithm. The algorithm requires a color image as input. The algorithm of K-means clustering is as follows

Step 1   Compute the distribution of the intensity values.

Step 2   Using k random intensities initialize the centroids.

Step 3   Repeat the step 4 and step 5 until the labels of the cluster do not change any more.

Step 4   Cluster the image points based on the distance of their intensity values from the centroid intensity values.

$$c^{(i)} := \arg \min_{j} \left\| x^{(i)} - \mu_j \right\|^2 \quad (2)$$

Step 5   Compute new centroid for each cluster.

$$\mu_i := \frac{\sum_{i=1}^{m} 1\{c_{(i)}=j\} x^{(i)}}{\sum_{i=1}^{m} 1\{c_{(i)}=j\}} \quad (3)$$

Where k is the number of clusters, i iterates over all the intensity values, j iterates over all the centroids (for each cluster) and μ_i are the centroid intensities.

IV. DEFECT SEGMENTATION

Image segmentation using k-means algorithm is quite useful for the image analysis. An important goal of image segmentation is to separate the object and background clear regardless the image has blur boundary. Defect segmentation of fruits can be seen as an instance of image segmentation in which number of segmentation is not clearly known. Figure 1 shows the framework for the fruits defect segmentation.

The basic aim of the proposed approach is to segment colors automatically using the K-means clustering technique and L*a*b* color space. The introduced framework of defect segmentation operates in six steps as follows

Step 1.   Read the input image of defected fruits.

Step 2.   Transform Image from RGB to L*a*b* Color Space. We have used L*a*b* color space because it consists of a luminosity layer in 'L*' channel and two chromaticity layer in 'a*' and 'b*' channels. Using L*a*b* color space is computationally efficient because all of the color information is present in the 'a*' and 'b*' layers only.

Step 3. Classify Colors using K-Means Clustering in 'a*b*' Space. To measure the difference between two colors, Euclidean distance metric is used.

Step 4. Label Each Pixel in the Image from the Results of K-Means. For every pixel in our input, K-means computes an index corresponding to a cluster. Every pixel of the image will be labeled with its cluster index.

Step 5. Generate Images that Segment the Input Image by Color. We have to separate the pixels in image by color using pixel labels, which will result different images based on the number of clusters.

Programmatically determine the index of each cluster containing the defected part of the fruit because K-means does not return the same cluster index value every time. But we can do this using the center value of clusters, which contains the mean value of 'a*' and 'b*' for each cluster.

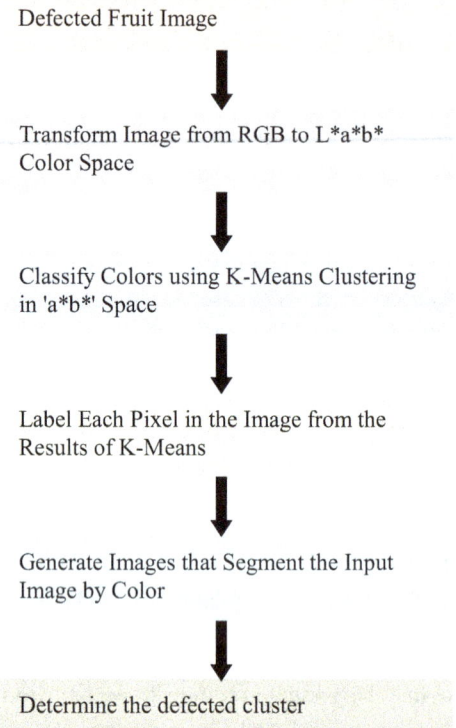

Defected Fruit Image

↓

Transform Image from RGB to L*a*b* Color Space

↓

Classify Colors using K-Means Clustering in 'a*b*' Space

↓

Label Each Pixel in the Image from the Results of K-Means

↓

Generate Images that Segment the Input Image by Color

↓

Determine the defected cluster

Fig. 1. Fruits defect segmentation.

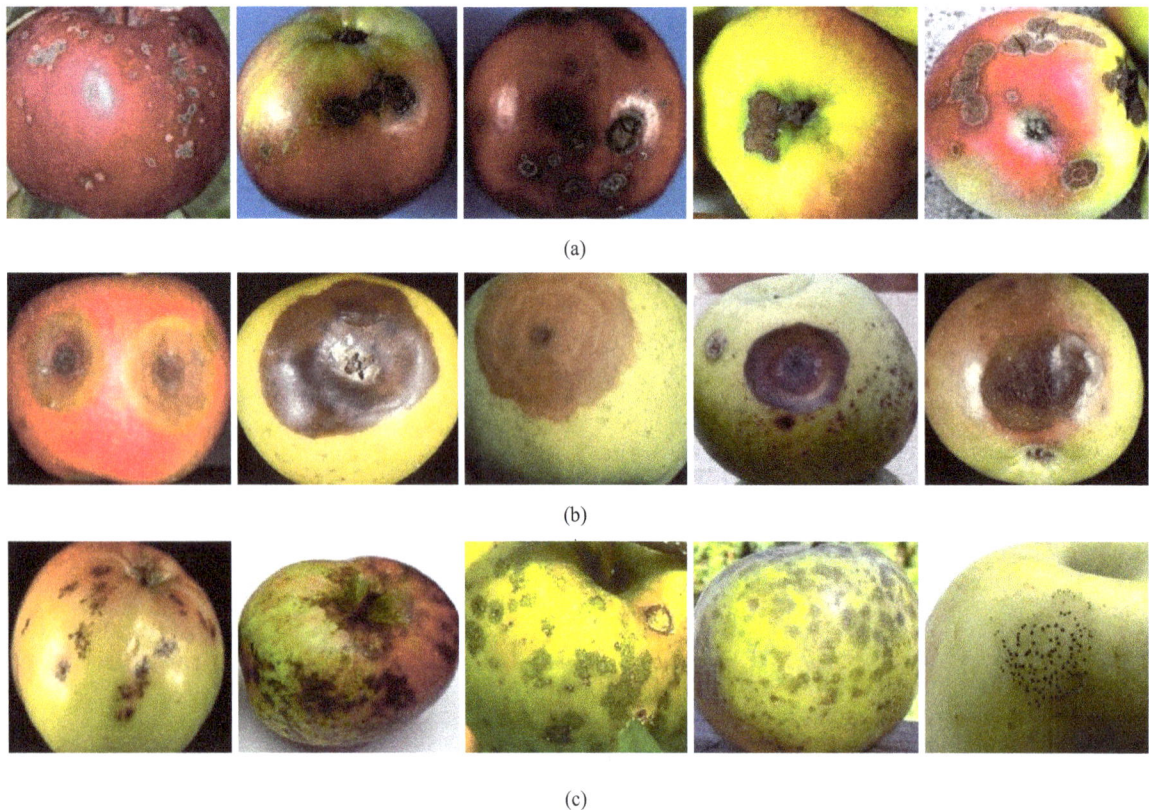

(a)

(b)

(c)

Fig. 2. Sample images from the data set infected with (a) apple scab, (b) apple rot, and (c) apple blotch diseases.

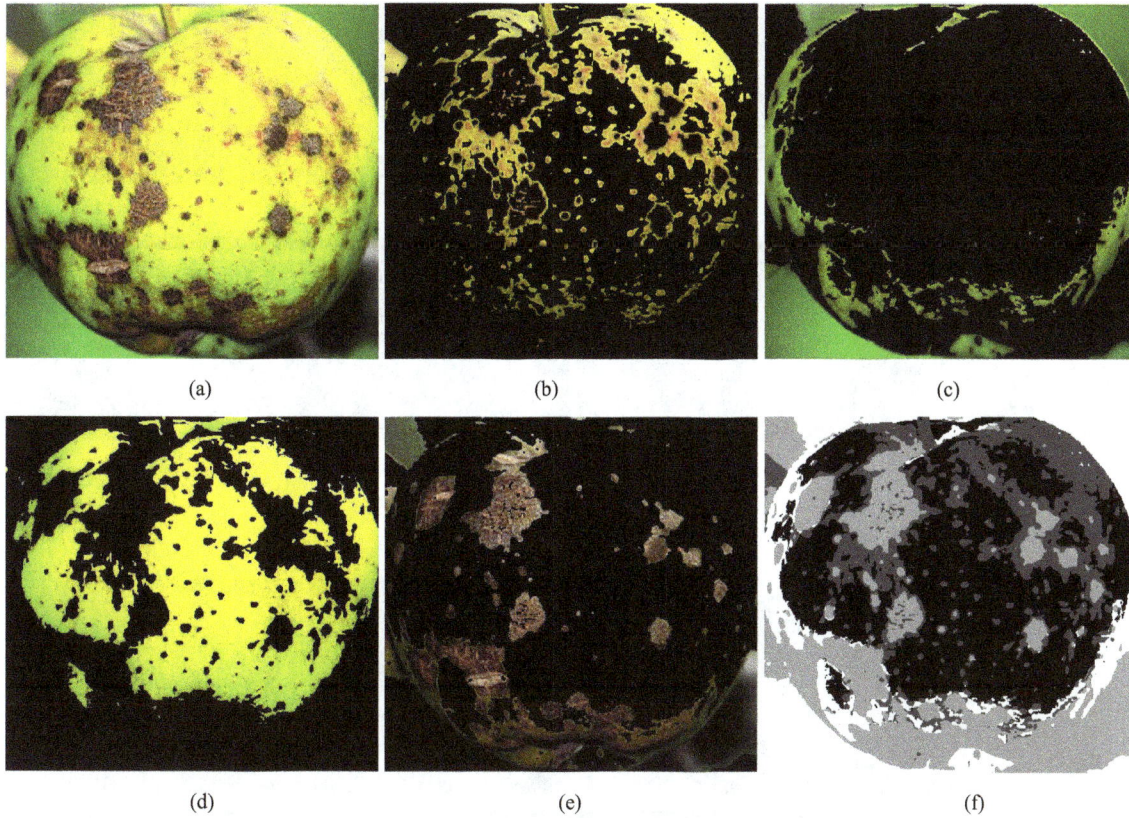

Fig. 3. K-Means clustering for an apple fruit that is infected with apple scab disease with four clusters (a) The infected fruit image, (b) first cluster, (c) second cluster, (d) third cluster, and (e) fourth cluster, respectively, (f) single gray-scale image colored based on their cluster index.

V. EXPERIMENTAL RESULT

To demonstrate the performance of the proposed approach, we have taken apples as a case study. The introduced method is evaluated on the defected apples. We have taken some of the diseases of the apples such as apple scab, apple rot and apple blotch for the defect segmentation. Figure 2 shows some images of the data set infected with various diseases. Presence of a lot of variations in the data set makes it more realistic.

Figure 3 shows the defect segmentation result of an apple fruit infected with the apple scab disease using K-means clustering technique. We have segmented the input image into four clusters in Figure 3 and it is clear that fourth cluster correctly segment the defected portion of the image. From the empirical observations it is found that using 3 or 4 clusters yields good segmentation results. So, in this experiment input images are partitioned into three or four segments as per requirement.

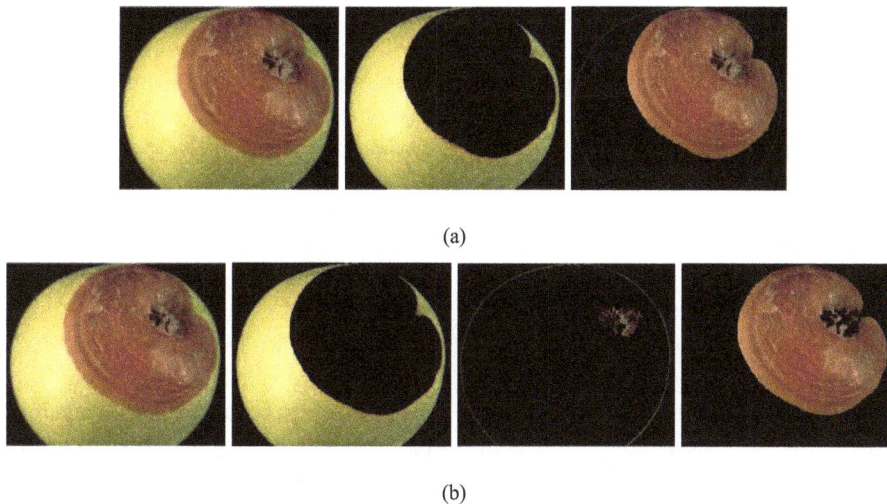

(a)

(b)

(c)

(d)

Fig. 4. Defect detection results when number of cluster is set to (a) 2, (b) 3, (c) 4, and (d) 5 respectively.

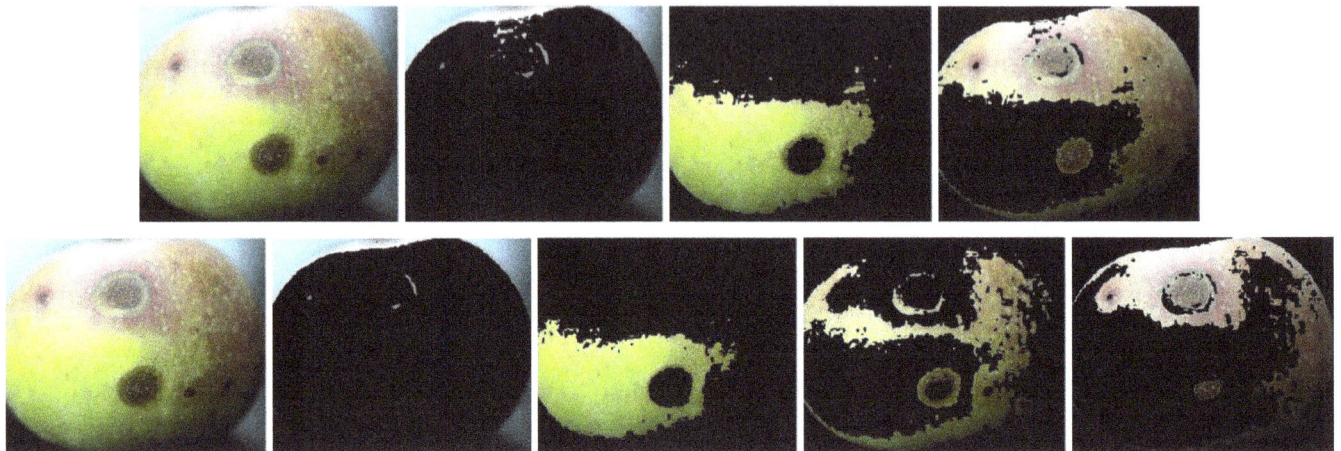

Fig. 5. Comparison of result while using 3 and 4 number of clusters.

Figure 4 shows the detection result on an image infected with apple rot while considering different number of clusters for K-Mean clustering. When number of cluster is set to 2, one cluster contains fruit part while other one contains defected part and background. But if we increase the number of cluster to 3, then defected part is separated with background. If we further increase number of clusters to 4 and 5, then we are not able to segment all defected portion in a single cluster (i.e. single segment) as shown in figure 4. But, in some cases using 3 clusters is not sufficient such as in the example taken in figure 5. In this figure, we have considered an image of apple infected with apple scab.

In figure 5, segmentation result is better for 4 clusters than 3 clusters because the area of infected portion in apple scab is less than the apple rot and the color of infected part is quite similar to the color of fruit part. Only 3 clusters is sufficient in the case of figure 4 because it is infected with apple rot which have generally larger area of infected portion. So, we can say that if defected area is larger, fewer clusters will be required

while if defected area is smaller more clusters will be needed. It means number of clusters required for the defect detection from the infected image is invertionally proportion to the defected area.

Figure 6 shows the results of defect segmentation of two defected apple fruits using K-means clustering method with only three clusters. Whereas, in Figure 3, we have used four clusters because using three clusters was not sufficient in that case due to the natural variability of skin color in the input apple fruit image. In the first case of Figure 7, there is the presence of the stem/calyx in the input image of defected apple, and using only three clusters our proposed approach are able to segment the defected portion with the stem/calyx of the image. Figure 7 shows more segmentation results using proposed approach of defect segmentation of fruits using K-means clustering technique. The experimental results suggest that the introduced method for defect segmentation in this paper is robust because it can accurately segment the defected part with the fruit region, background and stem/calyx.

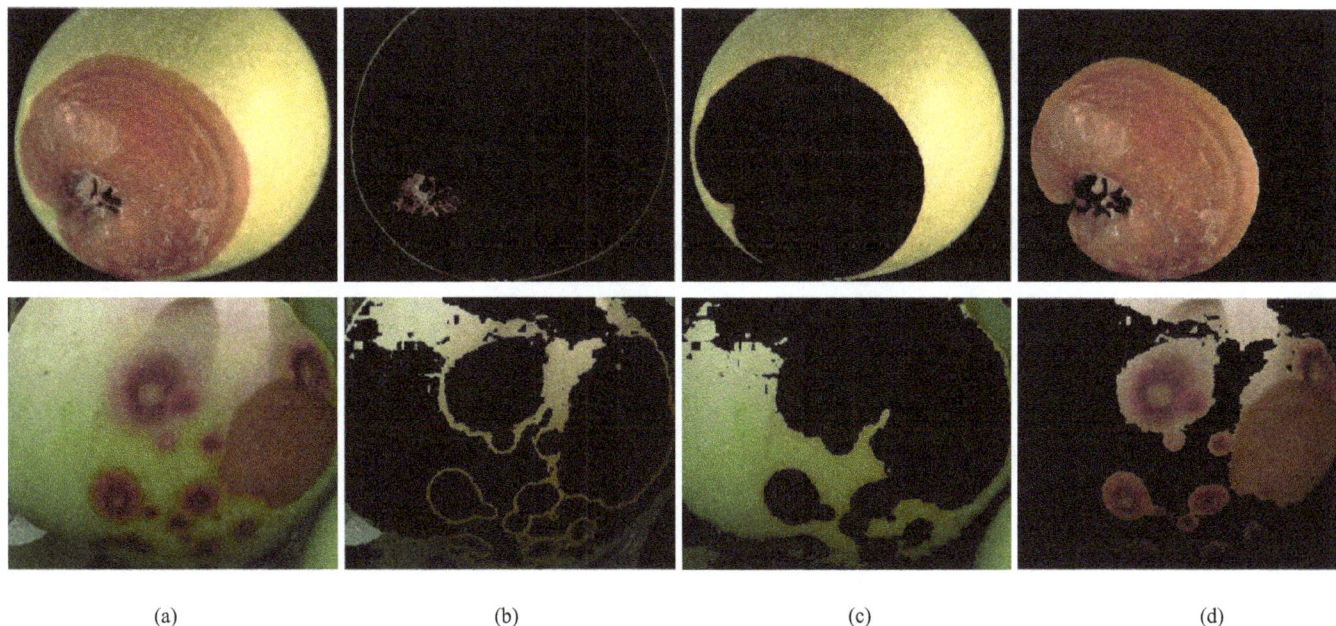

Fig. 6. K-Means clustering for defected apples with three clusters (a) The infected fruit images, (b) first cluster, (c) second cluster, and (d) third cluster

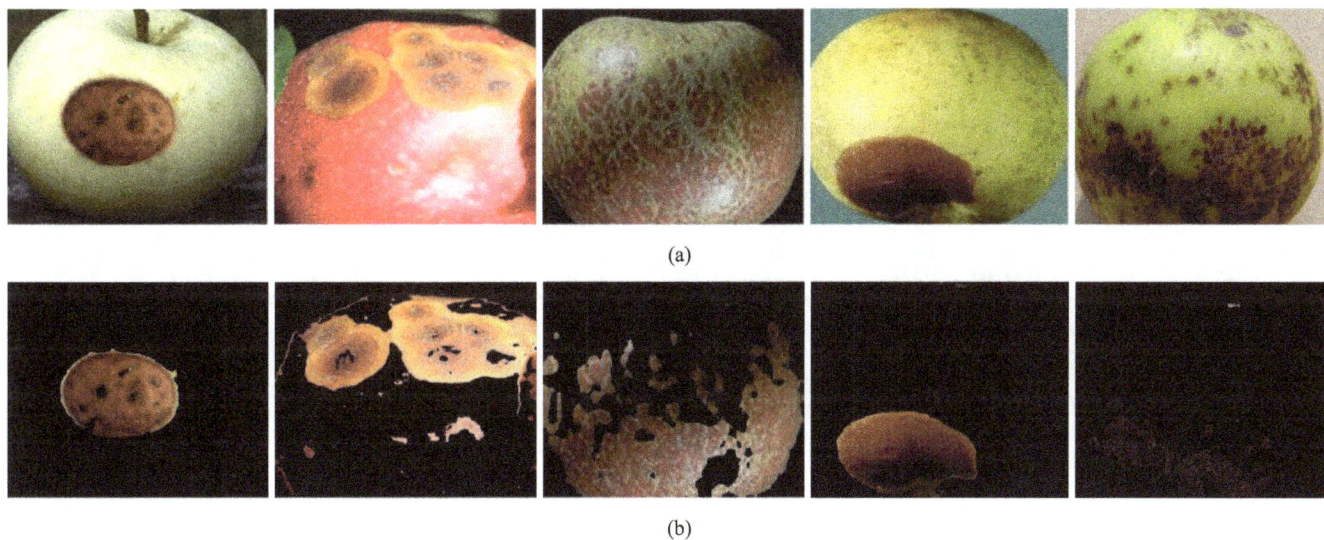

Fig. 7. Defect segmentation results of apples (a) Images before segmentation, (b) Images after segmentation.

VI. CONCLUSION

A framework for the defect segmentation of fruits using images is proposed and evaluated in this paper. The proposed approach used K-means clustering technique for segmenting defects with three or four clusters. We have used defected apples for the experimental observations and evaluated the introduced method considering apples as a case study. Experimental results suggest that the proposed approach is able to accurately segment the defected area of fruits present in the image. K-means based defect segmentation approach is also segment defected area with the stem and calyx of the fruits. The future work includes automatic determination of number of clusters required to segment the defects more accurately.

REFERENCES

[1] Y. Li and Y. Shen, "Robust image segmentation algorithm using fuzzy clustering based on kernel-induced distance measure," in *Proceedings of the International Conference on Computer Science and Software Engineering*, 2008, vol. 1, pp. 1065-1068.

[2] D. E. Ilea and P. F. Whelan, "Color image segmentation using a spatial k-means clustering algorithm," in . *Proceedings of the Irish Machine Vision & Image Processing Conference 2006 (IMVIP 2006),* Dublin City University, 2006, pp. 146–53.

[3] R. D. Silva, R. Minetto, W. R. Schwartz, and H. Pedrini, "Satellite image segmentation using wavelet transforms based on color and texture features," in *Proceedings of the 4th International Symposium on Advances in Visual Computing*, Part II, 2008, pp. 113-122.

[4] H. C. Chen and S. J. Wang, "Visible color difference-based quantitative evaluation of colour segmentation," in *Proceedings of the IEEE*

Conference on Vision Image and Signal processing, 2006, vol. 153, no. 5, pp. 598-609.

[5]   N. R. Pal and S. K. Pal, "A review on image segmentation techniques," *Pattern Recognition*, vol. 26, no. 9, pp. 1227-1294, 1993.

[6]   H. Frigui and R. Krishnapuram, "Clustering by competitive agglomeration," *Pattern Recognition*, vol. 30, no. 7, pp. 1109-1119, 1997.

[7]   Y. Boykov, "Graph cuts and efficient N-D image segmentation," *International Journal of Computer Vision* (IJCV), vol. 70, no. 2. pp. 109–131, 2006.

[8]   B. Sowmya and B. Sheelarani, "Colour image segmentation using soft computing techniques," *International Journal of Soft Computing Applications*, vol. 4, pp. 69-80, 2009.

[9]   A. Borji, M. Hamidi, and A. M. E. Moghadam, "CLPSO-based fuzzy color image segmentation," in *Proceedings of the North American Fuzzy Information Processing Society*, 2007, pp. 508-513.

[10]   H. D. Cheng, C. H. Chen, H. H. Chiu, and H. Xu, "Fuzzy homogeneity approach to multilevel thresholding," *IEEE Transactions On Image Processing* (TIP), vol. 7, no. 7, pp. 1084-1086, 1998.

[11]   P. M. Birgani, M. Ashtiyani, and S. Asadi, "MRI segmentation using fuzzy c-means clustering algorithm basis neural network," in *Proceedings of the 3rd International Conference on Information and Communication Technologies: From Theory to Applications*, 2008, pp. 1-5.

[12]   J. F. David, K. Y. Yau, and A. K. Elmagarmid, "Automatic image segmentation by integrating color-edge extraction and seeded region growing," *IEEE Transactions On Image Processing* (TIP), vol. 10, no. 10, pp. 1454-1466, 2001.

[13]   R. Adams and L. Bischof, "Seeded region growing," *IEEE Transaction on pattern analysis and machine intelligence* (PAMI), vol. 6, no. 6, pp. 641-647, 1994.

[14]   F. Y. Shih and S Cheng, "Automatic seeded region growing for color image segmentation," *Image and Vision Computing* (IVC), vol. 23, no. 10, pp. 877-886, 2005.

[15]   J. B. MacQueen, "Some methods for classification and analysis of multivariate observations," in *Proceedings of the 5th Berkeley Symposium on Mathematical Statistics and Probability*, Berkeley, University of California Press, 1967, pp. 281-297.

[16]   T. Kanungo, D. M. Mount, N. Netanyahu, C. Piatko, R. Silverman, and A. Y. Wu, "An efficient k-means clustering algorithm: Analysis and implementation," in *Proceedings of the IEEE Conference on Computer Vision and Pattern Recognition* (CVPR), 2002, pp. 881-892.

[17]   S. R. Dubey and A. S. Jalal, "Robust Approach for Fruit and Vegetable Classification", *Procedia Engineering*, vol. 38, pp. 3449 – 3453, 2012.

[18]   S. R. Dubey and A. S. Jalal, "Species and Variety Detection of Fruits and Vegetables from Images", *International Journal of Applied Pattern Recognition* (IJAPR), vol. 1, no. 1, pp. 108 – 126, 2013.

11

Open Data as a key factor for developing expert systems: a perspective from Spain

Luz Andrea Rodríguez Rojas[1], Juan Manuel Cueva Lovelle[1], Giovanny Mauricio Tarazona Bermúdez[2], Carlos Enrique Montenegro[2],

[1]*University of Oviedo, Asturias, Spain*
[2]*Francisco José de Caldas District University, Bogotá, Colombia*

Abstract — **The open data movement is relatively new but very significant, and potentially powerful. The overall intention is to make local, regional and national data available in a form that allows for direct manipulation. This paper is based on analyzing the current context of the Open Data initiative in Spain, from its origins and concepts, the legal framework, current initiatives and challenges that must be addressed for effective reuse of public sector information.**

Keywords — **Open Data, Interoperability, linked data, e-government.**

I. INTRODUCTION

THE new technologies of information and communication technologies (ICT) have dramatically changed the ways of access to public sector information facilitating information processing and its dissemination. The main producer of information are the governments, which, in the exercise of their functions, create, collect, treat, store, distribute and disseminate large amounts of information of various fields.

The e-government, the reuse of public sector information, with open data initiative has led to the spread of numerous campaigns opening public sector data and administrative transparency. Spain has not been unresponsive to these initiatives. However, despite the good position in the world ranking for the development of Open Data initiatives, the efforts made so far are still insufficient.

Artificial intelligence contributes to the advancement of government management models, where issues relating to electronic government and open, ensure transparency of data online and public accountability, and citizen participation.

This paper shows the current situation of open data initiative from a Spanish perspective, its importance, challenges and opportunities.

II. BACKGROUND

In recent years has begun a revolution called "Open Data" caused by the wide array of government data that are significant not only because of the quantity and centrality but also because most government data are public by law [1] and that entails access and reuse of public information [2].

The Open Data initiative is related to the e-government and the web presence administration and is about to publish the Public Sector Information in standard formats, open and interoperable, facilitating their access and enabling reuse.

Many central governments are making government information more easily available on the Web to the public and in formats that citizens can reuse. This occurs in response to the needs of society for transparency, participation, collaboration, innovation, accountability, economic development, job growth, cost reduction, interoperability, among other [3].

The open data value chain including the source which can be public or private, infomediaries which are responsible for data processing and end users who may be enterprises or citizens (Fig. 1).

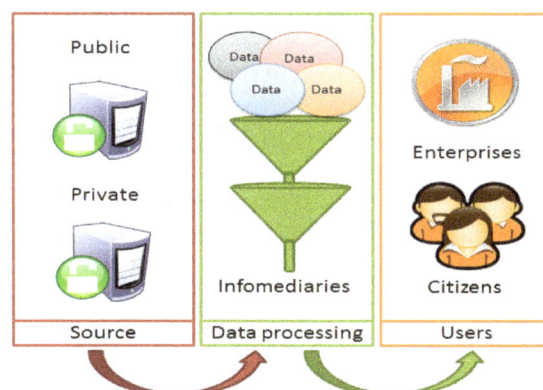

Fig. 1. Open Data value chain

Technical semantic and organizational interoperability difficulties are evident throughout the value chain of open data. From the lack of a clear legal framework and political leadership[4–6] until socio-technical impediments from the perspective of the open data user [7].

Reuse of Public Sector Information (RPSI) is the main objective of Open Data initiatives. RPSI means making public information generated by the public sector available to individuals and business. Although its great economic potential is not yet possible to quantify the benefits of this initiative.

A study, of use of open government data (OGD) from data.gov.uk [8] puts forward five processes of OGD use (fig. 2): Data to fact, information, interface, data or service.

For improving the open data process research should be about the impediments in users groups with the objective to propose strategies allowing for the accomplishment of the desired impact.

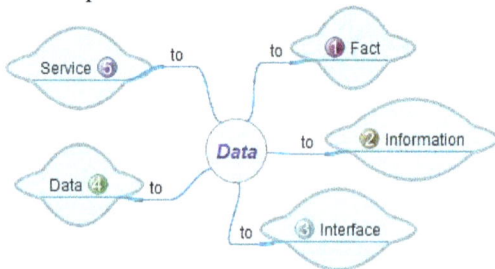

Fig. 2. Five processes of OGD

Moreover, linking information from different sources is a key for further innovation. The ultimate goal of Linked Data is to enable the use of the Web like a single global database [9] with easy accessible information obtained from different sources [10]. If data can be placed in a new context, more and more valuable applications will be generated [11]. It is clear that interoperability and standards are a key to fully benefit from OGD. Therefore, Linked Data is essential to generate an information management through standards to ensure interoperability. This needs to be considered at three levels [12]:

Organizational level: coordinated processes in which different organizations achieve a previously agreed and mutually beneficial goal.

Semantic level: precise meaning of exchanged information which is preserved and understood by all parties, enables organizations to process information from external sources in a meaningful manner.

Technical level: planning of technical issues involved in linking computer systems and services, It includes aspects such as interface specifications, interconnection services, data integration services, data presentation and exchange, etc.

Not only users, but also other stakeholders, may have greater freedom to make choices when the level of interoperability increases, there are other benefits to make interoperable government systems, for example, better data quality, the greatest opportunity in the delivery of information and finally non-tangible benefits such as gaining greater citizen trust[13]. Furthermore using the right architectural model to deploy interoperable e-government services results in a direct, positive impact on Gross domestic product (GDP) [14].

III. OPEN DATA IN WORLD

Open data is closely related to the maturity models as the e-government. The World Bank described e-government as the use of information technologies by government agencies that have the   ability to transform relations with citizens, businesses, and other arms of government [15].

Governments around the world are at various stages of e-Government readiness and implementation, European countries generally taking the top spots [16]. According to the 2012 United Nations E-government Survey rankings, the Republic of Korea is the world leader (0.9283) followed by the Netherlands (0.9125), the United Kingdom (0.8960) and Denmark (0.8889), with the United States, Canada, France, Norway, Singapore and Sweden close behind [17].

Data.gov is an initiative to put hundreds of thousands of public datasets in an easily usable and retrievable form in one location. It means democratizing access to data, it is one of the first initiatives, which since 1999, is driving the U.S. government.

The United Kingdom repository data.gov.uk is a large repository of public information that the British government launched to provide transparency to the management, data and financial issues on the UK government are an example of transparency.

The European Union (EU) recently released the beta version of its open data portal. This web page contains all public information generated by EU institutions. A study conducted ten portals, edited by public bodies member countries of the European Union in 2012 showed that most have less than two years and are in a phase of organization and implementation, so that consistency and data coverage is very uneven [18].

In Latin America, in Chile, there are three open data initiatives: an open data portal which publishes, on a single website, public information sets in more than one format, Chile Library of Congress the first Latin American state institution validated with five stars Open Data by CTIC Foundation and Open Data catalogue by the Chile's Transparency Board. In Brazil, there are open data portal and Federal Senate Legislative Open Data as national initiatives and other regional such as "Minas em Numbers" data, statistics and indicators of mines.

Several countries in Latin America are studying and making experiments with Open Data, the same is happening, on a much smaller scale, in a few parts of Asia and Africa. Kenya is the first developing country to have an open government data portal, the first in sub-Saharan Africa and second on the continent after Morocco[19].

IV. OPEN DATA IN SPAIN

A. legal framework

In 1998, the European Commission launched the "Green Paper on Public Sector Information in the Information Society" highlighting the need to improve synergies between the public and private sectors in the information market. In this Green Paper lies the foundation for the regulation of access to public information, transparency, reuse and trading it. The Green Paper proposals inspired the Directive 2003/98 of The European Parliament and of The Council of 17 November 2003 on the re-use of public sector information.

Meanwhile in Spain the law 37/2007 was passed for the regulation of the legal regime applicable to the reuse of documents produced or held by the Administrations and public sector bodies and the law 11/2007 which recognizes the right of citizens to interact with the public administrations by electronic means and regulates basic aspects of using information technologies. The Royal Decree 1495/2011 and the Royal Decree 1671, respectively, partially implemented such laws (fig. 3).

Recently, the EU Council approved the amendment of Directive 2003/98/EC which regulates the reuse of Public Sector Information. Among the innovations introduced is expanding its scope, the obligation to publish a standard format, open and processable by automatic means. On the other hand, member states will be held accountable for the implementation of the Directive every 36 months and the Commission should develop guidelines for licensing unique rates and publication preferred datasets.

Spain is working on transparency law and executing projects and initiatives in order to fulfill The European eGovernment Action Plan 2011-2015.

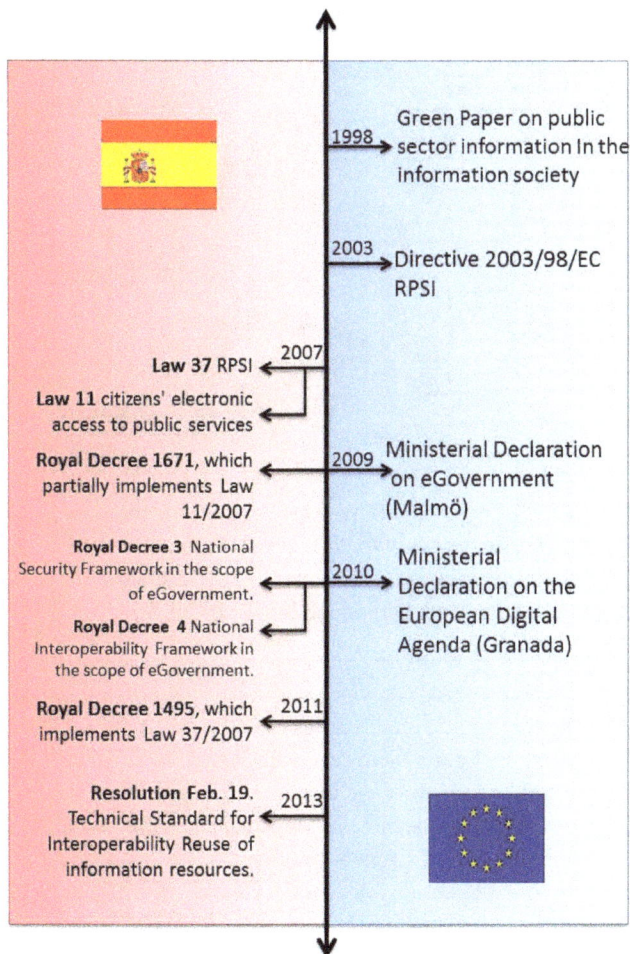

Fig. 3. Legal framework 1998-2013.

B. National and Regional initiatives

In 2005 Spain's government approved through the Avanza Plan, the first national action line dedicated specifically to promote the opening of public sector information. In 2009 it created the project Aporta, fruit Avanza2 Plan. Two of his main achievements were the launch of www.aporta.es and the

publication of the Aporta Guide. The Guide offers a series of recommendations and best practices in open public sector information and is a good starting point for any work related to Open Data. Through the Aporta Program, what has been working to boost the provision of public information by the different administrations, with several initiatives for effective implementation of the culture of reuse of information within the administration.

In March 2010 (fig. 4), the Public Information catalogue was launched, the first single access to available data sources in Spain's public sector. In June 2011, the "Characterization infomediary sector 2011" was published a study, which showed the situation of open data in Spain. In that same year, the initiative datos.gob.es was launched, taking over from project "Contribute" and assumed its commitment to open public sector information. Its main objective is to promote the publication, improve access and encourage reuse of public information corresponding to the General State Administration.

Fig. 4. Initiatives open data in Spain

Other examples of policies reuse of information is opening data from the General Directorate for Cadastre and National Geographic Institute.

The CTIC Foundation (Center for the Development of Information and Communication Technologies in Asturias) is a organization, constituted by a group of firms from the Information and Communication Technologies field, and the Government of the Principality of Asturias. The datos.fundacionctic.org is the CTIC website dedicated to Open Data. Its goal is to help government and public bodies in the publication of your information in accordance with current regulations.

There are also some regional projects that have been developed by several Public administrations to share their data, together with CTIC Foundation. Table I shows the main initiatives undertaken in provinces of Spain.

Although the government is working on the reuse of data and important steps have been taken in recent years, there is still much to be done in the field of open data.

Open data Euskadi was born in 2010 as the first open data

portal in Spain. It has been awarded the "Project Transparency Referent in central, regional or Local Government" in 2013 and shows a genuine opening of public data for reuse. In Asturias on the other hand, despite linked data, the number of published datasets is very low and not updated frequently. Gencat data catalogue uses different formats according to the data type, this variety makes it hard access. Most of the data sets published in the Open Data Junta de Castilla y León are in proprietary formats o Comma Separated Values (CSV). It particularly emphasized the launch of Open data Córdoba in 2011, with an investment of 382,000 [21] euros, which is not currently operating and for which the reasons are unknown.

TABLE I
OPEN DATA REGIONAL INITIATIVES IN SPAIN

Region	Open data initiatives	Catalogue status [20]
Andalucía	Andalucia Open Data Catalog	Open formats
	Open Data Universidad Pablo de Olavide - Sevilla	Open formats
	Open Data Córdoba	Extinct
Asturias	Asturias Public Data	Linked Data
	Gijón Public Data Catalogue	RDF Data
País Vasco	Open Data Euskadi	RDF Data
Navarra	Open Data Navarra	Open formats
	Pamplona Open Data	Open formats
Aragón	Aragon Open Data	RDF Data
	Zaragoza Public Data Catalogue	Linked Data
Galicia	Abert@s (Galicia Open Data)	RDF Data
Castilla La Mancha	Portal de Datos Abiertos de JCCM	RDF Data
Cataluña	Lleida Open Data	Open formats
	Gencat data catalogue	RDF Data
	Open Data Terrassa	RDF Data
	OpenData Sabadell	Open formats
	Badalona Open Data	Open formats
	OpenData Barcelona	RDF Data
	Observatori de la Ciutat Open Data Sant Boi	Open formats
Islas Baleares	Balearic Islands Open Data	RDF Data
Castilla y León	Open Data Junta de Castilla y León	RDF Data

Only 60% of open data catalogs of the initiatives mentioned above announced the update frequency and none makes visible downloaded datasets registration number. Also it is difficult to know the true value of such data. Furthermore Lleida and Sant Boi portals are only available in Catalan.

V. CHALLENGES AND OPPORTUNITIES

The impact of open data initiatives is emerging, while it is true, Spain has consolidated initiatives, a legislative framework to promote the reuse of public information, steady growth in electronic government there are many challenges still to face (fig. 5).

Technically speaking it is necessary to establish common parameters for all the autonomous communities regarding the information published, quality, format, licensing, processing and loading at each site. Regarding the semantic the challenge is to standardize formats and establish a standard reference vocabulary that various administrations can use to facilitate the linking of data and broaden the scope to nationwide solutions.

Finally, from an organizational perspective strong leadership from General State Administration is required in order to identify what are the data that generate added value, define indicators to measure and evaluate initiatives and recognize the importance of open data as an engine for the generation of economic benefit, transparency and improving vertical and horizontal interoperability in public administrations.

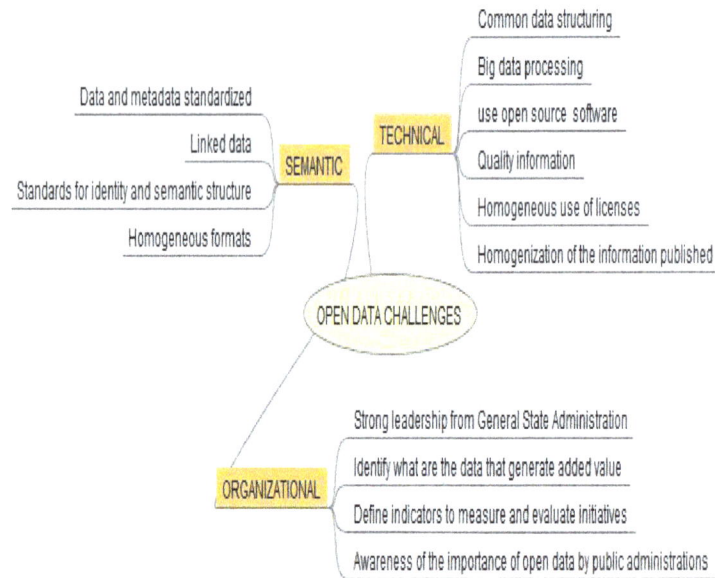

Fig. 5.Challenges open data

The actual citizen participation is essential, the frustration and lack of trust in institutions in many countries are high, so it's no surprise when people express skepticism that opening government data won't help much in fixing things.

VI. CONCLUSIONS

Despite the significant advances in the reuse of public sector information, in Spain there are still things to be done. The progress has been slow if we consider that for over 15 years, the European Commission has been working in this scope.

Probably the most difficult problems to overcome are not technical or semantic, but rather those relating to the understanding that the information belongs to everyone and that the empowerment of the public administrations is essential in order to share its information with others, and to raise awareness of the potential of opening data for reuse.

In this sense, research on new expert systems applied to the generation of an assistant to the search, extraction and understanding of information in large volumes of data is completely innovative, and allows for a tremendous value to projects that leverage the potential of Open Data.

REFERENCES

[1]   Open Knowledge Foundation, "Manual de los Datos Abiertos," *Portal de Datos Abiertos del Gobierno de Misiones*, 2012. [Online]. Available: http://www.datos.misiones.gov.ar/directorio/documentos/Manual_de_ Datos_Abiertos.pdf.

[2]   A. Naser and G. Concha, "Datos abiertos: Un nuevo desafío para los gobiernos de la región," *Naciones Unidas*, 2012. [Online]. Available: http://www.eclac.org/publicaciones/xml/7/46167/DatosAbiertos_17_0 4_2012.pdf.

[3]   Proyecto Aporta, "Guía Aporta sobre Reutilización de la Información del sector Público," 2011.

[4]   N. Huijboom and T. Van Den Broek, "Open data: an international comparison of strategies," *European Journal of ePractice*, pp. 1–13, 2011.

[5]   M. Fioretti, "Open Data : Emerging trends , issues and best practices administration," 2011. [Online]. Available: http://www.lem.sssup.it/WPLem/odos/odos_report_2.pdf.

[6]   E. D. Rico and R. M. Calenti, "Open Data y RISP: generando valor social y económico. Transparencia e innovación en la eAdministración," *Santiago de Compostela: Colexio Profesional de Enxeñaría en Informática de Galicia*, 2011.

[7]   A. Zuiderwijk, M. Janssen, and S. Choenni, "Socio-technical Impediments of Open Data," *Electronic Journal of eGovernment*, vol. 10, no. 2, pp. 156–172, 2013.

[8]   T. Davies, "Open data, democracy and public sector reform. A look at open government data use from data.gov.uk," University of Oxford, 2010.

[9]   C. Bizer, T. Heath, and T. Berners-Lee, "Linked data - The story so far," *International Journal on Semantic Web & Information Systems*, vol. 5, pp. 1–22, 2009.

[10] J. L. Redondo-García, V. Botón-Fernández, and A. Lozano-Tello, "Linked Data Methodologies for Managing Information about Television Content," *International Journal of Interactive Multimedia and Artificial Intelligence*, vol. 1, no. 6, p. 36, 2012.

[11] F. Bauer and M. Kaltenböck, "Linked Open Data: The Essentials," 2011.

[12] F. B. Vernadat, "Technical, semantic and organizational issues of enterprise interoperability and networking," *Annual Reviews in Control*, vol. 34, no. 1, pp. 139–144, 2010.

[13] G. Concha and A. Naser, "El desafío hacia el gobierno abierto en la hora de la igualdad," 2012.

[14] L. Madrid, "The Economic Impact of Interoperability Connected Government," *Microsoft Corporation*, 2012.

[15] The World Bank, "e-Government," 2011. [Online]. Available: http://go.worldbank.org/M1JHE0Z280. [Accessed: 15-Feb-2013].

[16] W. Olatokun and B. Adebayo, "Assessing E-Government Implementation in Ekiti State, Nigeria," *Journal of Emerging Trends in Computing and ...*, vol. 3, no. 4, pp. 497–505, 2012.

[17] United Nations Department of Economic and Social Affairs, "E-Government Survey 2012," 2012.

[18] F. Ramos, R. Arquero, I. Botezán, S. Cobo, A. Sala, R. Sánchez, and F. del V. Gastaminza, "De la reutilización de información del sector público a los portales de datos abiertos en Europa," *BiD: textos universitaris de biblioteconomia i documentació*, vol. 29, 2012.

[19] "Kenya openData," 2013. [Online]. Available: https://www.opendata.go.ke/.

[20] Fundación CTIC, "Public Dataset Catalogs Faceted Browser," 2013. [Online]. Available: http://datos.fundacionctic.org/sandbox/catalog/faceted/.

[21] Diario Córdoba, "Nace la plataforma ´Open data Córdoba´ para buscar empleo o crear empresas," 2011. [Online]. Available: http://www.diariocordoba.com/noticias/cordobalocal/nace-plataforma-open-data-cordoba-para-buscar-empleo-o-crear-empresas_625578.html.

12

An Analysis Architecture for Communications in Multi-agent Systems

Celia Gutiérrez
Complutense University in Madrid, Spain

Abstract — Evaluation tools are significant from the Agent Oriented Software Engineering (AOSE) point of view. Defective designs of communications in Multi-agent Systems (MAS) may overload one or several agents, causing a bullying effect on them. Bullying communications have avoidable consequences, as high response times and low quality of service (QoS). Architectures that perform evaluation functionality must include features to measure the bullying activity and QoS, but it is also recommendable that they have reusability and scalability features. Evaluation tools with these features can be applied to a wide range of MAS, while minimizing designer's effort. This work describes the design of an architecture for communication analysis, and its evolution to a modular version, that can be applied to different types of MAS. Experimentation of both versions shows differences between its executions.

Keywords — Analysis, architecture, bullying, communications, multi-agent systems.

I. Introduction

COMMUNICATIONS become complex to design in huge systems which interact frequently. In MAS, interactions among agents must be designed correctly to avoid behaviors that may collapse communications. The overall result of these behaviors is high response times, among other problems. Within this context, communication analysis techniques become relevant to evaluate the correct performance of the MAS. These techniques inspect the communications among agents in executions, to detect undesirable patterns of communications, like agents that are overloaded with the reception of too many messages. Once the undesirable situation is detected, the re-design the MAS communications is a straightforward task [1]. Other non desirable situations appear when there are not expected sequences of agents that interact in a conversation [2].

The effect of overloading can be compared to bullying, as explained in [3]. There are agents that play the bully role, when they send too many messages; other agents play the mistreated role when they received too many messages; other agents that play both roles, mistreated and bully; other ones that are considered as isolated because they neither send nor receive messages; and there are regular agents that behave correctly because they send and receive messages in a balanced way. There are metrics to measure the proportion of sent and received messages; these metrics are the values to classify agents into the mentioned patterns. The detection of non desired patterns in certain conversations can help the designer to modify the interactions, obtaining better response times and higher QoS results, [1], [3].

Previous frameworks for the analysis of these behaviors have been designed embedding the evaluation and debug tools within the execution of the MAS. Results can be inspected after the execution, and in consequence a straightforward re-design can be made.

Despite the satisfactory results obtained with this approach, reusability for other types of MAS becomes a difficult task, that involves re-codification of the evaluation and debug functionality. An efficient architecture is basic for the designer/tester, not only to obtain satisfactory results, but also to reuse the analysis tool in other type of MAS.

This work represents one step forward in architectures for MAS analysis. We provide a new framework for the MAS execution and evaluation in order to reach complete independence of both tasks. The result is a new architecture with two modules: one for the execution and another for the evaluation and debugging.

This research is presented in the following order: Section 2 describes the related work. The description of the new architecture is within Section 3. The results of the execution of the new architecture are included in Section 4. Finally, conclusions and future work can be found in Section 5.

II. Related Work

Literature regarding load balancing in MAS is relevant and plentiful. This problem has been focused using different strategies. [4] apply learning techniques in MAS load balancing. The task of the agent is to choose the correct resource using local information. Its objective is to optimize the resource usage. Unlike our work objective, they are not concerned in the scalability and adaptability of their solution to other problems or platforms.

The work in [5] resembles ours because they also use classification techniques and metrics to analyze the organization of MAS. They also relate their metrics and the response time, which is used as indicator of QoS. But it differs our work in the use of their metrics, which are used just to evaluate architectures; instead, we present an architecture to evaluate the communications in MAS.

AntNet [6], Challenger [7], and DIET (Decentralised Information Ecosystem Technologies) [8] use mobile agents to use their respective resources equitably, but they do not identify the cause of the overloading/bullying problem. DIET overcomes multi-agent platforms limitations in terms of adaptability and scalability, providing a foundation for an open, adaptive and scalable agent organization. In this way, they share the same interests as we do, but they are focused on supporting basic mobile agent capabilities.

Messor [9] uses adaptive system approach. It uses an algorithm that emulates the ant behavior to distribute workload among distributed nodes. In this case, they are specially focused in peer-to-peer systems.

Other work, the Anticipate Agent Assistance (AAA) [10] also uses an agent-based metric for testing and managing the resource information of the wireless points, choosing the less overloaded access points. They are also concerned in achieving high QoS indicators of communications. However, they have confined their solution to the wireless networks.

Finally, [11] perform debugging process on recorded data of the MAS execution, like in the current work. Their analysis helps understand the behavior of the system and can reveal undesirable social behaviors. So their testing and debugging of complex MAS remains just at social level.

In summary, there are works that are concerned in achieving equitable behaviors of agents in MAS executions. All of them differ in the way they make the analysis, design, or evaluation, and their purpose: ones are focused on load balancing in general, others on load balancing in communications, and others in social behaviors. But neither of them has the purpose of building a scalable architecture of MAS to evaluate its communications. This architecture can integrate the elements which are present in MAS communications, as the following section describes.

III. DESIGN OF THE NEW ARCHITECTURE

The new architecture, called IDKAnalysis 2.0, is based on a previous version, IDKAnalysis 1.0.

Both architectures follow the Ingenias methodology [12] and have been executed on Ingenias Development Kit (IDK) case studies, although they use different versions of IDK (IDKAnalysis 1.0 uses IDK 2.7, whereas IDKAnalysis 2.0 uses IDK 2.8). IDK versions use a template (build.xml) to detail the agent deployment of the case study one wants to run. At the same time, user inputs can be necessary to start the case study activity, although these inputs vary on each case study. Further details on this framework can be seen at [13]. Both versions of IDK are available at http://sourceforge.net/projects/ingenias/files/INGENIAS%20D evelopment%20Kit/Aranjuez/, on their corresponding option.

Fig. 1 shows the differences of both versions of the architectures:

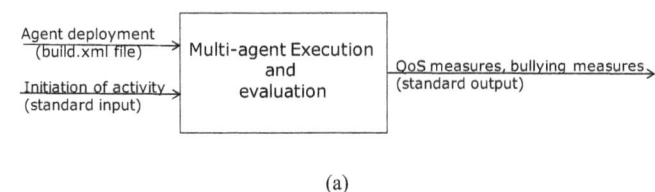

(a)

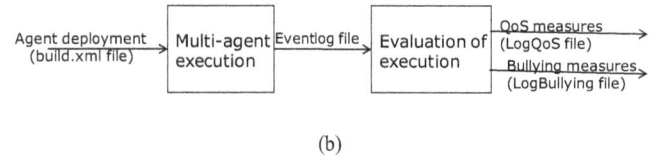

(b)

Fig. 1. Block diagram of the IDKAnalysis, version 1.0 (a) and 2.0 (b).

The first version performs the MAS execution and evaluation at the same time. The outputs only refer to the analysis, and extract the analysis measures and QoS measures. The second version is based on an architecture with a front-end that executes the main functionality of the MAS, and a back-end that analyzes the communications generated by the front-end.

There are also differences in the inputs and outputs of both architectures:

- In the first one, apart from the agent configuration, it is necessary human interaction to start the activity, whereas in the second one, the execution starts automatically (without the user input).
- The outputs of the first version are shown at the same time. In the second version, the front-end outputs a log file with the events recorded; the back-end receives as input the event log file, and produces the two outputs physically separated in two files.

Inputs and outputs of the back-end are described and analyzed in the following subsections.

A. Event log file

The event log file registers the main events of the MAS execution with certain format that corresponds to the main features of these events. The generation of this file is a characteristic functionality of IDK2.8.

The standard format of a line is as follows:

```
Timestap(hours:minutes:seconds:milesecond)
;Name of the event;Additional fields
```

Additional fields depend on the type of event it represents. Below there is an example of the event that represents *A new entity is added to the agent mental state*:

```
23:53:47:187;MEAddedToMentalState;BuyerAge
nt_0multipleBuyers@viriato:60000/JADE!Curr
entAssistedAgent!ME0
```

where the content of the additional fields are:

```
involved agent ->
BuyerAgent_0multipleBuyers@viriato:60000/J
ADE
```
kind of entity -> `CurrentAssistedAgent`
entity id -> `ME0`

To register all communication information, the types of

events of this version include message shipping and reception events, and others that are necessary to measure response times. Even more, with the intention of using the event log file for other purposes than communication analysis, a wide range of types of events is included:

1) A new entity is added to the mental state.
2) An agent was initialized completely.
3) A task was scheduled within the agent.
4) A task was executed.
5) An agent is starting collaboration as initiator.
6) An agent has accepted to participate in an interaction as collaborator.
7) An agent has received a request to participate in collaboration.
8) A mental entity has been removed from the conversation.
9) An agent received a message.
10) An agent sent a message.
11) An entity was added to a conversation.

In IDK 2.8 the name of the event log file is generated in such a way that it contains the day, month, year, hour and minute of its creation.

An excerpt of an event log file can be found in the Appendix section.

B. Outputs of the Evaluation Module

As Fig. 1 (b) shows, the outputs of the second module are the QoS measures (in this case response times) and the bullying measures. This module is coded in Java, JDK1.7.0_04. For this purpose, there are two types of events selected from the event log file.

The first output depends on each case study and basically is the time elapsed since a service is requested until an offer of that service is proposed. For MAS with a lot of service responses (as a consequence of having many agents offering services), it is may be useful to establish a number of iterations or responses until a response time is recorded.

It is necessary to choose the task when the time measuring process initiates and the task when it finishes. The response time is the elapsed time between them. This depends on each case. In the experimentation of Section 4, the initiating task is ChooseMovie, and the finishing task is ChooseCinema. The type of event that records the executed task is *TaskExecuted*. In the example below, the log refers to the starting time of execution of ChooseMovie task.

```
18:22:02:355;TaskExecuted;InterfaceAg
ent_3expInterfaceAgentwithprofile!ChooseMo
vie!ME103705
```

where the additional fields mean:

```
involved agent ->
InterfaceAgent_3expInterfaceAgentwithprofi
le
task type -> TaskExecuted
```

```
task name -> ChooseMovie
task id -> ME103705
```

The second output is the bullying measures, which are described in detail in a previous work [3]. In this case, *MessageReceived* event is used each time a message is received by an agent, as in the following example:

```
23:53:48:885;MessageReceived;BuyerAssignme
nt!0.InterfaceAgent_9multipleInterfaceAgen
tsvir1225148028355!RejectBecomingAssistant
!BuyerAgent_4multipleBuyers!InterfaceAgent
_9multipleInterfaceAgents
```

where the additional fields are:

```
protocol -> BuyerAssignment
conversation id ->
0.InterfaceAgent_9multipleInterfaceAgentsv
ir1225148028355
protocol state from which the message is sent ->
RejectBecomingAssistant
sender -> BuyerAgent_4multipleBuyers
receiver ->
InterfaceAgent_9multipleInterfaceAgents
```

In this way, information about senders and receivers is enough to compute the measures of [3] and start the evaluation process.

Although the measures are standard for any type of MAS with agents playing different roles, the designer must also specify which the role is going to be analyzed as bully, and which one as the mistreated. Besides, he must tune a threshold. As explained in [3], the computed measures are compared with the indicated values for each pattern, although a margin between both values is established as threshold.

Considering that all these features must be customizable for executions of other types of MAS, this module contains the following parameters:

1) Path of the Eventlog file, LogBullying file, QoS file.
2) Name of the LogBullying file
3) Name of the Qos file.
4) Role that is suspected to be the Bully in the conversations.
5) Role of that is suspected to be the Mistreated in the conversations.
6) Threshold for the bullying metrics.
7) Number of iterations that a task must be executed to calculate the response time.
8) The initial task that must be executed to start the response time counting.
9) The final task that must be executed to end the response time counting.

C. Advantages of IDKAnalysis 2.0 over IDKAnalysis 1.0

Case studies built under IDKAnalysis 2.0 offer several aspects of the executions that make it applicable to other case studies. These features appear on each module:

1) The event log file generated by the first module does not only record communication related to events, but also other events that can be analyzed for different purposes.

2) The second module produces two different files, so bullying measures and response times can be analyzed separately. Besides, this module contains some parameters that can be tuned, so it can be adapted to other methodology case studies.

Figs. 4 and 5 (in the Appendix section) show the running architectures of both versions using the experimentation described in Section 4. Fig. 4 (a) shows the architecture of the first version, where the distinction between the front-end and the back-end does not exist. The second version in Fig. 4 (b) contains the *srceclipse* package, which is the back-end, whereas the rest of the packages compose the front-end. The *srceclipse* package, which does not appear in Fig. 4 (a), is also composed of the *bullying* package and the *logs* package, as Fig. 5 shows. The first one contains the source and binary files for the evaluation process, and the second one is the directory where the log files (inputs and outputs) are placed. As explained in the previous subsection, this directory is the first parameter the designer/tester can customize.

IV. EXPERIMENTATION RESULTS

Executions of both versions have been carried out using the Cinema case study, pursuing the objective of acquiring cinema tickets according to certain user's preferences. The participant roles are the following:

- *Interface agent*, which represents the customer.
- *Seller agent*, which represents the cinema.
- *Buyer agent*, which represents the intermediary between the Seller and the Interface.

The hardware of the experimentation has been a machine with 2 GHz and 2GB RAM, using 32-bit Windows 7 Professional.

The Cinema case study uses Java Agent DEvelopment (JADE) platform. JADE framework uses the Foundation for Intelligent Physical Agents (FIPA) standard for communications among agents.

As table 1 shows, configurations with different numbers of agents for each role have been run:

TABLE I
CONFIGURATIONS FOR CINEMA CASE STUDY

Configuration	Number of Interface Agents	Number of Seller Agents	Number of Buyer Agents
Serious	10	5	10
Simple	20	4	20
FullSystem	100	8	100

The following subsections include examples of executions on both versions of the tool.

A. Execution using IDKAnalysis 1.0

The Cinema case study begins with two possible options for the use, as Fig. 2 shows.

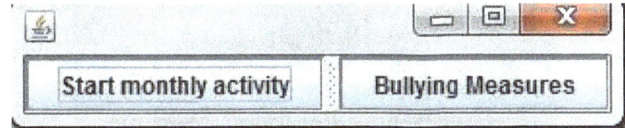

Fig. 2. Initial GUI of the Cinema case study built with IDKAnalysis 1.0

It is necessary to start running by selecting *Start monthly activity*. This will produce the conversations between the agents, in order to get the proposed tickets. As this is not the relevant part of this work, no output has been extracted. Then, *Bullying Measures* can be selected, to obtain the values for the bullying metrics and response times from the generated communications.

A snapshot of this execution on console can be seen in Fig. 3, where the metrics and classification for IntergaceAgent_16, IntergaceAgent_19, IntergaceAgent_18 agents, and the corresponding values for the roles and the system, can be seen alongside the extraction of a response time.

B. Execution using IDKAnalysis 2.0

Mentioned parameters in subsection 3.B, numbered from 4 to 9, have been tuned as follows:

- Role that is suspected to be the Bully in the conversations: Interface
- Role of that is suspected to be the Mistreated in the conversations: Buyer
- Threshold for the bullying metrics: 1.0
- Number of iterations that a task must be executed to calculate the response time: 10
- The initial task that must be executed to start the response time counting: ChooseMovie
- The final task that must be executed to end the response time counting: ChooseCinema

In this way, the response time which is recorded, is the elapsed time between the ChooseMovie task and the tenth occurrence of the ChooseCinema task.

In the Appendix section, there are examples of the two outputs generated by the IDKAnalysis 2.0 using the FullSystem configuration. They are generated in two separate files, to facilitate the designer analysis.

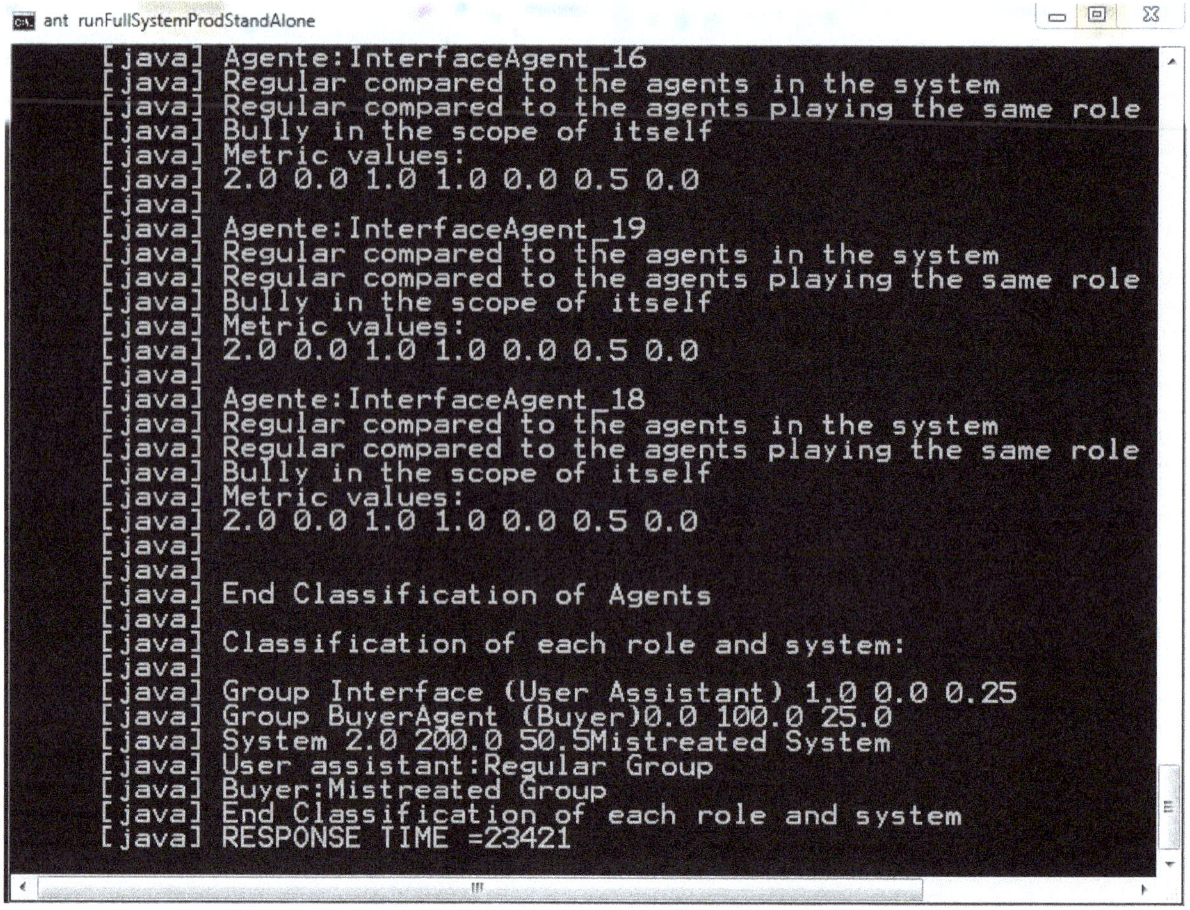

Fig. 3 Output of the IDKAnalysis 1.0 for a FullSystem configuration.

V. FUTURE WORK AND CONCLUSIONS

In this work we have presented a new framework which separates the multi-agent system execution and the evaluation of the communication among agents.

This perspective provides several advantages from the Software Engineering point of view:

- To work on the functionality or the evaluation process directly, by introducing changes in the front-end (for the first purpose), or the back-end (for the second purpose).
- To inspect bullying behaviors and QoS measures separately, by the analysis of the LogBullying file (in the first case), or the QoS file (in the second case).
- To reuse the evaluation module in other case studies, by tuning some parameters accordingly to each multi-agent
- system circumstances. The range of events generated by IDK8.0 (and IDKAnalysis2.0 in consequence) offers different possibilities to record QoS, which does not necessarily use the TaskExecuted event, but other ones.

This architecture offer several possibilities of future work. It is thought to use the evaluation module in MAS with different purposes and frameworks:

- ADELFE methodology [14] for Adaptive MAS.
- ICARO-T framework [15] for agent organizations. Available at http://icaro.morfeo-project.org/
- Agent Based Social Simulation frameworks.

The combination of IDKAnalysis 2.0 with the above methodologies will provide experimentation outputs with two purposes:

1) Validate and enlarge the evaluation framework with the experimentation results. In particular, it is necessary a previous extraction of the event logs. These logs must accomplish the basic format of the log file mentioned in subsection 3.A. Even more, as log extraction is used for other purposes, an ontology may be parsed to get the correct parameters for each purpose. This new component and other ones will be incorporated in a new version of the tool, IDKAnalysis 3.0.
2) Enlarge the mentioned methodologies and frameworks from the AOSE point of view, with a complete module that provides testing and debugging tools.

ACKNOWLEDGMENT

C. Gutiérrez thanks Jorge Gómez-Sanz for his continuous help on the use of IDK releases, and also for providing the infrastructure for the Cinema case study.

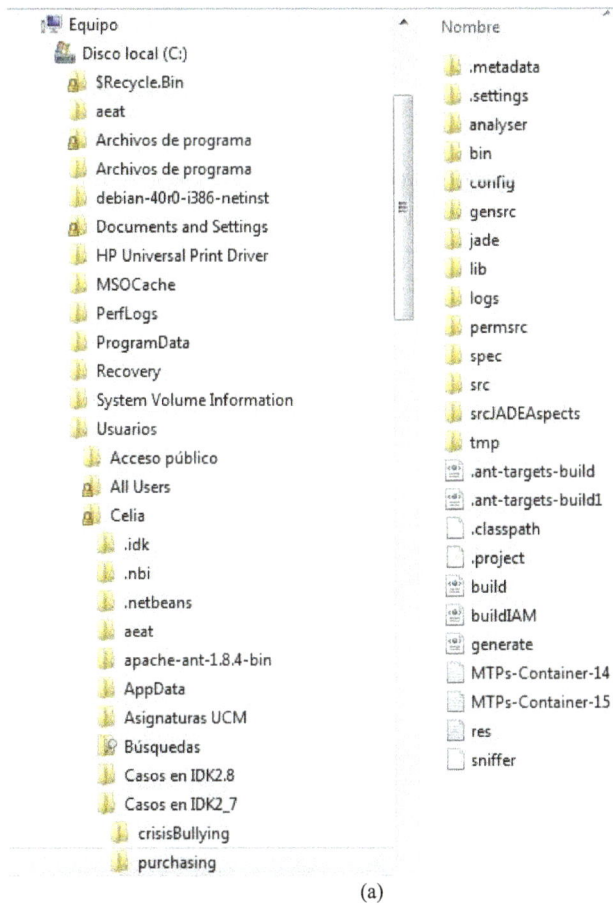

(a)

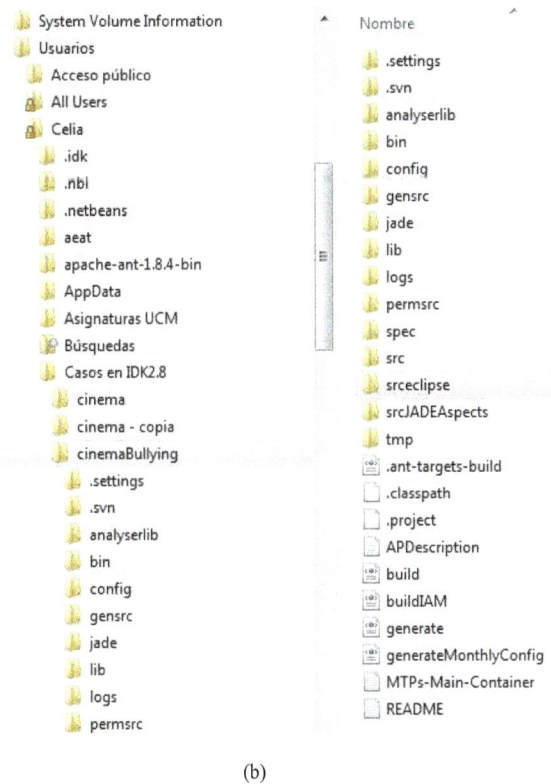

(b)

Fig. 4 A snapshot of the running architecture top level in IDKAnalysis1.0 (a) and IDKAnalysis2.0 (b).

APPENDIX

This section contains two types of information:

1) Snapshots of the running architecture of the Cinema case study using IDKAnalysis1.0 and IDKAnalysis2.0. In the first snapshot, belonging to IDKAnalysis1.0, the package deployment does not show the distinction between the front-end and the back-end. This fact is reflected in the second snapshot, belonging to IDKAnalysis2.0. The third snapshot shows the content of the back-end. Further explanations are provided in subsection 3.C.

2) Samples of the input and outputs of the evaluation module for the execution of the Cinema case study, using the parameter configuration in subsection 4.B.

- This is an excerpt of an event log file. Each line contains the information of an event, according to the syntax described in subsection 3.A:

```
18:21:26:770;TaskExecuted;InterfaceAgent_6
5expInterfaceAgentwithprofile!Look_for_an_
assistant!ME1044
18:21:26:770;MessageSent;BuyerAssignment!0
.InterfaceAgent_67expInterfaceAgentwithpro
filePC-
1227028885694!enable!InterfaceAgent_67expI
nterfaceAgentwithprofile!BuyerAgent_6expBu
yerAgentWithProfile@PC-
sheilacg:60000/JADE,
18:21:26:770;TaskScheduled;InterfaceAgent_
7expInterfaceAgentwithprofile!Look_for_an_
assistant!ME1167![ME19:GetAssignments]
18:21:26:770;MessageSent;BuyerAssignment!0
.InterfaceAgent_67expInterfaceAgentwithpro
filePC-
1227028885694!RequestBeingAssistant!Interf
aceAgent_67expInterfaceAgentwithprofile!Bu
yerAgent_6expBuyerAgentWithProfile@PC-
sheilacg:60000/JADE,
18:21:26:786;MessageSent;BuyerAssignment!0
.InterfaceAgent_1expInterfaceAgentwithprof
ilePC-
1227028886568!enable!InterfaceAgent_1expIn
terfaceAgentwithprofile!BuyerAgent_6expBuy
erAgentWithProfile@PC-sheilacg:60000/JADE,
```

- This is an excerpt of the LogBullying file. It reflects the classification values and measures for one of the Interface agents, both roles and the whole system:

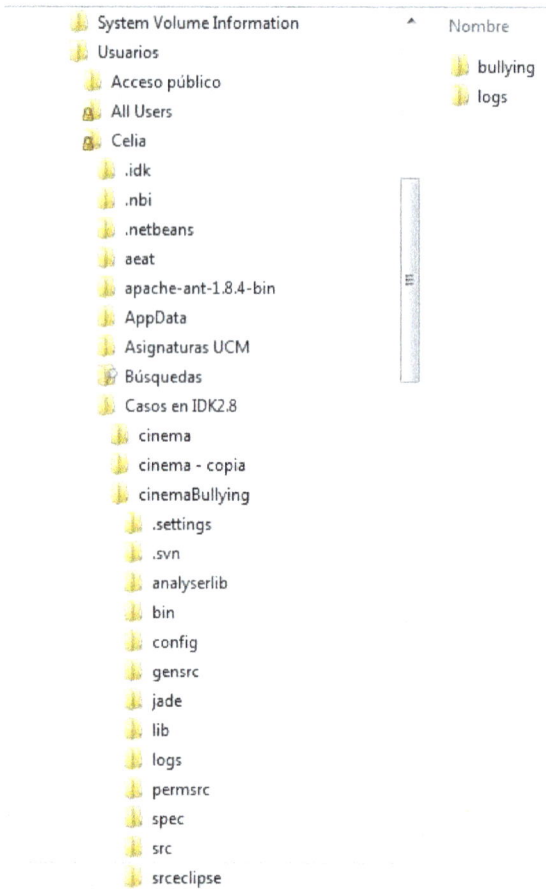

Fig. 5 A snapshot of the running architecture second level (back-end) in IDKAnalysis2.0.

```
Agente:InterfaceAgent_7expInterfaceAgentwi
thprofile
numOutputAgent =26.0 NumAgent =40
numOutput = 514.0 Bully proportionally to
the bully agents in the system
Regular compared to the agents playing the
same role
Bully in the scope of itself
Metric values:
2.0233462 0.0 1.0116731 0.50583655 0.0
0.25291827 0.0

End Classification of Agents

Classification of each role and system:

Group CoordA 0.0 0.0 0.0
Group NetworkA 0.28266892 4.5903044
1.2182432
System 2.2613513 36.722435
9.745946Mistreated System
Bully System
CoordA:Regular Group
NetworkA:Mistreated Group
End Classification of each role and system
```

- This is an exerpt of the LogQoS file. Each line contains the response times (in milliseconds) obtained with a frequency of 10 iterations:

```
10 iterations 6755
20 iterations 8106
30 iterations 8994
40 iterations 9511
```

REFERENCES

[1] C. Gutiérrez, I. García-Magariño, and R. Fuentes-Fernández, "Detection of undesirable communication patterns in multi-agent systems," *Engineering Applications of Artificial Intelligence,* vol. 24, no. 1, pp. 103-116, 2011.

[2] C. Gutierrez and I. García-Magariño, "Extraction of execution patterns in multi-agent systems," *IEEE Latin America Transactions,* vol. 8, no. 3, pp. 311-317, 2010.

[3] C. Gutiérrez and I. García-Magariño, "Revealing bullying patterns in multi-agent systems," *Journal of Systems and Software,* vol. 84, no. 9, pp. 1563-1575, 2011.

[4] S. Andrea, Y. Shoham, and M. Tennenholtz, "Adaptive Load Balancing: A Study in Multi-Agent Learning," *Journal of Artificial Intelligence Research,* vol. 2, no. 1, pp. 475-500, 1994.

[5] S. K. Lee and C. S. Hwang, "Architecture modeling and evaluation for design of agent-based system," *Journal of Systems and Software,* vol. 72, no. 2, pp. 195-208, 2004.

[6] R. Schoonderwoerd, O. Holl, J. Bruten, and L. Rothkrantz, "Ant-based load balancing in telecommunications networks," *Adaptive Behavior,* vol. 5, no. 2, pp. 169-207, 1996.

[7] A. Chavez, R. Moukas, and P. Maes, "Challenger: A Multiagent System for Distributed Resource Allocation," *Autonomous Agents,* vol. 97, pp. 323-331, 1997.

[8] C. Hoile, F. Wang, E. Bonsma, and P. Marrow, "Core Specification and Experiments in DIET: A Decentralised Ecosystem-inspired Mobile Agent System," in *Proc. 1st Int. Conf. Autonomous Agents and Multi-Agent Systems,* Bologna, 2002, pp. 623-630

[9] A. Montresor and H. Meling, "Messor: Load-Balancing through a Swarm of Autonomous Agents," in *Proc. 1st Workshop on Agent and Peer-to-Peer Syst.,* Bologna, 2002, pp. 125–137.

[10] Y. C. Chen and W. Y. Chen, "An agent-based metric for quality of services over wireless networks," *Journal of Systems and Software,* vol. 81, no. 10, pp. 1625-1639, 2008.

[11] E. Serrano, A. Quirin, J. Botia, and O. Cordón, "Debugging complex software systems by means of path finder networks," *Information Sciences,* vol. 180, no. 5, pp. 561-583, 2010.

[12] J. Pavón and J.J. Gomez-Sanz, "Agent Oriented Software Engineering with INGENIAS," in *Proc. 3rd Int. Central and Eastern European Conf. on Multi-Agent Systems,* vol. 2691, Prague, pp. 394-403, 2003.

[13] J. J. Gomez-Sanz, R. Fuentes-Fernández, J. Pavón, and I. García-Magariño, "INGENIAS Development Kit: a visual multi-agent system development environment," in *Proc. 7th Int. Conf. Autonomous Agents and Multiagent Systems,* 2008, pp. 1675-1676.

[14] G. Picard and M-P. Gleizes, "The ADELFE Methodology," *Designing Adaptive Cooperative Multi-Agent Systems,* chapter 8, pp. 157-176, Kluwer Publishing, 2004.

[15] J. M. Gascueña, A. Fernández-Caballero, and F. J. Garijo, " Programming Reactive Agent-based Mobile Robots using ICARO-T Framework," in *2010 Proc. ICAART Conf.*, vol. 2, pp. 287-291.

13

Accessing Wireless Sensor Networks Via Dynamically Reconfigurable Interaction Models

Maria Cecília Gomes, Hervé Paulino, Adérito Baptista, Filipe Araújo,
CITI/Departamento de Informática, Faculdade de Ciências e Tecnologia, Universidade Nova de
Lisboa, Caparica, Portugal

Abstract — The Wireless Sensor Networks (WSNs) technology is already perceived as fundamental for science across many domains, since it provides a low cost solution for environment monitoring. WSNs representation via the service concept and its inclusion in Web environments, e.g. through Web services, supports particularly their open/standard access and integration. Although such Web enabled WSNs simplify data access, network parameterization and aggregation, the existing interaction models and run-time adaptation mechanisms available to clients are still scarce.

Nevertheless, applications increasingly demand richer and more flexible accesses besides the traditional client/server. For instance, applications may require a streaming model in order to avoid sequential data requests, or the asynchronous notification of subscribed data through the publish/subscriber. Moreover, the possibility to automatically switch between such models at runtime allows applications to define flexible context-based data acquisition. To this extent, this paper discusses the relevance of the session and pattern abstractions on the design of a middleware prototype providing richer and dynamically reconfigurable interaction models to Web enabled WSNs.

Keyword — *Web Enabled* Wireless Sensor Networks, Dynamic Interaction Models, Design Patterns

I. Introduction

Simulation applications for unexpected but extreme events like large-scale flooding, hurricanes, severe droughts, etc., demand the access to different types of data collected across wide scale geographic areas, and for long periods of time. Only large amounts of diverse data support more precise information extraction and knowledge, concerning a better evaluation of complex events of this kind.

Wireless Sensor Networks (*WSNs*), in particular, offer a good low cost solution for such large-scale environmental monitoring since they comprise a high number of sensor devices deployed throughout the geographic area to be evaluated. Current WSNs may include different types of sensors, spanning from simple and static devices to increasingly complex mobile devices. WSNs allow hence the development of more or less elaborated applications [1] to which the interaction with the real world is a pressing requirement. Such includes not only more traditional

applications like the ones mentioned above, but WSNs also allow the surge of novel ones. This is the case of the *Participatory Sensing* area [2] where applications like urban traffic management or virtual communities' support typically rely on data acquisition and dissemination through mobile devices (e.g. using sensors embedded in private cars and mobile phones).

Nevertheless, one disadvantage of WSNs is still their low-level limited interfaces. To this concern, high-level abstractions have been used to simplify WSNs access, allowing their representation as data streams, databases, or through mobile agent models, for instance. Likewise, abstracting WSNs as Web services [3][4] allows their inclusion in Web environments, e.g. in the context of business processes. Namely, the service paradigm via standard Web technologies supports a uniform and simple access to WSNs, their parameterization and aggregation, and the systematic access to collected data.

A service-based access to WSNs also allows their integration with very different systems, since the service paradigm provides a uniform access to, and aggregation of, distinct entities. One example may be the seamless integration of WSNs providing online, almost real-time, data acquisition with Cloud-based applications consuming that data. In fact, and considering the perceivable trend on making everything accessible as a service (*XaaS*), the service concept may provide a powerful but simple abstraction for heterogeneous systems' access, interaction, and integration, may those systems be *Web enabled WSNs*, *Internet of Things* entities (*IoT*) [5][6], *Grid* or *Cloud* computing services (for standardization efforts in this area see [7]), etc.

Nevertheless, the access to those types of services may have requirements behind the traditional request/response interaction, demanding therefore dynamic/richer interaction models [8]. For instance IoT entities having one single client (the owner) may be interfaced through a stateful Web service. Cloud computing services, in turn, may interface stateful resources or long running activities which need to be inspected in terms of resource consumption, dynamic requirements, or overall cost [9][10]. Considering specifically Web enabled WSNs, sensors may have to be inspected/interrogated (e.g. in terms of sensor autonomy evaluation and sensing frequency) and also be modified (e.g. sensor parameterization).

Additionally, sensing data may have to be acquired with different *QoS* depending on contextual information (e.g. sensing data streaming on an emergency situation versus periodic data notification for sensors' autonomy preservation).

WSNs accesses may consequently be modeled as Web services interfacing *stateful resources* [11][12] requiring the realization of a dynamic/variable state which has to be kept consistent along several message exchanges between a service and each one of its clients [13]. This is captured in the *Web Services Resource Framework* (*WSRF*) norm [14], which had its origin in the context of Grid computing [13] in order to represent the access to typical, long running, *High Performance Computing* applications. Consequently, such Web enabled WSNs may benefit from richer/dynamic interaction models for sensor data acquisition and dissemination that however are not generally available in current solutions.

The following sections describe the dimensions concerning such limitations and propose a solution towards richer interactions for Web enabled WSNs access. Subsequent Sections IV and V describe, respectively, the implementation architecture and an application scenario. The conclusions are described in the final section as well as future work.

II. PROBLEM DIMENSIONS

Consider an emergency application for a critical area prone to cyclic wild fire situations. In order to more accurately calculate a fire ignition probability [15], simulation applications in this domain benefit from consuming almost real-time/online sensing data provided by different types of WSNs deployed in the area, e.g. temperature, humidity and wind characteristics' monitoring. Under normal conditions, temperature data acquisition from a single type of sensors may be enough. However, in the presence of draught weather conditions, more precise temperature data may be needed, e.g. collected from different sensors at different heights. Moreover, if a fire ignition does occur, different types of data like wind velocity and direction are also needed.

In case a client uses a traditional request/reply interaction model to collected sensing data, several independent client requests are necessary in order to process enough quantities of different data. One solution is to support the collective data processing and dissemination as a single interaction action, similarly to what happens in the *mashups* concept. Moreover, the supporting system should allow the dynamic selection of those data sources at runtime. Additionally, and due to the low autonomy of typical sensing devices, the *QoS* in terms data acquisition rate and delivery should be low under normal situations, e.g. winter time for the fire application, and high in emergency situations.

Other requirements may also be considered. For instance, critical data may be needed not only to the fire simulation's execution, but also to the firemen deployed in the area. Namely, these may be using mobile devices for their coordination and relevant data may now depend upon their geographical location (e.g. data collected at the vicinities of the firemen's position). Likewise, if the mobile devices have already a low battery level, a data stream cannot be processed anymore, but sporadic data delivery is required instead.

Whatever the clients' perspective, the most adequate solution is to provide flexibility on data sources' dynamic selection and aggregation, and also in terms of the data acquisition rate. Such may be supported through selecting an adequate interaction model between the service and its clients, at some point in time. The supporting system should also provide their dynamic modification based on context data. For instance, a *Streaming* model is preferable for a continuous sensing data delivery; a *Producer/Consumer* is necessary if there are data delivery requirements; and a Publish/subscriber model is more adequate whenever low rate data transmission is enough.

Having defined such a (more or less) complex monitoring scenario for different Web enabled WSNs data sensing, its reuse for related clients/applications may also be useful. For example, the described scenario could be used in the context of a similar tornado simulation application for the area. Likewise, in case additional firemen corporations are deployed into the affected location, the contextual information perceived by the former firemen should be quickly and easily shared to the new ones.

Contextual information sharing may also support the coordination of relevant agents, e.g. considering that emergency protocols have to be precisely defined and known both by the actors in the field and authority entities. Based on a common context, emergency support systems may hence enforce some forms of pre-defined automatic dynamic reconfiguration capabilities concerning the evolution of a critical event. Such rules may be incorporated in those systems and be automatically triggered in face of particular events, e.g. sensing data values collected in a problematic area may trigger a switch from normal to an emergency situation.

Therefore, it is our opinion that richer/dynamic interaction models are necessary on accessing Web enabled WSNs and that they should be captured allowing their sharing among different clients and reuse for similar situations. In the following, we propose a novel session-based abstraction to represent and contextualize such dynamic interaction models.

III. PROPOSED SOLUTION

The conceptual view of the proposed solution is depicted in Fig. 1. The middleware layer hides the details inherent to accessing Web enabled WSNs and provides an interaction context to clients, either individual or to a set (e.g. clients which may benefit from sharing a particular interaction). The solution is based on a) the *Session* concept to capture dynamic rich interactions with Web enabled WSNs, and on b) the *Pattern* concept to implement a confined, structured, and well defined mechanism for dynamic reconfiguration within a session context.

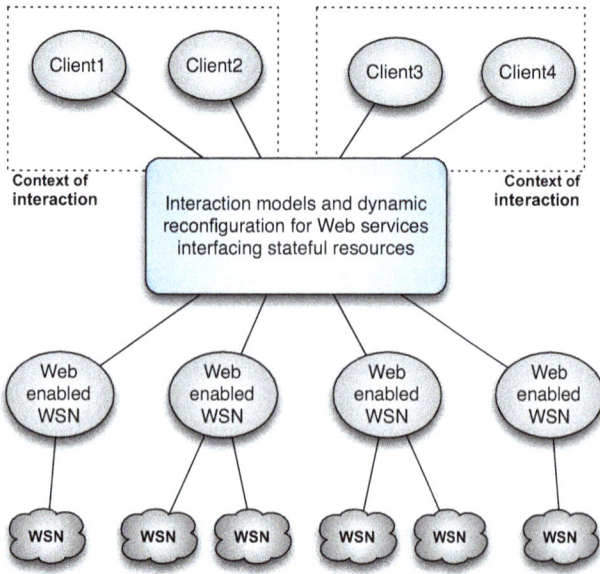

Fig. 1. Conceptual view.

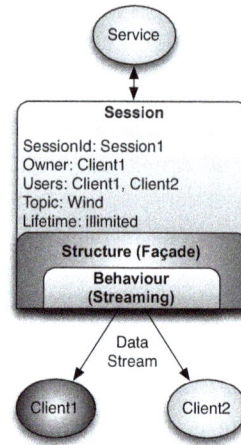

Fig. 2. Session abstraction with a Stream-based interaction model.

A. *Sessions Capturing Dynamic Interaction Models*

The *Session* concept represents the interaction context of a set of users accessing the same Web enabled WSN services, as well as the dynamic reconfiguration features possible within that context. A session includes:

1) The identification of the data sources plus the particular interaction model in use at some point in time for data dissemination. All client accesses within this session's context obey the semantics of that interaction model, which defines the service/users' data and control flow dependencies. Basic interaction models are Client/Server, Publish/Subscriber, Streaming, and Producer/Consumer. Fig. 2 depicts an example for a Wind data source, whose data is disseminated through a Streaming interaction model.

2) Management information, such as a unique session identifier used by new clients to join the session; the identifiers of the session's current members; the identifier of the *session's owner*, the sole that can perform explicit dynamic reconfigurations and terminate the session; and the session's life time limit which when expired causes the session's termination and the consequent notification of all its members. If this time is *unbounded*, the owner must explicitly request the termination. The session in Fig. 2 has two clients and an unbound lifetime limit.

3) The possible adaptation mechanisms consisting of structured and context-based dynamic reconfigurations. These depend on the characteristics of the WSNs service/users interaction context and may also be pre-defined:

- *The interaction's* context includes:
  i. The context of the service client (e.g. a mobile device with limited autonomy or progressing to a different geographic area).
  ii. The interaction medium between the Web enabled WSN service and its user (e.g. the characteristics of the

supporting communication networks).
  iii. The Web enabled WSN service's context (e.g. services representing relevant data sources like temperature or humidity sensing data whose critical values have to be acknowledged).

- *System evolution results from on-demand/pre-defined interaction models' dynamic modifications.* Users may explicitly require dynamic reconfigurations, or these may be automatically triggered by the runtime system based on pre-defined rules and upon change detection of the cited interaction context.

B. *Pattern-based Dynamic Interaction Models*

Within a session's context, the *pattern concept* is used both

- To implement the interaction model in use by all clients belonging to the session at some point in time; and
- To provide a structured dynamic adaptation mechanism ruling a session's evolution.

Implementing the Session's Interaction Model

Patterns underlie an interaction model's implementation in the context of a session. Such is accomplished following the ideas in [16] where pattern abstractions in the form of parameterized *Pattern Templates* capture structure and behavior with separation of concerns, allowing their flexible composition.

The implementation of a particular interaction model is based on the composition of one or more structural patterns with a behavioral pattern. *Structural Patterns* capture a session's "static view" in terms of the structural dependencies/relations among its members (e.g. a Façade or a pipeline) without specifying any restrictions in terms of data or control flows.

The "dynamic view" is defined, on the other hand, by *Behavioral Patterns* like *Producer/Consumer*, *Streaming*, *Publish/Subscriber*, and so on. These characterize the dependencies in terms of data and control flows among a session's members, as well as their role concerning the behavioral patterns' semantics (e.g. roles of *producer* and *consumer* when considering the Producer/Consumer pattern).

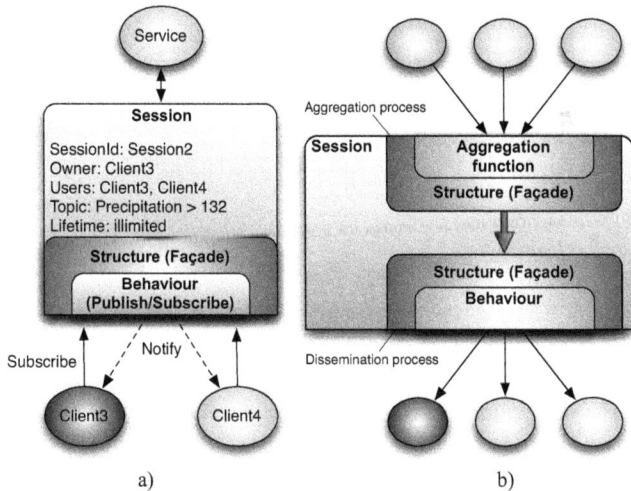

Fig. 3. Session's Interaction Models implemented as the composition of structural and behavioral patterns.

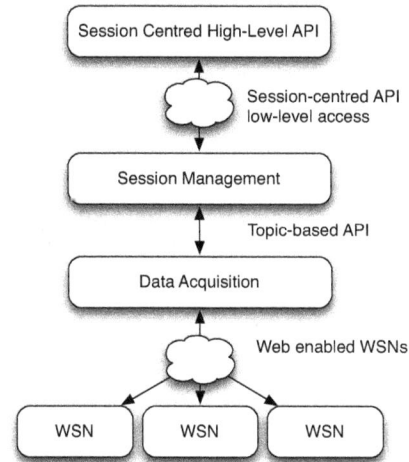

Fig. 4. Overall architecture.

The left-hand side of Fig. 3 (a) presents the composition of a Façade structural pattern with the Publish/Subscriber behavioral pattern. The Façade captures the common interface for data dissemination to all clients in the session, and the behavior defines how that data is disseminated to session's clients.

Different interaction models enable the presentation of data flows with distinct quality services at different points in time. This allows diversity on accessing Web enabled WSN services/data sources, as well as for their modification when convenient. For example, the use of a Client/Server model to inspect a data source versus a Publish/Subscriber model to receive asynchronous event notifications.

The right-hand side of Fig. 3 (b), in turn, presents the implementation of an *Aggregation* model in the context of a session, which consists on the aggregation, and possible processing, of multiple data sources, and their dissemination. Such is supported by a hierarchical structure, namely a two-staged process (a two stage pipeline structure) consisting of an aggregation phase and a dissemination phase. Both phases must present the same behavior, for instance, an aggregation of streams must be disseminated according to the Streaming behavior. The logic used to combine the multiple data sources is defined in the form of an *aggregation function* parameterized upon the model's definition. This approach accommodates the definition of application-specific stream processing techniques to filter the data, compute statistics, and so forth.

Structured Dynamic Adaptation

The pattern abstraction also supports a structured dynamic adaptation mechanism dependent on the current state of the interaction's context. As a result

a) Each pattern can be directly reconfigured at runtime, both in the dimensions of structure and/or behavior (e.g. to replace a behavior by another one);

b) The adaptation/evolution of the system may be represented as a pre-defined sequence of patterns captured as a state machine (see Section IV).

IV. A MIDDLEWARE FOR WSNs ADAPTABLE ACCESS

The proposed middleware implements the concepts described in the previous section providing rich and dynamic interaction models for Web enabled WSNs. It is implemented as a Web accessible platform upon which sessions can be shared by multiple geographically dispersed users.

The middleware's architecture, depicted in Fig. 4, follows a multi-tier model that cleanly separates the multiple concerns of the system, such as presentation, logic and data access. From a bottom-up perspective, the layers that compose the middleware are:

- **Data Acquisition:** interacts with Web enabled WSNs, the data sources, providing a topic-based API. Upper layers can hence associate topics to data sources or define restrictions on those same sources. For example, a topic may refer to a stream of data produced by a given service or only to the items of the stream that obey a given condition (e.g. subscription of precipitation levels above 132 units, as depicted in the left-hand side of Fig. 3 (b)).

- **Session Management:** implements the session abstraction, supplying tools for session creation/termination; session management, ranging from membership accounting to parameter configuration (e.g. lifetime specification); and possible dynamic reconfiguration mechanisms. Since a session may comprise geographically dispersed members, this layer exposes a simple Web service interface intended to be used by higher-level language APIs.

- **Session-Centered High-Level API:** provides a high-level session-centric interface for the cited capabilities.

The remainder of this section will further detail the *Session Management layer*, the core of the middleware, and the *Session-Centered API* used in the example of Section V.

A. Session Management Layer

A session hosts a single behavior/interaction model to which all of its clients are automatically bound. This behavior must be defined when the session is created but may also change in time, as a response to a reconfiguration action. The client that creates a session is titled its *owner* and is the sole with

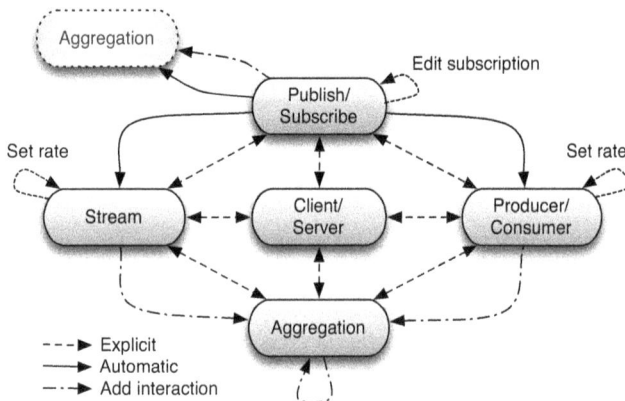

Fig. 5. Reconfiguration state machine: explicit reconfigurations.

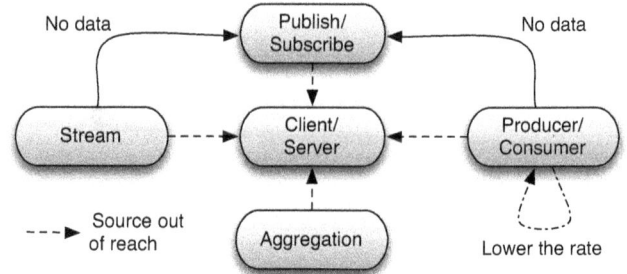

Fig. 6. Reconfiguration state machine: implicit reconfigurations.

permissions to perform reconfiguration actions that have a session-wide impact. The other members must comply with the session's current configuration, and adapt to any consummated reconfiguration or leave. The composition of one or more structural patterns with a behavioral pattern provides the framework upon which sessions are implemented, as described in Section III.B.

Pattern-based Dynamic Reconfiguration

The reconfiguration mechanisms featured in the middleware have the purpose of adapting, in the context of a session, the way a particular client or a set of clients (the session's members) interacts with a set of Web services.

The separation of the *session*, *structure*, and *behavior* concepts, and the way they are combined to support session execution, cleanly evidences the responsibility of each one. For instance, the session contextualizes the overall interaction; a new client joining an existing session is captured as a structural reconfiguration independent from the behavior (i.e. the new client has the same behavior as the other existing clients in the same session); the replacement of the session's interaction model in use is captured as a behavior reconfiguration independent from the structure (all clients in the session are notified of a new behavior ruling data dissemination).

Additionally, the reconfiguration actions can be characterized as implicit (automatically triggered by the middleware) and explicit (requested by a client). Orthogonally, their scope may be confined to the tuning of the current interaction model, or have a session-wide impact, replacing the current model altogether. The conjunction of all the reconfigurations supported by the middleware defines a state machine whose description follows.

Explicit Reconfigurations

Valid reconfiguration requests may be issued by any member of any session, at any moment in time. Their purpose is twofold: to tune or to replace the current interaction model. Tuning requests are model dependent, and must conform to the currently active reconfiguration interface. For instance, setting the data rate is only available in the *Streaming* and *Producer/Consumer* models.

The remainder requests have a broader impact and thus have their semantics bound to the role of the client in the session. Only a reconfiguration request issued by the session's owner may encompass the entire session. The other members are notified of such reconfiguration and will have to adapt to the new configuration or leave the session. Requests issued by some other member than the owner do not affect the target session. It is the client that is moved to another session fulfilling the required parameters. If no such session exists at the time, it is created on the fly.

Fig. 5 illustrates the transitions of the state machine that are triggered by explicit requests. The ones that actually perform a state transition have been divided into three categories:

- **Explicit**: an explicit *reconfigure* request.
- **Automatic**: reconfiguration actions that, when in the scope of a *Publish/Subscribe* model, can be programmatically associated to a particular topic subscription. As soon as the middleware receives a notification on that topic it automatically reconfigures the client, according to its role in the session (owner or regular member).
- **Add interaction:** addition of new data sources to the session. This reconfiguration forces the interaction model to become an aggregation, being that the dissemination model is inherited from the current configuration, e.g. adding a new source to a stream will result in the aggregation of two streams.

Implicit Reconfigurations

These constitute responses to changes in the context of the client, the service, or their communication channel. Their purpose is to ensure that the data flow between a session's sources and clients is adjusted according to the session configuration parameters and the ability of the sources to meet these requirements.

Fig. 6 presents the transitions of the state machine dedicated to this type of reconfigurations. Three scenarios are handled:

- **Session out of reach**: this transition is triggered whenever the data source is no longer reachable. The session's clients are notified of the incident and from that point on they will only able to interact with the source through the Client/Server model. Naturally, as long as the source is out of reach, any request will return an error message.
- **No data**: when in the context of the Stream and Producer/Consumer interaction models, the absence of

Session
Session(String topic, InteractionModel m, NotificationListener l, **int** duration)
Session(String[] topic, InteractionModel m, NotificationListener l, **int** duration)
boolean join(String SessionId, NotificationListener l)
void start()
void finish()
DataItem query()
void setListener(NotificationListener l)
boolean reconfigure(String topic, InteractionModel m, Listener l)
boolean reconfigure(String topic, InteractionModel m)
boolean reconfigure(InteractionModel m, Listener l)
boolean reconfigure(InteractionModel m)
boolean reconfigureCurrent(String operation, String[] parameters)
boolean addInteraction(String[] topics, NotificationListener l, String aggrFunctionClass)
boolean addInteraction(String[] topics, NotificationListener l)
String getClientId()
String getSessionId()
InteractionModelId getCurrentInteractionModel()
List<String> getAvailableTopics()
List<String> getActiveSessions()

InteractionModel
InteractionModelId getId()
List<String> getTopics()
NotificationListener getListener()

Client/Server
Client/Server()
DataItem query()

Stream
Stream(**int** rate)
Stream()
boolean setRate(**int** rate)

ProdCons
ProdCons(**int** rate)
ProdCons()
boolean setRate(**int** rate)

Pubsub
PubSub()
void onNotification(String[] topics, InteractionModel m, NotificationListerner l)

Aggregation
Aggregation(InteractionModelId disseminationModel, String aggrFunctionClass)
Aggregation(InteractionModelId disseminationModel)

NotificationListener
NotificationListerner()
void processMessage()
void sessionTerminated()
void sourceOutOfReach()
void noData()
void sessionReconfiguration()
void interactionModelReconfiguration()
void addedInteraction()

<uses>
<uses>

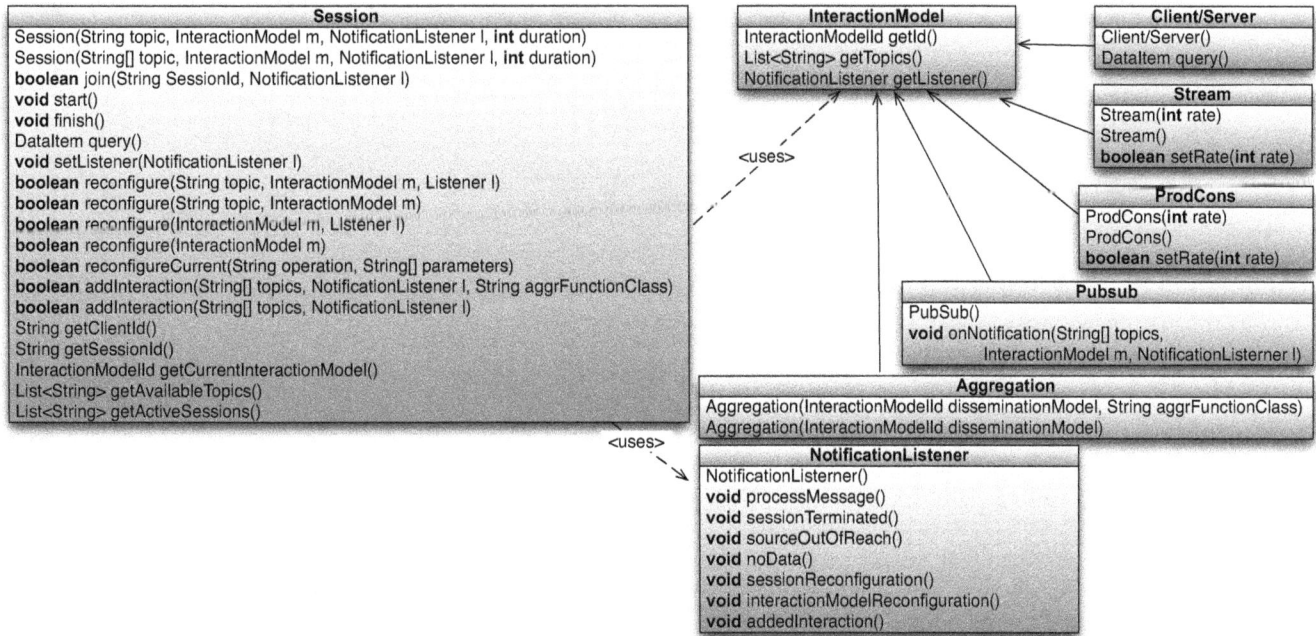

Fig. 7. API's simplified class diagram.

new data items causes the session to be reconfigured to Publish/Subscribe. Clients are notified of both the data stream's interruption and resuming.

- **Lower the rate:** the Producer/Consumer interaction model enables clients to consume data-streams at their own pace, which may be significantly slower or faster than their production rate. To support such feature, the middleware buffers data items on both ends of a client connection. In this context, the *Lower the rate* transition is triggered whenever the buffer that resides on the client end detects that it is no longer able to consume the data at the current pace. As the name implies, the reconfiguration lowers the rate to which the data items are sent to that particular client. Thereby, this reconfiguration targets a single connection, and not the whole session.

B. The Java Session-Centered API

A high-level session-centric API has been developed for the Java language. It exposes all of the middleware's features, providing the means for applications to create, destroy, join and reconfigure existing sessions. Moreover, it specifies how an application can process incoming data items and react to consummated reconfigurations. Fig. 7 showcases a simplified version of the API's class diagram.

Creating and Joining Sessions

Session are instances of the Session class that can be parameterized with the topic(s) of the data sources, an interaction model (the default is *Client/Server*), a listener to handle incoming data (more on this ahead), and a duration in minutes (the default is *unbound*). All interaction models share a common interface (InteractionModel) but provide specific reconfiguration interfaces (the methods of each class). The ability to join existing sessions is provided by the join() method. It requires the identifier of the session to be joined

and the listener to handle incoming data. The inquiry of which sessions and topics are currently active is possible through methods getActiveSession() and getAvailableTopics(), respectively.

Reconfiguration Requests

Three methods are provided for requesting explicit reconfigurations: reconfigureCurrent(), reconfigure() and addInteraction(). The first empowers the tuning of the current interaction model, while the remainder two instantiate the *explicit* and *add interaction* transitions of Fig. 5, respectively.

Handling Incoming Data and Notifications

A special handler that we refer as *listener* must process all the data received in the scope of a session. This handler must subtype abstract class NotificationListener and implement methods to process the reception of new application data items (processMessage()) and of all possible exceptions and reconfiguration notifications (the remainder methods).

V. APPLICATION SCENARIO

The application scenario chosen to illustrate some capabilities of our proposal belongs to the domain of the *Data Driven Applications and Systems* (DDDAS) [21]. These applications are characterized by the need to dynamically incorporate sensing data into a running simulation. Inversely, the simulation should also be able to dynamically parameterize how such sensing data is collected (e.g. restricting data acquisition to the most affected areas in order to reduce data processing). Our example describes only a partial scenario in the context of a fire monitoring and simulation application, as introduced in section II. Namely, a session contextualizes the dynamic aggregation of sensing data collected in a critical area, and typical clients to this session are fire workers and a fire evolution simulation. These clients may hence share the

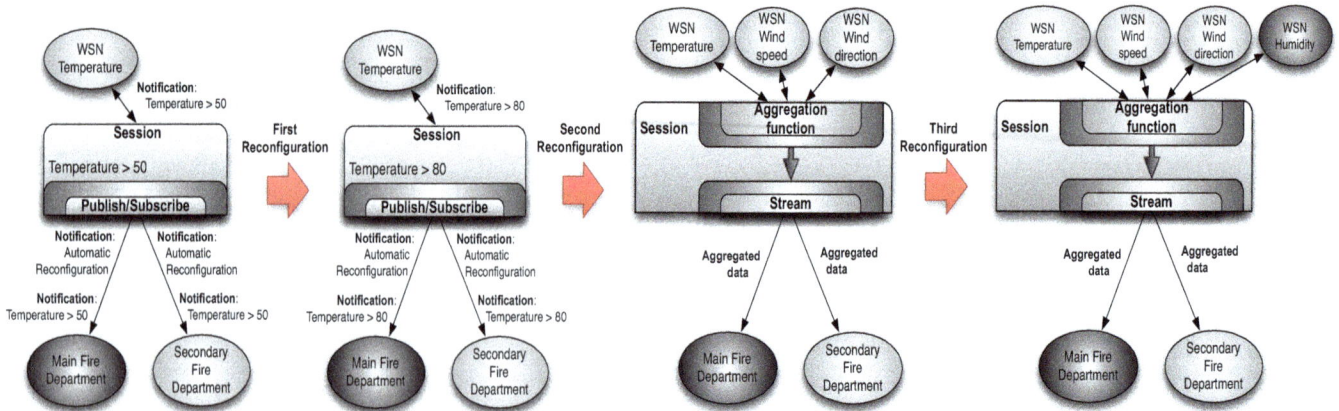

Fig. 8. Dynamic reconfigurations in a session context.

same context both in terms of collected data and the used interaction model for that data dissemination; additionally, all clients in the session are notified of the same dynamic reconfiguration events. The example in the next sub-section describes only the perspective of the fire workers.

A. Wildfires notification application

Consider a fire detection application supporting a fire department responsible for a critical geographical area prone to recurrent wild fire events. The department is interested on receiving a notification whenever the temperature in the area rises to values above 50° Celsius. Furthermore, when this happens, a dynamic reconfiguration should cause a switch from an *alert state* scenario to a *critical state* contemplating the raise of the temperature above 80°. Based on this last notification indicating a probable imminent fire ignition, the next step requires on-line (almost real time) data acquisition on wind-speed and direction, besides temperature. Such different data types should also be aggregated according to user's defined criteria.

In case a secondary fire department is appointed to fight a fire in the same area, the application should provide them with access to the same data as the main fire department. Furthermore, if during the fire fighting period the main fire department decides to add another source of data, e.g. "Humidity", in order to gain more precise information about the conditions in the terrain, this has to be acknowledged by the secondary fire corporation as well. Fig. 8 represents such modifications in the context of a session capturing this application scenario.

B. System dynamic evolution

The first image in Fig. 8 (on the left-hand side) depicts a session created by the middleware including:
1) The interaction context between the session clients, namely the *Main Fire Department* (the session's owner) and the *Secondary Fire Department* (the auxiliary corporation).
2) The available data sources accessible in the session, i.e. a Web enabled WSNs acquiring temperature data.
3) The interaction model in use is the *Publish/Subscribe* being the subscription topic: temperature values above 50,

which defines an alert state.
4) The dynamic reconfiguration rules.

Namely, if such temperature value of 50 is observed, a user-defined dynamic reconfiguration takes place (*First reconfiguration* in 8) modifying the subscription topic. The fire departments are now interested in being notified when the temperature reaches 80° or above which indicates a critical situation. Note that the interaction model is left unaltered, and thus both departments are notified of this event.

On such scenario, another automatic dynamic reconfiguration (*Second reconfiguration*) is triggered to build an aggregation of multiple data sources. In the face of a critical situation, temperature data inspection is not enough, and new data sources on wind speed and direction are dynamically added to the session context. Data collected from different types of Web enabled WSNs may hence be aggregated in the context of the session and processed according to a user-defined aggregation function. Moreover, for a precise evaluation of the fire situation (e.g. if a fire ignition is imminent or has already occurred), a continuous data flow from the sensor devices monitoring the area is now mandatory. Such is also depicted in the new configuration, where the interaction model used for both the aggregation and dissemination stages is the *Streaming* model.

Finally, the case when additional data sources are still needed, e.g. on humidity values, is illustrated by the *Third reconfiguration*. The aggregation model remains as the underlying interaction model but a new Web service interfacing WSNs has been added, allowing the definition of a different aggregation function for processing all the types of incoming data.

```
// Listeners
Listener ownerListerner1 = new TemperatureListener50();
Listener ownerListerner2 = new TemperatureListener80();
Listener ownerListerner3 = new
TemperatureWindSpeedWindDirectionListener();
// Interaction models
Aggregation fire =
         new Aggregation(InteractionModel.STREAM,
                  "fire.HPAggregationFunction");
PubSub critical = new PubSub();
critical.onNotification({"Temperature", "WindSpeed",
                  "WindDir"}, fire, ownerListerner3);
PubSub alert = new PubSub();
alert.onNotification( "Temperature>80", critical,
                  ownerListerner2);
// Session creation
Session s = new Session("Temperature", alert,
                  ownerListerner1);
s.start();
…
// Later, during the fire fighting
Listener ownerListerner4 = new HumidityListener();
s.addInteraction("Humidity", ownerListerner4);
```

Listing 1: Main Fire Department's session

Note that this session's context captures the subordination, in the field, of the *Secondary Fire Corporation* to the *Main Fire Department* in terms of relevant collected data and the associated response. For instance, Listing 1 sketches the creation of a session with a *Publish/Subscribe* interaction model used to notify temperature values. When those values exceed a minimum threshold, the above critical situation is established and a pre-defined dynamic modification takes place ("Second reconfiguration" in Fig. 8). This reconfiguration is defined by the session owner (*Main Fire Department*) and consists on an *Aggregation* of streams on temperature, wind direction, and wind speed values, as depicted in Listing 1.

To share the same session context and subsequently be notified of the same events as the owner - including dynamic reconfigurations - the *Secondary Fire Corporation* has to know this session's identifier and join it as specified in Listing 2.

```
Session s = Session.join(sessionId, new ClientListener1());
```

Listing 2: A new fire corporation joins the session

In order to acknowledge and handle the events occurring in the context of the session, the above *Secondary Fire Corporation* (or other novel clients joining the session at some point in time) has to implement the ClientListener1 handler as it is sketched in Listing 3. The disclosed methods handle the reception of data items, displaying them in a user interface (gui), and define a new listener able to process the

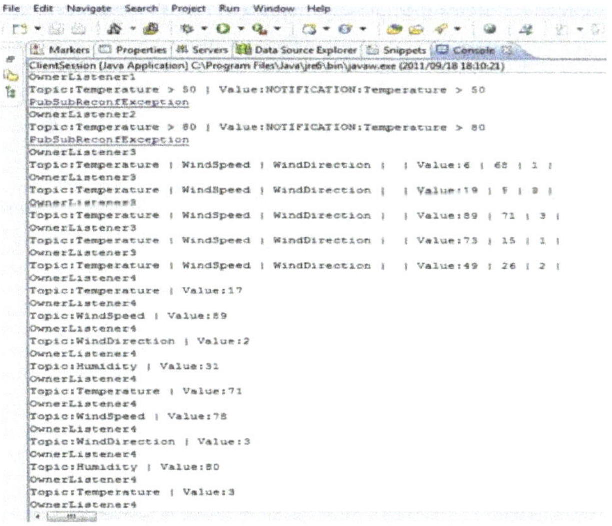

Fig. 9. Output of the main fire department - owner.

reconfigurations possible in a new session's state.

```
public class ClientListener1 extends AbstractListener {

  public void processMessage(DataItem msg) {
    gui.display(msg.getTopic() ,msg.getContents());
  }
  public void InteractionModelReconfiguration(
            ReconfException n) {
    gui.displayNotification(n.getTopic(), n.getReason());
    getSession().setListener(new ClientListerner2());
  }
}
```

Listing 3: Sketching the implementation of ClientListener1

C. Example output

In the context of the previous particular scenario Figs. 9 and 10 illustrate the reception, on both fire departments, of the data values and notifications disseminated in the context of the session. The display of data values complies with the following format:

Topic*: subscribed_topic* | Value*: received_value*

while the display of notifications adheres to format:

Topic*: subscribed_topic* | Value*:*NOTIFICATION*:reason*

Reconfiguration notifications are disseminated to all members of a session, including its owner, which pre-defined the reconfiguration request. This approach entails a uniform way to react to a given notification, regardless of the member's role in the session.

Fig. 10 also shows the situation when the Main Fire Department requests a novel stream on humidity values (third reconfiguration on Figure 7) but an aggregation function is not supplied in the invocation of addInteraction() (last line of Listing 1). As a consequence, the messages are no longer aggregated by the middleware, who simply forwards them. Such behavior can be observer on both figures from the point the fourth listener takes action.

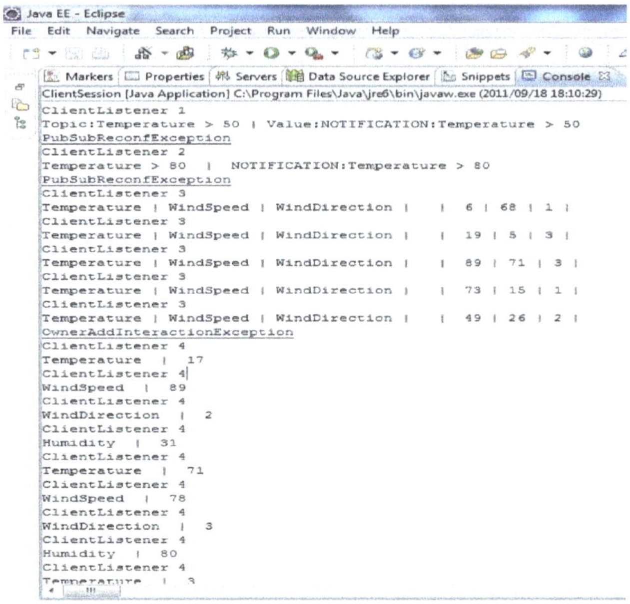

Fig. 10. Output of the second fire department – participative user.

VI. Related Work

To the best of our knowledge, existing middleware platforms for Web enabled WSNs do not address client-WSN interaction model's dynamic reconfiguration concerns nor provide a session abstraction to capture and reuse such dynamicity. Among those platforms we highlight:

52º North [12], the most known implementation of the Sensor Web Enablement (SWE) [3], a set of models and Web service interfaces proposed by the Open Geospatial Consortium for the Web integration of sensor systems. The models focus on the description of sensor systems and their capabilities to collect and process observations, while the services address the collection, storing, and dissemination of sensor reading and alerts (notifications).

Global Sensor Network (GSN) [20], which aims at building a sensor Internet by connecting virtual sensors, abstracting data-streams originating from either a WSN or from another virtual sensor. SQL queries can be performed on top of these virtual sensors.

SenSer [4], a generic middleware for the remote access and management of WSNs, being the latter virtualized as Web services, in a way that is programming language and WSN development platform independent. Its distinguishable properties include the ability to filter the acquired data and to submit WSN reprogramming requests.

As for the presence of the pattern concept on system's dynamic adaptations, the work in [17] presents one solution for self-adaptability of service-generated data streams targeting problems such as data loss or delays associated with communication networks disruptions. However, interaction models are not present as explicit configuration options

considering service interactions as described in our proposal. Although the cited approach does implement a (sophisticated) *Producer/Consumer* interaction model, such is restricted to the support system (i.e. it is not explicitly visible at the point-to-point interaction level between a service and its user). Furthermore, in [17] there is no reference to the possibility of dynamically adding new data flow consumers or additional data sources, as we proposed in the session's context.

Some other works use reconfigurable *Architectural Patterns* for adaptable system's definition [18]. The architecture of the *Publish/Subscriber* pattern, for instance, allows the reconfiguration of publishers, subscribers, and subscribed events. The *Master/Slave* pattern also allows the addition of new slaves to optimize task execution [19]; such is also incorporated in our solution within the context of a session. In spite of such reconfigurable system architecture definition, these works do not provide a session capturing an interaction's context, nor a pattern-based system evolution based on pre-defined rules conform to those pattern's semantics.

Finally, our solution is based on the work by [16] which, however, does not provide a session abstraction to contextualize and reuse dynamic interaction models, nor implements a state machine for pattern-based system evolution.

VII. Conclusions and Future Work

Current applications relying on WSNs for large-scale environment monitoring require adequate abstractions for network access and parameterization, and sensing data acquisition. However, applications also increasingly request the seamless integration of WSNs in heterogeneous and dynamic complex systems, what is possible via the service concept. Moreover, the access to sensing data requires richer interaction models besides the traditional synchronous request/reply model, for example the *Publish/Subscribe* and *Streaming* models. Based on such Web enabled WSNs this work proposes a *session* abstraction in order to capture, contextualize, and reuse diverse richer dynamic interaction models to those services.

A session embodies the common interaction characteristics relating a set of users accessing the same service at some point in time, and all perceive the same events occurring meanwhile in the session's context. A session also contextualizes the possible dynamic adaptations both in terms of the service, the communication medium, or the clients' contexts. For instance the sensing data, the data transfer rate, or a client's mobile device autonomy, may all trigger the modification of the interaction model. Furthermore, both the interaction models and the rules for their dynamic adaptation rely on the *pattern* concept and depend on individual pattern semantics. The system's evolution is captured in a state machine based on pre-defined pattern-based rules. Being well defined, such per-pattern reconfigurations allow adaptation automation and contribute to limiting, to some extent, the impact of the dynamic reconfiguration upon the overall system.

The performance evaluation in terms of the overhead of one additional middleware layer between a Web enabled WSN and its users (SenSer platform [4]) is one point that unfortunately is missing in this paper but which will be studied in the near future. Likewise, more application scenarios are needed in order to evaluate the expressiveness of the model.

Nevertheless, it is our opinion that such novel session-based abstraction opens several interesting further developments concerning the inclusion and aggregation of diverse WSNs sensing data in different domains. For instance the aggregation of session-based interactions may be captured in the form of workflow dependencies and be used in ambient intelligence contexts and participatory sensing applications. Furthermore, the proposed middleware's deployment in a Cloud computing platform may provide clients a ubiquitous and reliable access to sessions. These cases are already under development.

ACKNOWLEDGMENT

The authors would like to thank Professors Omer Rana and José Cardoso e Cunha for the initial ideas on pattern and service abstractions, which ultimately conduced to this work.

REFERENCES

[1] C. F. García-Hernández, P. H. Ibargüengoytia-González, J. García-Hernández, and J. A. Pérez-Díaz, "Wireless sensor networks and applications: a survey," International Journal of Computer Science and Network Security, vol. 17, no. 3, pp. 264–273, 2007.

[2] A. T. Campbell, S. B. Eisenman, N. D. Lane, E. Miluzzo, R. A. Peterson, H. Lu, X. Zheng, M. Musolesi, K. Fodor, and G.-S. Ahn, "The rise of people-centric sensing," IEEE Internet Computing, vol. 12, pp. 12–21, July 2008.

[3] M. E. Botts, G. Percivall, C. Reed, and J. Davidson, "OGC Sensor Web Enablement: Overview and high level architecture," in GeoSensor Networks, Second International Conference, GSN 2006, Revised Selected Papers, Lecture Notes in Computer Science, S. Nittel, A. Labrinidis, and A. Stefanidis, Eds, vol. 4540, Springer, 2008, pp. 175–190.

[4] H. Paulino and J. R. Santos, "A middleware framework for the Web integration of sensor networks," in Sensor Systems and Software - 2nd International ICST Conference, S-CUBE 2010, Revised Selected Papers, Lecture Notes of the ICST, Springer-Verlag, 2011, pp. 75–90.

[5] T. Kindberg, J. J. Barton, J. Morgan, G. Becker, D. Caswell, P. Debaty, G. Gopal, M. Frid, V. Krishnan, H. Morris, J. Schettino, B. Serra, and M. Spasojevic, "People, places, things: Web presence for the real world," MONET, vol. 7, no. 5, pp. 365–376, 2002.

[6] ITU, "ITU Internet report 2005: The Internet of Things," International Telecommunication Union, Tech. Rep., 2005 [Online]. Available: http://www.itu.int/osg/spu/publications/internetofthings/

[7] L. Badger, T. Grance, R. Patt-Corner, and J. Voas, "Cloud computing synopsis and recommendations (draft), NIST special publication 800-146," Recommendations of the National Institute of Standards and Technology, Tech. Rep., 2011 [Online]. Available: http://www.nist.gov/itl/csd/20110512_cloud_guide.cfm

[8] A. Baptista, M. C. Gomes, and H. Paulino, "Session-based dynamic interaction models for stateful Web services," in Exploring Services Science - Third International Conference, IESS 2012, Lecture Notes in Business Information Processing, Springer-Verlag, 2012.

[9] M. Creeger, "Cloud computing: An overview," ACM Queue, vol. 7, no. 5, p. 2, 2009.

[10] M. Armbrust, A. Fox, R. Griffith, A. D. Joseph, R. Katz, A. Konwinski, G. Lee, D. Patterson, A. Rabkin, I. Stoica, and M. Zaharia, "A view of cloud computing," Commun. ACM, vol. 53, pp. 50–58, April 2010.

[11] T. Kobialka, R. Buyya, C. Leckie, and R. Kotagiri, "A Sensor Web middleware with stateful services for heterogeneous sensor networks," in Intelligent Sensors, Sensor Networks and Information, 2007, ISSNIP 2007, 3rd International Conference, 2007, pp. 491–496.

[12] C. Stasch, A. C. Walkowski, and S. Jirka, "A geosensor network architecture for disaster management based on open standards." in Digital Earth Summit on Geoinformatics 2008: Tools for Climate Change Research, 2008, pp. 54–59.

[13] I. F. at al., "Modeling stateful resources with web services v. 1.1," Computer Associates International, Inc., Fujitsu Limited, Hewlett-Packard Development Company, International Business Machines Corporation and The University of Chicago, Tech. Rep., 2004 [Online]. Available: http://www-106.ibm.com/developerworks/library/wsresource/ws-modelingresources.pdf

[14] OASIS, "Oasis web services resource framework (WSRF) TC," [Online]. Available: http://www.oasis-open.org/committees/tc_home.php?wg_abbrev=wsrf

[15] F. Darema, "Dynamic data driven applications systems: New capabilities for application simulations and measurements," in International Conference on Computational Science (2), 2005, pp. 610–615.

[16] C. Gomes, O. F. Rana, and J. Cunha, "Extending grid-based workflow tools with patterns/operators," Int. J. High Perf. Comput. Appl., vol. 22, pp. 301–318, August 2008.

[17] V. Bhat, M. Parashar, M. Kh, N. K, and S. Klasky, "A self-managing wide-area data streaming service using model-based online control," in in Proc. 7th IEEE Int. Conf. on Grid Computing, 2006, pp. 176–183.

[18] M. C. Huebscher and J. A. McCann, "A survey of autonomic computing degrees, models, and applications," ACM Comput. Surv., vol. 40, pp. 2–25, August 2008.

[19] M. Aldinucci, M. Danelutto, and P. Kilpatrick, "Towards hierarchical management of autonomic components: A case study," in Proceedings of the 17th Euromicro International Conference on Parallel, Distributed and Network-Based Processing, PDP2009, IEEE Computer Society, 2009, pp. 3–10.

[20] K. Aberer, M. Hauswirth, and A. Salehi, "A middleware for fast and flexible sensor network deployment," in Proceedings of the 32nd International Conference on Very Large Data Bases, Seoul, Korea, September 12-15, 2006, 2006, pp. 1199–1202.

[21] F. Darema, "Dynamic data driven applications systems: A new paradigm for application simulations and measurements," In Int. Conf. on Computational Science, volume 3038, Springer LNCS, 2004.

Dr. Maria Cecília Gomes, PhD, is an Assistant Professor at the Departamento de Informática (Computer Science Department) of the Faculdade de Ciências e Tecnologia/Universidade Nova de Lisboa. She is a member of the Research Center for Informatics and Information Technologies (CITI), and member of the management committee and national PI of *ARTS: Towards Autonomic Road Transport Support Systems* (COST Action TU1102). Her current research interests include autonomic systems particularly applied to road transportation support systems, service-oriented computing, and models.

Dr. Hervé Paulino, PhD, is an Assistant Professor at the Departamento de Informática (Computer Science Department) of the Faculdade de Ciências e Tecnologia/Universidade Nova de Lisboa. He is a member of the Research Center for Informatics and Information Technologies (CITI), on which he is the principal investigator of the SABLE (Service Abstractions for Parallel Computing) research team, and also a collaborator member of the Research Center for Research in Advanced Computing Systems (CRACS). His current interests are on the fields of service-oriented computing and, concurrent and parallel programming.

Adérito Baptista, MsC, was a computer science master student at Faculdade de Ciências e Tecnologia/Universidade Nova de Lisboa. He is currently a software developer at Novabase, Portugal.

Filipe Araújo, MsC, was a computer science master student at Faculdade de Ciências e Tecnologia/Universidade Nova de Lisboa. He is a record management systems and R&D developer at Quidgest, Portugal.

14

GPGPU Implementation of a Genetic Algorithm for Stereo Refinement

Álvaro Arranz, Manuel Alvar

Zed Worldwide

Abstract — **During the last decade, the general-purpose computing on graphics processing units Graphics (GPGPU) has turned out to be a useful tool for speeding up many scientific calculations. Computer vision is known to be one of the fields with more penetration of these new techniques. This paper explores the advantages of using GPGPU implementation to speedup a genetic algorithm used for stereo refinement. The main contribution of this paper is analyzing which genetic operators take advantage of a parallel approach and the description of an efficient state- of-the-art implementation for each one. As a result, speed-ups close to x80 can be achieved, demonstrating to be the only way of achieving close to real-time performance.**

Keywords — **Parallel processing, GPGPU, genetic algorithm, stereo.**

I. Introduction

RECENTLY, custom GPU programming has become one of the most popular tools for increasing the efficiency of parallel algorithms thanks to the computational capacity of the Graphics Processing Unit (GPU) compared to serial CPU programs.

Traditionally, GPUs appeared in the computer market as hardware products specialized on rendering tasks and, more specifically, for improving the gaming experience. Given that most of the rendering pipeline's steps of were parallel, these products rapidly evolved to machines capable of efficiently running highly parallel algorithms. In the last decade, the flexibilization of the GPU hardware and tools has enabled the use of these parallel-processing units for general scientific purposes.

Stereo analysis is a Computer Vision research area that has been widely studied in the literature. However, it remains an unsolved problem and many algorithms are still proposed every year. The aim of the stereo analysis is to obtain depth information from a couple of stereo images, simulating how the human's can perceive the depth using just two eyes. Solving this problem is very computationally demanding, especially when dealing with high-resolution images. GPGPU techniques have been recently used for speeding up these tasks and great results have been reported in the literature.

GPGPU primarily aims to improve the program's performance. It has been demonstrated that using these techniques could result in a speed-up of up to x100, depending on the algorithms' nature. This paper proposes to study the speed-up achieved by GPGPU programming applied to an evolutionary algorithm. A genetic algorithm for stereo refinement is implemented in both CPU and GPU and its performance analyzed and compared.

Improving the accuracy and performance of stereo algorithms is crucial for many real applications. Robotics has been traditionally a research area that has used these techniques, but new fields are arising. The digitalization of the automotive sector is leading to the incorporation of new sensors such as high definition cameras to high-end cars. Fast stereo algorithms are needed to provide accurate information about the car's environment. Other applications of stereo algorithms are biomedicine, virtual reality, automation or the entertainment industry. However, note that any optimization problem solved with evolutionary algorithms might benefit from the work herein proposed.

The paper is structured as follows: Section II is a brief overview of the GPGPU implementations found in the literature, Section III explains the stereo refinement genetic algorithm implemented, Section IV describes the details about the GPGPU implementation, in Section V some results are presented and finally in Section VI some conclusions are drawn.

II. GPGPU Overview

GPGPU has been widely used in the literature by the computer vision community. Its main role has been to enable real-time performance on many demanding algorithms.

First works on stereo GPU processing were proposed in [11]. SSD dissimilarity techniques, a multi-resolution approach and a very primitive GeForce4 were used to obtain performance equivalent to the fastest CPU commercial implementations available. Later, [3] proposed a multi-view plane- sweep-based stereo algorithm for handling correctly slanted surfaces applied to urban environments. Assuming a highly structured scene with buildings, they used a planar prior for estimating disparity maps. The algorithm was successfully implemented in an Nvidia GPU obtaining real-time frame rates.

In [6] a high-performance stereo-matching algorithm both fast and accurate is proposed. Using a parallel designed AD-census and scanline optimization implemented in CUDA in an NVIDIA GeForce GTX 480 they achieve near to real- time frame-rates. They report an impressive 140x speed up compared to the CPU implementation. Another GPU stereo matching algorithm using adaptive windows can be found in

Fig. 1 Disparity map examples

[13].

In [4] and [7] a real-time camera tracking and mapping using RGB-D cameras is proposed, obtaining quite impressive results. Their implementation relays heavily on the use of GPGPU, both for tracking and TSDF mapping. Depending on the voxel's resolution, they achieve execution times from 10 to 25ms.

GPGPU has also been applied in other fields, such as in feature detection and tracking, as proposed in [9]. Their KLT GPU implementation achieves real-time 30Hz on 1024768 resolution images, which is a 20x speed-up compared to their CPU implementation. A 10x improve is also reported for the SIFT [5] detector implemented in GPU. A CUDA implementation of the famous graph cuts algorithm [2, 14, 15]

is presented in [10], obtaining a 12x performance enhancement.

A similar system to the one herein proposed is presented in [8]. In this work, a genetic algorithm for stereo matching is also implemented in GPU. However, the genetic algorithms have quite different approaches, and their parallel implementation does not seam to provide any performance boost compared to the CPU one. This paper shows that, with the proper GPU implementation, a 50x speed-up can be achieved.

III. GENETIC ALGORITHM FOR STEREO REFINEMENT

The genetic algorithm for stereo refinement implemented in this paper is based on the work proposed in [1]. The implementation minimizes a fitness function that is related to a Markov Random Field (MRF) and is equivalent to minimizing a global energy function. Due to the flexibility of genetic algorithms, this function is able to include occlusion handling.

This algorithm uses a guided search approach with new crossover and mutation operators adapted to the stereo refinement problem. Each operator will be explained briefly in this section. An example of the results that can be achieved using these techniques is shown in Figure 1.

A. Genome representation

Each individual includes the whole disparity map estimate and the occlusion map for both left and right images.

$$\bar{g}\begin{cases} \bar{f}_L = \{f_{L_1}, f_{L_2}, ..., f_{L_N}\} \\ \bar{f}_R = \{f_{R_1}, f_{R_2}, ..., f_{R_N}\} \\ \bar{O}_L \\ \bar{O}_R \end{cases}, \quad f_i \in \Lambda, \Lambda = \{1, 2, ..., L\} \quad (1)$$

where g is the genome, gL and gR are the representation of the left and right disparity images respectively, XiL and XiR are the disparities estimated for pixel i on the left and right disparity images, N the total number of pixels in each image and Li the set of different disparity labels.

Occlusion maps are defined as:

$$O(p) = \begin{cases} 0 & \text{if not occluded} \\ 1 & \text{if occluded} \end{cases} \quad (2)$$

where O(p) is the occlusion map and p is the pixel.

B. Initialization

For the initialization process two different window-based algorithms with different window sizes, the adaptive support-weight approach [12] with random parameters and the census based with window-cost aggregation have been used. This variation aims to provide a wide range of initial solutions.

$$E(\bar{g}) = E_{data}(\bar{g}_L) + E_{smooth}(\bar{g}_L) \tag{3}$$

$$E_{data}(\bar{g}_L) = \begin{cases} \lambda_d & \text{if } i \text{ is occluded} \\ \sum_{i \in \bar{g}_L} |I_L(x_i, y_i) - I_R(x_i - X_i, y_i))| & \text{otherwise} \end{cases} \tag{4}$$

$$E_{smooth}(\bar{g}_L) = \sum_{\{p,q\} \in N} min\left(\frac{\beta_s}{\varphi_s}|X_p - X_q|, \lambda_{st}\right) \tag{5}$$

$$\beta_s = max(\lambda_s, \gamma_s - |I_L(p) - I_L(q)|) \tag{6}$$

C. Fitness function

An energy function that considers discontinuities and occlusions is used for the fitness function:

where g is a certain individual, gL is the left disparity image, Il and Ir stand for the left and right stereo pair, xi and yi are the image coordinates of pixel i, V{p,q} is a smoothing function and λs, γs and φs are constant parameters for every pixel.

Before any fitness function evaluation, a occlusion management process is triggered for classifying pixels correctly before any energy evaluation.

D. Occlusion management

The process of handling the occluded areas is a two-step operation: occlusion detection followed by an occlusion management.

The following operations are defined for calculating the left occlusion map:

$$O_L(p) = \begin{cases} 0 & \exists i / \begin{pmatrix} x(i) + \bar{g}_R(i) \\ y(i) \end{pmatrix} = \begin{pmatrix} x(p) \\ y(p) \end{pmatrix} & p, i \in P \\ 1 & \text{otherwise} \end{cases} \tag{7}$$

being OL the left occlusion map, x(p) and y(p) the x and y coordinates of point p respectively and P the set of disparity image points. A similar expression can be deduced for the right occlusion map. This occlusion map identifies which areas of the image are classified as occluded regions.

For the occlusion management, an iterative process based on neighboring disparities of the occluded pixels is applied. For the left image, each occluded pixel is assigned the disparity value of the most photo-consistent non-occluded neighbor from left to right and afterwards it is marked as non-occluded. If no non-occluded neighbors exist, it maintains its occluded status for the next iteration. Special status have the occluded pixels whose x(p) coordinate is less than the number of disparities analyzed. In this case the iteration is made from right to left and bottom-up. The iteration is finished when no occluded pixels are left on the left occluded map.

For the right image it is similarly done but vice versa (right to left for common pixels and left to right for pixels whose x(p) is at a distance of the number of disparities analyzed from the right image border).

E. Crossover

The crossover is based on comparing parent's blocks of different sizes and assign the best ones to the same son. This operator can be summarized in the following steps:

1) Parents are divided into blocks (random sizes)
2) The fitness function of each block is evaluated
3) Best block is selected to persist in the same child

F. Mutation

Three different mutation operations may occur to each individual. Firstly, one possible mutation operation is to initialize again a group of pixels following the steps explained in Subsection III-B with a probability PMa. Secondly, a bilateral filter operation with a random window size with a probability PMb. Finally, a morphological operation such as erode or dilate may occur with a probability PMc.

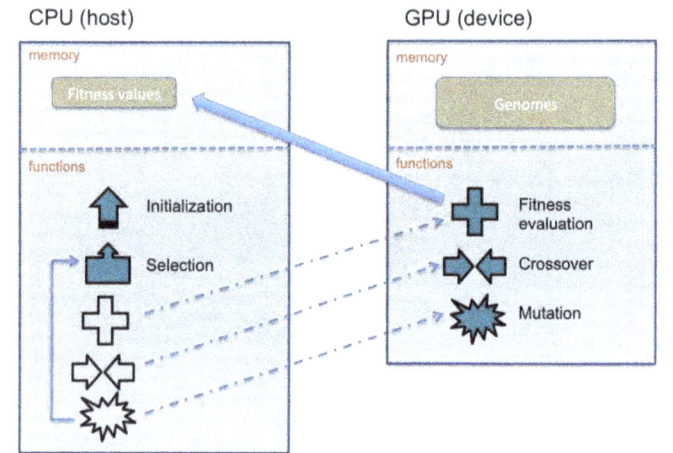

Fig. 2: Assignment of genetic operators to GPU and CPU

IV. ALGORITHM'S GPGPU IMPLEMENTATION

After analyzing the performance of the serial version of the genetic algorithm, it is easy to conclude that the most computationally demanding functions are the genetic operators and not the genetic algorithm itself. This result is straightforward because each genome includes a lot of data and information inside (whole four images: two disparity maps and two occlusion maps). For example, each genome evaluation implies evaluating the energy function for each pixel and neighborhood individually. Besides, each genome operator is naturally parallel, which suggests that implementing these operators in CUDA will have a dramatic impact on the genetic algorithm performance.

In Figure 2 is shown where is computed each genetic operation. The left side of Figure 2 represents data information is stored and which functions are implemented and executed in the CPU. The fitness values are stored in the CPU because they are needed for the selection operator in order to decide which individuals of the actual population will survive to the next one. The right side of the diagram represents which information is stored and which functions are evaluated in the

GPU. All the genomes are stored in the GPU in order to enable fast access to the data from the functions evaluated in the device. The only memory transaction between the CPU and GPU needed is the copy of the fitness value of each individual from device to host and is represented by the big blue arrow from the fitness evaluation function icon to the fitness value memory in the CPU. Remember that this device-host and vice-versa transactions are very costly and must be minimized for achieving the best performance.

The genetic algorithm has been implemented in the CPU using the GAlib library. For the image processing and allocation it has been used the OpenCV library, specifically the GPU module, which facilitates the memory allocation and transaction and has quite a lot processing algorithms built-in in the GPU already. Finally, evaluation, crossover and mutation operators have been implemented in CUDA in several kernels. The next sections describe in detail the strategy used for implementing efficiently each operator in CUDA language.

1) CUDA evaluation kernel

Although the title may suggest that the evaluation of a genome is carried out just by one kernel, the reality is that it is a process composed by three steps. The first two are solved using a single kernel each while the third has to be solved by two kernels. The first two steps could be executed in parallel by two different CUDA streams but the last have to be executed after the firsts have finished. This parallel capability has not been implemented and all four kernels have been

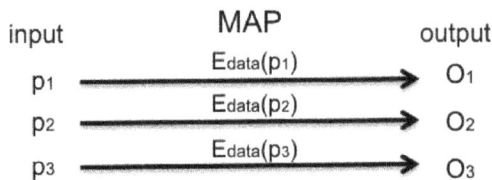

Fig. 3: Parallel MAP operation

programmed to run in the same stream.

The first step in the evaluation process is the data term evaluation of the energy function. The result is one value for each pixel and its calculation is independent from the values of the neighboring pixels. Thus, the relation is one to one and its parallel implementation is very efficient and straightforward. This type of operation is also called MAP, and it has been implemented using one thread per pixel in the disparity image. A simple diagram of MAP is shown in Figure 3.

The data term only depends on the values of the left and right stereo image and on the disparity image evaluated. Left and right stereo images have been allocated in the device as 2D textures, which are very efficient for interpolation. Note that in this case, using shared memory does not make much sense because the number of memory accesses needed per thread would not bee minimized. The result is saved in a floating-point structure of the same size as the original image, and here will be referred as memData.

The second step is the evaluation of the smoothing term. If one thread per pixel is used, it requires to access to its own disparity value and the neighbouring disparities. This

operation can be considered a type of Stencil operation, in which many reads are needed as input while only one write is performed. An illustrative example is shown in Figure 4.

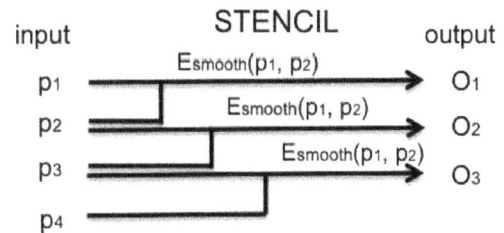

Fig. 4: Parallel STENCIL operation

In order to maximize the performance, shared memory is used for first loading all the disparities in a block and then using that shared memory in all the threads of the same block. Remember that the access to shared memory is much faster than the access to global memory. Each thread is related to each pixel in the disparity image and it is in charge of evaluating the smoothing function that relates itself with the right and bottom neighbors. As happened in the data kernel, the result is saved into a floating point vector with the same size of the disparity image and here will be called memSmooth.

The third and last step is composed of two kernels, one executed after the other. It is in charge of performing the summation over all pixels of the memData and memSmooth structures calculated in the two first steps. This type of summation is an operation also known as Reduce. Although at first glance this operation might seem difficult to parallelize, actually it is fairly simple. Figure 5 shows the two-step reduction implemented. Besides, this step sums memData and memSmooth individually for each pixel and saving it in memTotal in order to facilitate the crossover task explained in subsection IV-2.

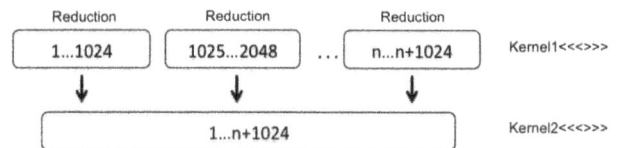

Fig. 5: Reduction operation in two kernels

For enhancing the performance of the Reduce operation, the data has been divided in groups of 1024 addends, each being processed by a CUDA block. All the data in each group is loaded in shared memory to improve its read and write speed. The first kernel performs the summation over each group, obtaining one result per group. Finally, the last kernel performs the last summation over all the results of the previous kernel, and obtains the final value for the fitness function.

Finally, an asynchronous memory copy is performed from device to host to copy the final fitness value calculated for that genome. This operation is recommended to be asynchronous because the memory copy can be performed at the same time as other kernels are executed in other streams, instead of waiting until the memory copy has finished.

Note that in spite of running all the evaluation kernels in the same stream, different individuals are able to run their evaluation in different streams, which enables copying memory from device to host at the same time as other kernels and a higher level of parallel exploitation.

The evaluation process is performed over the left and right disparity map, but for the right one, it is not necessary to perform the final Reduction and memory copy. This optimization can be achieved because de fitness function of a genome is just the fitness function of its left disparity image.

2) CUDA crossover kernel

As explained in Subsection III-E, the crossover operation consists of three steps:
- Divide the two disparity maps into blocks. In this case, the number of pixel of each block will not be greater than 1024.
- Sum up the memTotal for each pixel inside the blocks, compare them by pairs (one for each parent) and keep the block with the best fitness function.
- Copy the best block to the children.

The limit of 1024 pixels per block is related to the maximum number of threads per CUDA block available by the GPU. All the pixels in a block must be part of the same CUDA block be- cause the summation can be performed using shared memory, which is much more efficient than global memory. Therefore, the CPU realizes the first step, and the following two are done by the GPU, the first one as a Reduction operation very similar to the one in subsection IV-1 and the second one as a very simple copy operation as a MAP.

3) CUDA occlusion handling kernel

Occlusion handling encompasses two different tasks: occlusion estimation and occlusion management. The occlusion estimation is calculated through an image warping, where each pixel of the other disparity image is displaced a number of pixels equal to its disparity level. Pixels left without any assignation are considered to be occluded pixels. Thus, each pixel operation is independent from the rest, but several threads can output their result to the same piece of memory. This operation is also known as Scatter and can be solved using, for example, atomic operations. In our case it is not necessary because the function aims only to output a boolean value, more precisely a zero to indicate that the pixel is not occluded.

The second task is the occlusion management, where the

main objective is to re-estimate the disparity value for the pixels that where labelled as occluded. For this parallel implementation the horizontal fast occlusion filling algorithm explained in Subsection III-D was used. Given that a horizontal search for the closest non-occluded pixel has to be performed, the occlusion information was loaded in shared memory, being each block responsible for each independent scan-line. Each thread is in charge of estimating the new depth for each occluded pixel. Figure 8 shows the per-thread operations and the memory accesses incurred.

4) CUDA mutation kernel

The mutation kernel comprises three different operations: bilateral filtering, erosion and dilation. These morphological operations are already efficiently implemented in the OpenCV library using CUDA. A problem that may rise using a third party library is the performance penalty incurred while parsing from the data-types used in your application to the data-types used in the library and vice- versa. However, in the implementation herein proposed, the data types are compatible with those from OpenCV, so this transformation is trivial. Thus, this library has been used for this purpose.

Fig. 7: Nvidia Visual Profiler tool

V. RESULTS

In this section the parallel capabilities of the genetic algorithm are discussed. Both the serial implementation and the parallel one using CUDA are compared. Given the stochastic nature of the algorithm and the various types of mutations that are likely to happen, the algorithm was run for different images during five hundred generations and an average per individual and generations was calculated. The Middlebury dataset will be used for comparison, as it is a standard and well-known test-bed.

For the tests, an Intel i7-2600 at 3.4 GHz CPU and an Nvidia GeForce GTX 770 were used. As operating system, Ubuntu Linux 14.04LTS was used given the CUDA performance improvement compared to Windows. The measuring tool used was the Nvidia Visual Profiler, obtaining valuable data such as timing, occupancy, optimizations, et. A capture of the profiler is shown in Figure 7.

The parameters used in the experiments carried out along this section are shown in Table I and Table II.

A comparison between the performances of the GPGPU versus the CPU implementation for four Middlebury's common test images is shown in Table III.

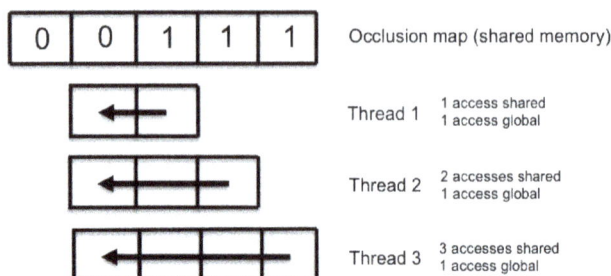

Fig. 6: Horizontal fast occlusion filling implementation example

TABLE I
PARAMETERS FOR THE NEW ENERGY FUNCTION

λ_d	λ_s	γ_s	φ_s	λ_{st}
10.0	50.0	2.0	10.0	$ndisp/2$

TABLE II
PARAMETERS FOR THE GENETIC ALGORITHM

Population	Generations	P_{cross}	P_{Ma}	P_{Mb}	P_{Mc}
50.0	500	0.9	0.01	0.033	0.066

TABLE III
MEAN TIME SPENT FOR EACH INDIVIDUAL AND GENERATION IN CPU
AND CUDA IMPLEMENTATION

	CPU (ms)	CUDA(ms)	Speed-up
Tsukuba	20.84	0.359	58.05
Venus	31.6	0.52	60.77
Teddy	45.63	0.579	78.8
Cones	46.77	0.574	81.48

The first column of Table III shows the mean total time spent for the CPU implementation for one genome. Note that not all genetic operations always occur in each genome and, therefore, these results are obtained dividing the total time spent by the algorithm by the number of genomes and generations. The increment in the CPU execution time from Tsukuba to Cones is explained due to the increment in the size of the test images.

The second row shows the same measure, but using now the GPU implementation. It is shown that parallelization of the genetic algorithm provides a great performance improvement compared with the serial one. The speed-up comparison between the two algorithms is shown in the third column. Note the increment in the performance improvement when the images get bigger, suggesting that with more pixels the GPU performs more efficiently. However, both CPU and GPU implementation still depend highly on the number of pixels in the image analyzed.

TABLE IV
MEAN TIME SPENT BY EACH GENETIC OPERATION FOR EACH INDIVIDUAL
AND GENERATION IN TSKUBA AND CUDA IMPLEMENTATION

	time (ms)
Evaluation	0.155
Crossover	0.061
Occlusion handling	0.062
mutation	0.033
CPU	0.0467
Total	0.359

In order to study the impact of each genetic operation, Table IV shows in detail how the time is divided for each genome. It shows that evaluating the genome is the most demanding operation. Given that it is an operation that has to be run always in every gnome and that it is quite complex (energy function composed by several complex terms), this result is comprehensible. In comparison, the other operation that is run

always and has a lot less impact in the total time is the occlusion handling. The percentage of the impact is shown in Figure 8.

As a result, it can be said that adding the occlusion handling to the algorithm implies a 17% impact on the performance. This result does not account for the impact of the occlusion variable in the evaluation operator, which here will be considered negligible.

TABLE V
MEAN TIME FOR EACH OPERATOR IN TSKUBA AND CUDA
IMPLEMENTATION

	time (ms)	times per genome
Evaluation	0.155	1
Crossover	0.0663	0.9018
Occlusion handling	0.062	1
mutation	0.277	0.121

TABLE VI
MUTATION OPERATION EXECUTION TIMES

	time (ms)	times called (%)
Bilateral filter	0.273	36.78
dilate	0.283	32.19
erode	0.275	31.03

Maybe the result that was unexpected was the efficiency of the crossover function. However, although being a demanding operation, a lot of information from the evaluation process could be reused, leading to an efficient implementation. Bear in mind that the crossover it is run with a probability of Pcross, so this fact also has an impact on this measure. The same occurs with the mutation operation, that it is has a low impact due to it is rarely run.

Finally, a CPU entry in this table might seem strange at first. This time is attributed to the tasks of launching the CUDA kernels and managing the genetic algorithm itself, not the operators. As shown in Figure 2, this includes the selection operation, sorting, etc.

The measures presented in Table IV were calculated aggregating the occurrences of all the operations, but they do not occur in the same proportion. Therefore, those metrics do not represent the true performance penalty of each operation. In Table V the performance of each individual operator is shown.

These measures are the mean time spent value for each operation individually. It can be seen that, although the mutation operation has little impact on the total time spent on the algorithm, individually, it is by far the most demanding one. This is explained by the fact that a low mutation probability was set. Incrementing the mutation probability would have a great impact in the algorithm's performance. The second row of Table V shows statistically how many times each operation is called for each genome.

Finally, a more in-depth analysis of the mutation operation

is shown in Table VI.

The three different operations were configured to be triggered with the same probability, and this is represented in the second row of the table. It is shown that the three algorithms perform very similarly.

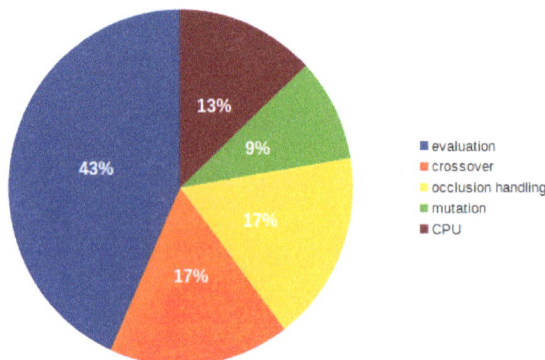

Fig. 8: Portion of time spent by each operation for Tsukuba

As a conclusion, it can be stated that approximately a 80x speedup can be achieved using a parallel implementation of the algorithm used on sufficiently big images.

VI. CONCLUSIONS

In this paper, a parallel GPGPU implementation of a genetic algorithm has been proposed. These evolutionary algorithms are very flexible and fit nicely in a parallel architecture given that the operators act independently on each individual genome. This quality suggests that parallelizing the main genetic operators would have a great impact in the algorithm's performance.

The most time-demanding genetic operators considered to be run in GPGPU were the fitness evaluation, the crossover and the mutation. However, the selection and the maintenance of the genetic algorithm itself was decided to be kept in the CPU. The main reason was that these tasks are negligible compared to the other operators; this assumption was supported by the results presented. Each operator was analyzed and a specific parallel implementation was proposed for each one.

A genetic stereo refinement algorithm with occlusion handling was selected for analysis. Using the standard Middlebury's stereo test-set, a comparison between a CPU and a GPGPU implementation was shown. As a conclusion, a great performance improvement can be achieved using GPGPU computation: a x80 speed-up has been achieved for some images. An analysis of the time spent by each operation and the impact of modifying the genetic parameters has been discussed. As a result, the most demanding operation was the fitness evaluation. This is reasonable due to the complexity of the energy function used for testing. However, considering individual function performance, the mutation operations are the most expensive, so an increment in the mutation probability would have a noticeable impact on the performance.

Evolutionary algorithms are generally not designed for real-time applications. Although a great performance improvement has been obtained, real-time performance is still not achievable for these applications. However, the GPGPU implementation improved the algorithm's performance from minutes to seconds order of magnitude.

In order to continue with this line of research, in future works, it would be interesting to try different genetic algorithm's formulations such as migrating or overlapping populations. These approaches might help avoiding local minima during the optimization. In [1] was demonstrated that this algorithm is very sensible to the fitness function. Therefore, trying different and new energy functions is likely to enhance its accuracy. Finally, for improving the algorithm's performance, trying double core GPGPUs and different platforms such as OpenCL is suggested.

REFERENCES

[1] A. Arranz, M. Alvar, J. Boal, A. S. Miralles, and A. de la Escalera. Genetic algorithm for stereo correspondence with a novel fitness function and occlusion handling. In *VISAPP (2)'13*, pages 294–299, 2013

[2] Y. Boykov, O. Veksler, and R. Zabih. Fast approximate energy minimization via graph cuts. *IEEE Transactions On Pattern Analysis And Machine Intelligence*, 23(11):1222–1239, 2001.

[3] D. Gallup, J. M. Frahm, P. Mordohai, Q. X. Yang, and M. Pollefeys. Real-time plane-sweeping stereo with multiple sweeping directions. *2007 IEEE Conference on Computer Vision and Pattern Recognition, Vols 1-8*, pages 2110–2117, 2007.

[4] S. Izadi, R. A. Newcombe, D. Kim, O. Hilliges, D. Molyneaux, S. Hodges, P. Kohli, J. Shotton, A. J. Davison, and A. Fitzgibbon. KinectFusion: real-time dynamic 3D surface reconstruction and interac- tion. In *ACM SIGGRAPH 2011 Talks*, SIGGRAPH '11, page 23:123:1, New York, NY, USA, 2011. ACM.

[5] D. G. Lowe. Distinctive image features from scale-invariant keypoints. *Int.J.Comput.Vision*, 60(2):91–110, 2004.

[6] X. Mei, X. Sun, M. Zhou, S. Jiao, H. Wang, and X. Zhang. On building an accurate stereo matching system on graphics hardware. In *Computer Vision Workshops (ICCV Workshops), 2011 IEEE International Confer- ence on*, pages 467 –474, nov. 2011.

[7] R. A. Newcombe, A. J. Davison, S. Izadi, P. Kohli, O. Hilliges, J. Shotton, D. Molyneaux, S. Hodges, D. Kim, and A. Fitzgibbon. KinectFusion: real-time dense surface mapping and tracking. In *Mixed and Augmented Reality (ISMAR), 2011 10th IEEE International Sympo- sium on*, pages 127–136, 2011.

[8] D.-H. Nie, K.-P. Han, and H.-S. Lee. Stereo matching algorithm using population-based incremental learning on gpu. In *Intelligent Systems and Applications, 2009. ISA 2009. International Workshop on*, pages 1–4, 2009.

[9] S. Sinha, J.-M. Frahm, M. Pollefeys, and Y. Genc. Feature tracking and matching in video using programmable graphics hardware. *Machine Vision and Applications*, 2007.

[10] V. Vineet and P. J. Narayanan. Cuda cuts: Fast graph cuts on the gpu. In *Computer Vision and Pattern Recognition Workshops, 2008. CVPRW '08. IEEE Computer Society Conference on*, pages 1–8, 2008.

[11] R. Yang and M. Pollefeys. Multi-resolution real-time stereo on com- modity graphics hardware, 2003.

[12] K. J. Yoon and I. S. Kweon. Adaptive support-weight approach for correspondence search. *Ieee Transactions On Pattern Analysis And Machine Intelligence*, 28(4):650–656, Apr 2006.

[13] Y. Zhao and G. Taubin. Real-time stereo on gpgpu using progressive multi-resolution adaptive windows. pages 420–432, 2011.

[14] Seoane, P., M. Gestal, and J. Dorado, "Approach for solving multimodal problems using Genetic Algorithms with Grouped into Species optimized with Predator-Prey", International Journal of Interactive Multimedia and Artificial Intelligence, vol. 1, no. 5, pp. 6-13, 06/2012

[15] Mey Rodríguez, M., and E. P. Gayoso, "EvoWild: a demosimulator about wild life", International Journal of Interactive Multimedia and Artificial Intelligence, vol. 1, issue Experimental Simulations, no. 1, pp. 25-30, 12/2008

15

GLOA: A New Job Scheduling Algorithm for Grid Computing

[1]Zahra Pooranian, [2]Mohammad Shojafar, [3]Jemal H. Abawajy,[4]Mukesh Singhal
[1]Graduate School, *Dezful Islamic Azad University, Dezful, Iran*
[2]Dept.of Information Engineering, Electronic and Telecommunication (DIET), *"Sapienza"*
University of Rome, Rome, Italy
[3]School of Information Technology, *Deakin University, Geelong, Australia*
[4]Computer Science & Engineering, *University of California, Merced, USA*

Abstract — **The purpose of grid computing is to produce a virtual supercomputer by using free resources available through widespread networks such as the Internet. This resource distribution, changes in resource availability, and an unreliable communication infrastructure pose a major challenge for efficient resource allocation. Because of the geographical spread of resources and their distributed management, grid scheduling is considered to be a NP-complete problem. It has been shown that evolutionary algorithms offer good performance for grid scheduling. This article uses a new evaluation (distributed) algorithm inspired by the effect of leaders in social groups, the group leaders' optimization algorithm (GLOA), to solve the problem of scheduling independent tasks in a grid computing system. Simulation results comparing GLOA with several other evaluation algorithms show that GLOA produces shorter makespans.**

Keywords — **Artificial Intelligence, Distributed Computing, Grid Computing, Job Scheduling, Makespan.**

I. INTRODUCTION

NEW technology has taken communication to the field of grid computing. This allows personal computers (PCs) to participate in a global network when they are idle, and it allows large systems to utilize unused resources. Like the human brain, modern computers usually use only a small fraction of their potential and are often inactive while waiting for incoming data. When all the hardware resources of inactive computers are collected as an all-in-one computer, a powerful system emerges.

With the help of the Internet, grid computing has provided the ability to use hardware resources that belong to other systems. "Grid computing" may have different meanings for different people, but as a simple definition, grid computing is a system that allows us to connect to network resources and services and create a large powerful system that has the ability to perform very complex operations that a single computer cannot accomplish. That is, from the perspective of the users of grid systems, these operations can only be performed through these systems. As large-scale infrastructures for parallel and distributed computing systems, grid systems enable the virtualization of a wide range of resources, despite their significant heterogeneity [1].

Grid computing has many advantages for administrators and developers. For example, grid computing systems can run programs that require a large amount of memory and can make information easier to access. Grid computing can help large organizations and corporations that have made an enormous investment to take advantage of their systems. Thus, grid computing has attracted the attention of industrial managers and investors in companies that have become involved in grid computing, such as IBM, HP, Intel, and Sun [2].

By focusing on resource sharing and coordination, managing capabilities, and attaining high efficiency, grid computing has become an important component of the computer industry. However, it is still in the developmental stage, and several issues and challenges remain to be addressed [3].

Of these issues and challenges, resource scheduling in computational grids has an important role in improving the efficiency. The grid environment is very dynamic, with the number of resources, their availability, CPU loads, and the amount of unused memory constantly changing. In addition, different tasks have different characteristics that require different schedules. For instance, some tasks require high processing speeds and may require a great deal of coordination between their processes. Finally, one of the most important distinctive requirements of grid scheduling compared with other scheduling (such as scheduling clusters) is scalability.

With more applications looking for faster performance, makespan is the most important measurement that scheduling algorithms attempt to optimize. Makespan is the resource consumption time between the beginning of the first task and the completion of the last task in a job. The algorithm presented in this paper seeks to optimize makespan. Given the complexity and magnitude of the problem space, grid job scheduling is an NP-complete problem. Therefore, deterministic methods are not suitable for solving this problem. Although several deterministic algorithms such as min-min and max-min [4] have been proposed for grid job scheduling, it has been shown that heuristic algorithms provide better solutions. These algorithms include particle swarm optimization (PSO)[5], genetic algorithms (GAs)[6], simulating annealing (SA)[7], tabu search (TS)[8], gravitational emulation local search(GELS)[9], ant colony optimization (ACO) [10], and recently Learning Automata (LA) [26]. Also, some researchers have proposed combinations of these algorithms, such as GA-SA[11], GA-TS[12], PSO-SA[13], GPSO[14], and GGA[15].

It is important that an optimization algorithm for optimization problems should converge to the optimal solution in a short period of time. The group leaders optimization algorithm (GLOA) [16] was inspired by the influence of leaders in social groups. The idea behind the algorithm is that the problem space is divided into several smaller parts (several groups), and each part is searched separately and in parallel to increase the optimization speed. Each separate space can be searched by its leader, who tries to find a solution by checking whether it is the closest member to the local and global minimum.

In this paper, we use GLOA for independent task/job scheduling in grid computing. In addition to the simplicity of its implementation, GLOA reduces optimization time. The remainder of this paper is organized as follows. Section II discusses related methods. Section III presents a general model for job/task scheduling. Section IV presents the GLOA method and modifies it based on our problem. Section V compares simulation results obtained with this algorithm and several other heuristic algorithms. Finally, the last section presents the conclusion of this study.

II. RELATED WORK

In [17], the TS algorithm, which is a local search algorithm, is used for scheduling tasks in a grid system. In [18], the SA algorithm is used to solve the workflow scheduling problem in a computational grid. Simulation results show that this algorithm is highly efficient in a grid environment. The TS algorithm uses a perturbation scheme for pair changing.

In [19], the PSO algorithm is used for job scheduling with two heuristic algorithms, latest finish time (LFT) and best performance resource (BPR), used to decide task priorities in resource queues. In [20], the critical path genetic algorithm (CPGA) and task duplication genetic algorithm (TDGA) are proposed; they modify the standard GA to improve its efficiency. They add two greedy algorithms to the GA so that the wait times for tasks to start and ultimately the makespan can be reduced. The proposed algorithms consider dependent tasks, so that computation costs among resources are considered as well. Chromosomes are divided into two parts, and the graph under consideration is transformed into a chromosome that performs mapping and scheduling. The mapping part determines the processors on which tasks will execute, and the scheduling part determines the sequence of tasks for execution. In the representation of a chromosome, task priorities are considered by examining the graph.

The CPGA algorithm combines the modified critical path (MCP) algorithm [21] and a GA. The MCP algorithm first determines critical paths, and if the parent of tasks being executed on a processor is executing on another processor, these tasks are transported to the parent's processor to reduce the cost of transportation between processors.

The TDGA algorithm combines the duplication scheduling heuristic (DSH) algorithm [22] and a GA. This algorithm first sorts tasks in descending order and then repeats the parent task on all processors so that the children can execute earlier, because the transportation cost between processors becomes zero. By repeating the parent task, overload and communication delays are reduced and total execution time is minimized.

The resource fault occurrence history (RFOH) [23] algorithm is used for job scheduling fault-tolerant tasks in a computational grid. This method stores resource fault occurrence histories in a fault occurrence history table (FOHT) in the grid information server. Each row of the FOHT table represents are source and includes two columns. One column shows the failure occurrence history for the resource and the other shows the number of tasks executing on the resource. The broker uses information in this table in the GA when it schedules tasks. This reduces the possibility of selecting resources with more occurrences of failures.

The chaos-genetic algorithm [24] is a GA for solving the problem of dependent task/job scheduling. This algorithm uses two parameters, time and cost, to evaluate quality of service (QOS), and chaos variables are used rather than randomly producing the initial population. This combination of the advantages of GAs and chaos variables to search the search space inhibits premature convergence of the algorithm and produces solutions more quickly, with a faster convergence.

The integer genetic algorithm (IGA) [25] is a genetic algorithm for solving dependent task/job scheduling that simultaneously considers three QOS parameters: time, cost, and reliability. Since these parameters conflict with one another and cannot be simultaneously optimized—as improvement of one reduces the quality of another—weights are assigned to each parameter, either by the user or randomly. If the user provides the weighting, the parameter that is more important to the user is given more weight than the others.

III. PROBLEM DESCRIPTION

The problem studied in this paper is independent task/job scheduling in grid computing. The proposed algorithm should be efficient in finding a solution that produces the minimum makespan. Thus, the problem is to assign a set of m input tasks $(T=T_1,T_2,...,T_m)$ to n resources $(R=R_1,R_2,...,R_n)$, with the minimum makespan.

IV. THE GLOA ALGORITHM

GLOA is an evolutionary algorithm that is inspired by the effect of leaders in social groups. The problem space is divided into different groups, and each group has its own leader. The members of each group don't necessarily have similar characteristics, and they have quite random values. The

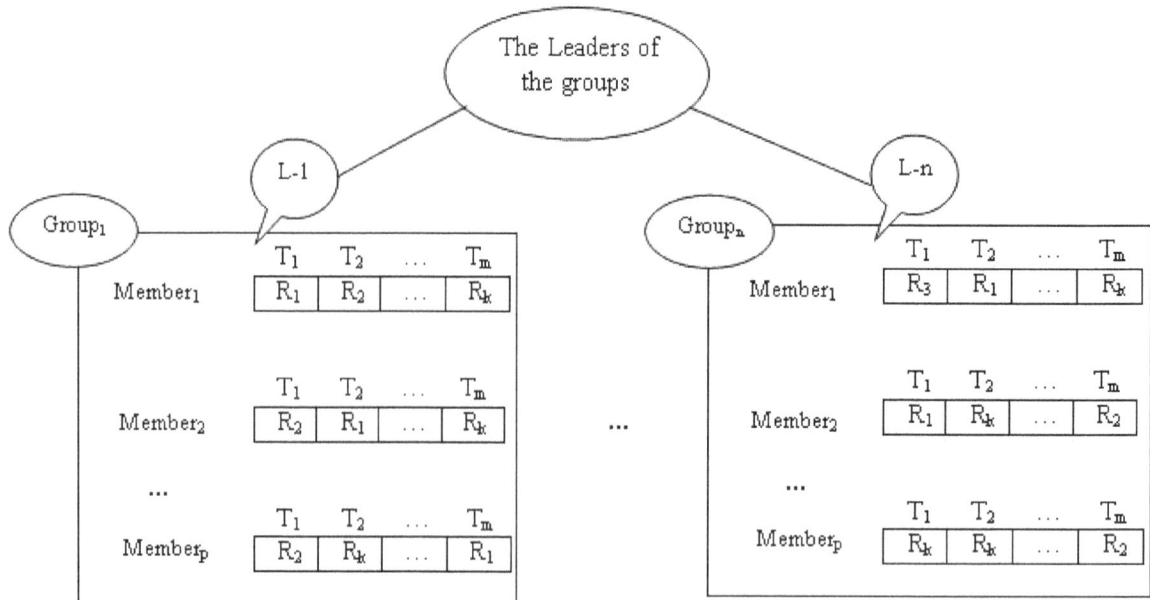

Fig. 1. Steps 1–3 of the algorithm: *n* groups consisting of *p* members are created, and their leaders are chosen based on their fitness values.

best member of each group is selected as the leader. The members of each group try to become similar to their leader in each iteration. In this way, the algorithm is able to search a solution space between a leader and its group members. It is obvious that after some iteration, members of a group may become similar to their leader. In order to introduce diversity within a group, one of its members is selected randomly and some of its variables are interchanged with a member of another group. In addition, a crossover operator helps a group come out of local minima, and the solution space can be searched again so to produce diversity. The algorithm steps are as follows:

A. Initial Population Production

A set of *p* members is produced for each group. The total population is therefore $n*p$, where *n* is the number of groups. Group and member values are produced randomly. Since the number of entering tasks is *m*, the members are represented as an *m*-dimensional array in which the stored values are resource numbers. For example, in Figure 1 we have *n* groups, each with *p* members.

B. Calculating Fitness Values of All Group Members

The fitness value is calculated for each member of each group. Since the purpose of task/job scheduling in a grid is to assign tasks to resources in a way that minimizes makespan, makespan has been chosen as the criterion for evaluating members. The less a member's makespan is, the greater is its fitness value, according to (1):

$$fitness(member_k) = \frac{1}{makespan(member_k)} \qquad (1)$$

C. Determining Leader of Each Group

In each group, after the fitness value is computed for each

member, the member with the best fitness value is selected as the group leader.

D. Mutation Operator

In this step, a new member is produced in each group from an older member, the leader of the group, and a random element, using (2). If the fitness value of the new member is better than the fitness value of the older member, it replaces the older member. Otherwise, the older member is retained.

$$new = r_1 * old + r_2 * leader + r_3 * random \qquad (2)$$

where r_1, r_2, and r_3 are the rates determining the portion of the older member, the leader, and the random element that are used to generate the new population, such that $r_1 + r_2 + r_3 \leq 1$. Pseudocode for this step follows:

for i=1 **to** n do {

for j=1 **to** p do {

$new_{ij} = r_1 * member_{ij} + r_2 * Li + r_3 * random$

if fitness (new_{ij}) **better than** fitness ($member_{ij}$)

then

$member_{ij} = new_{ij}$

end if

} **end for**

} **end for**

The value of r_1 determines the extent to which a member retains its original characteristics, and r_2 moves the member toward the leader of its group in different iterations, thus making the member similar to the leader. Careful selection of these two parameters plays an important role in the optimization of the results. The main characteristic of this algorithm is that it searches the problem space surrounded by

TABLE I
PARAMETERS FOR THE ALGORITHMS

Algorithm	Parameter	Value
GLOA	Number of groups	3
	Population in each group	10
	r_1	0.8
	r_2	0.1
	r_3	0.1
GA	P-Crossover	0.85
	P-Mutation	0.02

the leaders. This leads to very rapid convergence to a global minimum. Note that eq. (2) is similar to the update equation for the PSO algorithm. The difference is that here, unlike PSO, the best position value of each member is not stored and so there is no information about the past positions of members.

TABLE II
THE ALGORITHMS' MAKESPAN AFTER 100 ITERATIONS (IN SECONDS)

(No. Tasks, No. Resources)	SA	GA	GSA	GGA	GLOA
(50,10)	136.742	99.198	95.562	90	89
(100,10)	307.738	183.49	190.353	181.028	167
(300,10)	973.728	638.082	626.66	597	581.842
(500,10)	1837.662	1105.56	1087	1087.216	1072.362

E. One-way Crossover Operator

In this step, a random number of members are selected from the first group and some of their parameter values are replaced with those of a member of another group that is selected randomly. It should be noted that in each iteration, only one parameter is replaced. If any new member is better it replaces the old one; otherwise the old member remains in the group. An important issue here is selecting the correct crossover rate, for otherwise all members will rapidly become similar to each other. The transfer rate t is a random number such that $1 \le t \le (\frac{m}{2})+1$ for each group. The purpose of the crossover operator is to escape local minima.

F. Repetition of Steps C to V according to the Determined Number of Iterations

This algorithm is repeated according to the determined number of iterations. At the end, from the different groups, the leader with the best fitness value is chosen as the problem solution.

V. SIMULATION

This section compares simulation results for our proposed algorithm with the results of several other algorithms. All algorithms were simulated in a Java environment on a system with a 2.66 GHZ CPU and 4GBRAM. Table I lists the parameters used in the performance study of our proposed algorithm and the other algorithms.

Table II shows the five algorithms' makespans for various numbers of independent tasks and 10 resources. As can be seen, SA has the worst makespans and GLOA has the best. We

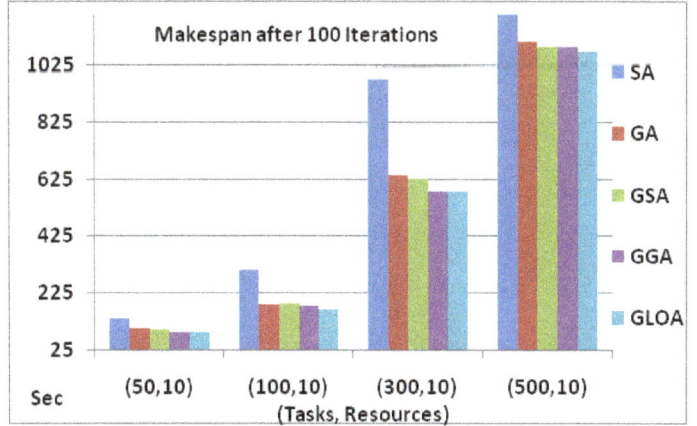

Fig. 2. Algorithms' makespan after 100 iterations, with 10 resources

provide more details in Fig. 2.

As we can see in Fig. 2, the SA algorithm's makespan increases rapidly as the number of tasks grows from 50 to

TABLE III
ALGORITHMS' MAKESPAN AFTER 300 ITERATIONS (IN SECONDS)

(No. Tasks, No. Resources)	SA	GA	GSA	GGA	GLOA
(100,10)	233.2	172.628	179.062	175.598	166.14
(100,20)	173.116	111.946	105.314	103.092	94.55
(100,30)	120.452	90.716	87.846	80.086	77.75

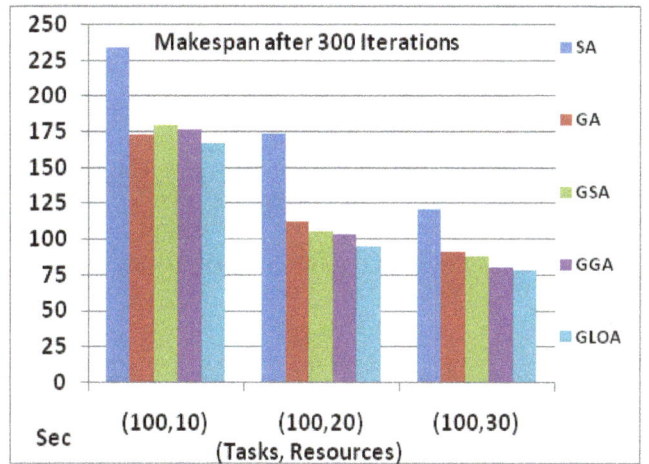

Fig. 3. Algorithms' makespan after 300 iterations, for various numbers of resources

500.

Hence, SA is the worst algorithm for minimizing makespan and GLOA is the best in every case. In the 50-task case, the difference between SA and GLOA is approximately 48 seconds, which is less than half of the SA makespan. Here GLOA has the least makespan. When there are only a few tasks, the makespans for all of the algorithms are low, and GLOA produces the minimum. For the 300-task and 500-task cases, GGA has a similar makespan to the GLOA algorithm. For example, in the 300-task case, GGA's makespan is approximately 597 seconds but GLOA's is approximately 582

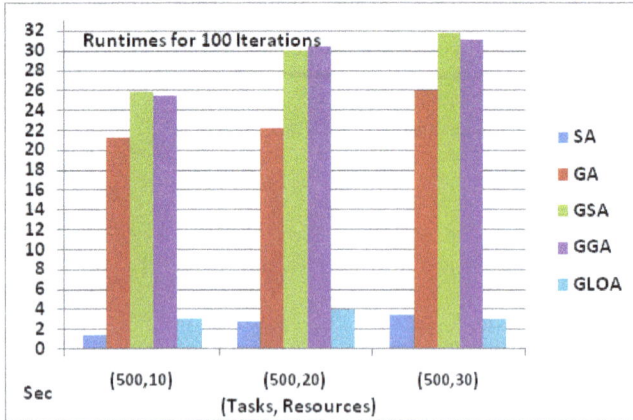

Fig. 4. Algorithm runtimes for 100 iterations with varying numbers of resources.

seconds.

Table III shows the makespans the algorithms produce for 100 fixed independent tasks for various numbers of resources. As can be seen, SA has the worst makespan in all of these cases and GLOA has the best. More details are shown in Fig. 3.

As can be seen in Fig. 3, as the number of resources increases, the makespan decreases for all algorithms, because when there are several ready resources with empty queues, tasks can be assigned to the new resources. The variation is the difference between the algorithms' structures. When the number of resources triples, the decrease for the makespan in SA is approximately 100 seconds and in GLOA it is approximately 90 seconds. As shown, GLOA has the minimum makespan in each case. Its structure provides it with the ability to be close to GGA, because like GGA, it can search the problem space both locally and globally. Hence, GLOA reaches the best solution more rapidly (e.g., in 95 seconds for 20 resources) than the other methods, particularly SA (which takes approximately 174 seconds). GLOA's makespan decreases up to 45% compared to SA, 15% compared to GA, 11% compared to GSA, and approximately 8% compared to GGA.

TABLE IV
ALGORITHMS' RUNTIME FOR 100 ITERATIONS (SECONDS)

(No. Tasks, No. Resources)	SA	GA	GSA	GGA	GLOA
(500,10)	1.4	21.2	25.8	25.4	3
(500,20)	2.8	22.2	30	30.4	4
(500,30)	3.4	26	31.8	31	3

Table IV shows the algorithms' runtime for job 500 independent tasks with varying numbers of resources. As shown, SA has the best runtime for 10 and 20 resources (because it considers only one solution, it can search more quickly than the other algorithms), and GLOA has the second best for 30 resources (because it divides the problem solutions into several groups that search in parallel, it reaches the optimum more quickly, but it takes some time to produce several groups). Fig. 4 provides more details.

As can be seen in Fig. 4, when the number of resources increases, all algorithm runtimes increase, because when there are several new resources with empty queues, these resources must be searched and tasks assigned to them. SA is the least time-consuming algorithm (except for the 30-resource case) and GSA is the worst (except for the 500-task and 20-resource case). When the number of resources increases to 30, GLOA's runtime decreases less than SA's, because the resources have sufficiently many empty queues to be able to respond to 500 tasks more quickly, and SA considers the entire problem while GLOA divides the problem into several groups and considers the queue sizes and makespans for the tasks. When there are only a few resources (10), GA executes in just under 22 seconds, GSA and GGA have similar runtimes (just under 26 seconds), and GLOA requires just over 2 seconds, but SA requires less than 2 second. Although SA is the best algorithm in terms of runtime, it cannot produce better makespan results (as seen in Figure 2), and therefore we exempt this algorithm from consideration. When the number of resources triples (from 10 to 30), SA's runtime increases by 80%, GA's by 24%, GGA's and GSA's by 26%, but GLOA's increases by less than 10%. Therefore, while GLOA's runtime increases with the number of resources, it does so at a very low rate.

VI. CONCLUSION

Grid technology has made it possible to use idle resources as part of a single integrated system. The main purpose of grid computing is to make common resources such as computational power, bandwidth, and databases available to a central computer. The geographic spread and dynamic states of the grid space present challenges in resource management that necessitate an efficient scheduler. This scheduler should assign tasks to resources in such a way that they are executed in the shortest possible time.

This paper used a new evolutionary algorithm, GLOA, for scheduling tasks/jobs in a computational grid. Simulation results for GLOA were compared with results for four other intelligent algorithms: GA, SA, GGA, and GSA, and it was shown that in addition to wasting less computation time than the other algorithms, GLOA is able to produce shortest makespans. Also, GLOA could be applied in the real world because its runtime and makspan is less than other AI methods and produce less overhead on resources while responding the independent tasks.

In the future, we will change GLOA structure and apply it into dependent tasks in Grid Environment to cover the current gap into scheduling of dependent tasks.

REFERENCES

[1] J. Kołodziej, F. Xhafa, "Meeting security and user behavior requirements in Grid scheduling," *Simulation Modeling Practice and Theory, Elsevier*, pp. 213–226, 2011.

[2] S. Garg, R. Buyyaa, H. Siegel, "Time and cost trade-off management for scheduling parallel applications on Utility Grids," *Future Generation Computer Systems, Elsevier*, pp. 1344-1355, 2010.

[3] F. Xhafa, A. Abraham, "Computational models and heuristic methods for Grid scheduling problems," *Future Generation Computer Systems, Elsevier*, pp. 608-621, 2010.

[4] M. Shojafar, S. Barzegar, M. R, Meibody, "A new Method on Resource Scheduling in grid systems based on Hierarchical Stochastic Petri net," *3rd International Conference on Computer and Electrical Engineering (ICCEE 2010)*, pp. 175-180, 2010.

[5] Q. Tao, H. Chang, Y. Yi, Ch. Gu, W. Li, "A rotary chaotic PSO algorithm for trustworthy scheduling of a grid workflow, " *Computers & Operations Research, Elsevier*, Vol. 38, pp.824–836, 2011.

[6] R. P. Prado, S. García-Galán, A. J. Yuste and J. E. M. Expósito, "Genetic fuzzy rule-based scheduling system for grid computing in virtual organizations," *Soft Computing - A Fusion of Foundations, Methodologies and Applications*, Vol. 15, No. 7, pp.1255-1271, 2011.

[7] S. Fidanova, "Simulated Annealing for Grid Scheduling Problem, International Symposium on Modern Computing," *IEEE John Vincent Atanasoff*, pp.41-45, 2006.

[8] B. Eksioglu, S. DuniEksioglu, P. Jain, "A tabu search algorithm for the flowshop scheduling problem with changing neighborhoods," *Computers & Industrial Engineering*, 54, pp.1–11, 2008.

[9] B. Barzegar, A. M. Rahmani, K. Zamani far, "Advanced Reservation and Scheduling in Grid Computing Systems by Gravitational Emulation Local Search Algorithm," *American Journal of Scientific Research*, No. 18, pp. 62-70, 2011.

[10] Y. Yang, G. Wua, J.Chen, W. Dai, "Multi-objective optimization based on ant colony optimization in grid over optical burst switching networks," *Expert Systems with Applications*, 37, pp.1769–1775, 2010.

[11] G. Guo-ning , "Genetic simulated annealing algorithm for task scheduling based on cloud computing environment," *International Conference of Intelligent Computing and Integrated Systems (ICISS)*, pp-60-63, 2010.

[12] J.M. Garibaldi, D. Ouelhadj, "Fuzzy Grid Scheduling Using Tabu Search," IEEE International Conference of Fuzzy Systems, pp.1-6, 2007.

[13] R. Chen, D. Shiau, Sh. Tang Lo, Combined Discrete Particle Swarm Optimization and Simulated Annealing for Grid Computing Scheduling Problem, Lecture Notes in Computer Science, Springer, Vol. 5755, 2009, pp.242-251.

[14] Z. Pooranian, A. Harounabadi, M. Shojafar and J. Mirabedini, "Hybrid PSO for Independent Task scheduling in Grid Computing to Decrease Makespan," *International Conference on Future Information Technology (ICFIT 2011)*, Singapore, pp.435-439, 2011.

[15] Z. Pooranian, A. Harounabadi, M. Shojafar, N. hedayat, " New Hybrid Algorithm for Task Scheduling in Grid Computing to Decrease missed Task," *World Academy of Science, Engineering and Technology 79*, pp. 262-268, 2011.

[16] A. Daskin, S. Kais, "Group leaders optimization algorithm, Molecular Physics," *An International Journal at the Interface Between Chemistry and Physics*, Vol. 109, No. 5, pp. 761–772, 2011.

[17] M. Yusof, K. Badak, M. Stapa, "Achieving of Tabu Search Algorithm for Scheduling Technique in Grid Computing Using GridSim Simulation Tool: Multiple Jobs on Limited Resource," *International Journal of Grid and Distributed Computing*, Vol. 3, No. 4, pp. 19-32., 2010.

[18] R. Joshua Samuel Raj, V. Vasudevan, Beyond Simulated Annealing in Grid Scheduling, *International Journal on Computer Science and Engineering (IJCSE)*, Vol. 3, No. 3, pp. 1312- 1318, Mar. 2011.

[19] Ruey-Maw Chen and Chuin-Mu Wang, Project Scheduling Heuristics-Based Standard PSO for Task-Resource Assignment in Heterogeneous Grid, *Abstract and Applied Analysis*, Vol. 2011, pp.1-20, 2011.

[20] F. A. Omaraa, M.M. Arafa, "Genetic algorithms for task scheduling problem," *Journal Parallel Distributed Computing, Elsevier*, Vol. 70, No. 1, pp. 13-22, 2010.

[21] M. Wu, D. D. Gajski, "Hyper tool: A programming aid for message-passing systems," *IEEE Transactions on Parallel and Distributed Systems*, Vol. 1, No. 3, pp. 330-343, 1990.

[22] H. El-Rewini, T. G. Lewis, H. H. Ali, *Task Scheduling in Parallel and Distributed Systems*, Prentice-Hall, 1994, ISBN:0-13-099235-6.

[23] L. Khanli, M. Etminan Far , A. Ghaffari, "Reliable Job Scheduler using RFOH in Grid Computing," *Journal of Emerging Trends in Computing and Information Sciences*, Vol. 1, No. 1, pp. 43- 47, 2010.

[24] G. Gharoonifard, F. Moeindarbari, H. Deldari, A. Morvaridi, "Scheduling of scientific workflows using a chaos- genetic algorithm," Procedia Computer Science, Elsevier, Vol. 1, No.1, pp. 1445- 1454, 2010.

[25] Q. Tao, H. Chang, Y. Yi, CH. Gu, "A Grid Workflow Scheduling Optimization approach for e-Business Application," *Proceedings of the 10th International Conference on E-Business and E-Government*, pp. 168-171, 2010.

[26] J. A. Torkestani, "A New Distributed Job Scheduling Algorithm for Grid Systems," An International Journal of Cybernetics and Systems, Vol. 44, Issue 1, pp.77-93, 2013.

16

Mining Web-based Educational Systems to Predict Student Learning Achievements

José del Campo-Ávila, Ricardo Conejo, Francisco Triguero, Rafael Morales-Bueno

Universidad de Málaga, Andalucía Tech, Departamento de Lenguajes y Ciencias de la Computación, Campus de Teatinos, Málaga, España

Abstract — Educational Data Mining (EDM) is getting great importance as a new interdisciplinary research field related to some other areas. It is directly connected with Web-based Educational Systems (WBES) and Data Mining (DM, a fundamental part of Knowledge Discovery in Databases).

The former defines the context: WBES store and manage huge amounts of data. Such data are increasingly growing and they contain hidden knowledge that could be very useful to the users (both teachers and students). It is desirable to identify such knowledge in the form of models, patterns or any other representation schema that allows a better exploitation of the system. The latter reveals itself as the tool to achieve such discovering. Data mining must afford very complex and different situations to reach quality solutions. Therefore, data mining is a research field where many advances are being done to accommodate and solve emerging problems. For this purpose, many techniques are usually considered.

In this paper we study how data mining can be used to induce student models from the data acquired by a specific Web-based tool for adaptive testing, called SIETTE. Concretely we have used top down induction decision trees algorithms to extract the patterns because these models, decision trees, are easily understandable. In addition, the conducted validation processes have assured high quality models.

Keywords — Data Mining, Decision Trees, Educational technology, Knowledge discovery.

I. INTRODUCTION

SINCE Internet opened a new way to communicate in many different forms, the educational sector adopted such technology and developed the Web-based Educational Systems (WBES). Firstly, they were static systems, mainly dedicated to divulgate contents. But progressively, they extended their capabilities with new characteristics in order to make the systems adaptive and intelligent [1].

At this moment there exist many different systems that combine different elements to achieve some level of intelligence. Therefore, we can find WBES with adaptive techniques [2], some other WBES with intelligent mechanisms [3] and more complex systems that combine both properties (a detailed review of AIWBES was presented by Brusilovsky and Peylo [4]).

What it is evident is the high volume of data that these systems are storing and processing continuously: relations between contents offered to students, interactions with students, number of visits, marks achieved in tests, time used to respond those tests, etc.

Knowledge discovery in databases (KDD) continues extending to almost every field where large amount of data are stored and processed (databases, system logs, activity logs, etc.), so WBES becomes another environment to apply KDD processes.

The data mining techniques are essential for one of the most important points of KDD: they are applied in data analysis phase and machine learning algorithms are used to produce the models that summarize the knowledge discovered [5]. Therefore, it is easy to see that educational tasks can benefit from the knowledge extracted by data mining.

This research field is called Educational Data Mining (EDM) and its main objective is to analyze data stored in WBES in order to resolve educational research issues [6]: validation of the educational system, prediction of students learning achievements, identification of misconceptions [7], assessment and feedback to the authors of courses [8], etc.

In this paper we try to determine that data mining techniques can help to predict students learning achievements, mainly oriented to find relations between continual assessment (or evaluation) and the final grade achieved.

This paper is organized as follow. In Section 2 we describe the materials used and the conducted methodology. Basically, our materials are data collected by SIETTE[8], a Web-based tool for adaptive testing [9] and the framework for data mining called Weka [10]. Then, in Section 3, we present the results and comment the patterns discovered by machine learning algorithms. Finally, in Section 4, we summarize the most relevant conclusions and propose new research lines for futures works.

II. MATERIALS AND METHODS

Considering the features offered by data mining in order to discover patterns in datasets, in this case extracted from Web-based Educational System, we propose to study the existence of different kinds of relations between the continuous

[8] http://www.siette.org

evaluation of students and their final achievements in the subject. For this purpose we work with the following materials and methodologies.

A. Materials

The raw materials of any process of knowledge discovery that uses data mining techniques are data, grouped in subsets called datasets. Every dataset is composed of examples described by attributes and labeled with a class (supervised learning). Values for these attributes can be numerical or nominal.

For this study we have focused in students that took the subject "Principles in Informatics" in two consecutive courses. The skills and competences to be achieved are varied: from basic concepts related with Computer Science (hardware, software, algorithms, etc.) to elementary abilities to develop computer programs using the C programming language.

The evaluation of this subject includes a continuous evaluation during the course (with a weight of 40% in the final grade) that ends with a final evaluation exam (60% weight). The continual assessment (or continuous evaluation) is compound of three tests (20%) and three practical exercises (20%). What we are using in this study are the marks achieved by the students in the tests that have been completed using the SIETTE Web-based Educational System [9]. First test (T1) is used to check how concepts related with Computer Science are assimilated, the second one (T2) focus on initial programming abilities with C (types, expressions, operators and control flow) and the third one (T3) check the knowledge about more advanced concepts in C (functions and structures). The final exam is mostly prepared to check the programming abilities; so basic concepts related with Computer Science are only evaluated with one test (T1).

In Table I we show some statistics related to the real marks achieved in the tests. The maximum value cannot be greater than 100.00, but minimum values can be lower than 0.00 because wrongly answered questions count negatively (if a student answers many questions incorrectly, the mark is lower than 0.00).

Taking this context in consideration, now we can describe the datasets that we have used. In our case the examples summarize the evaluation achieved by the students (116) that took the subject "Principles in Informatics". In a first approach we only consider the marks for every test, but in a second step we added the differences with respect to the average value, in order to establish a relative comparison between the results.

The class attribute is the final grade achieved in the global subject evaluation. We have used the numerical grade, defined in [0,10], and transformed it to the European ECTS grading scale (A for the best grades and F for the worst ones, F corresponds to students that fail) [11].

To carry out the mining process there exist different frameworks that implement multiple machine learning algorithms. We have used Weka [10] because it includes TDIDT (Top Down Induction Decision Trees) algorithms that represent the knowledge extracted in form of decision trees

TABLE I
MARKS ACHIEVED IN TESTS

	T1 (HW, SW, algorithms)	T2 (types, operators, control flow)	T3 (functions, structures)
Minimum	-20.00	-20.00	-18.33
Average	34.18 ± 19.30	35.50 ± 23.42	30.47 ± 27.84
Maximum	76.67	86.67	100.00

Minimum, average and maximum values observed in the tests answered by students. Maximum never can be greater than 100.00, but minimum values can be lower than 0.00 because wrongly answered questions count negatively.

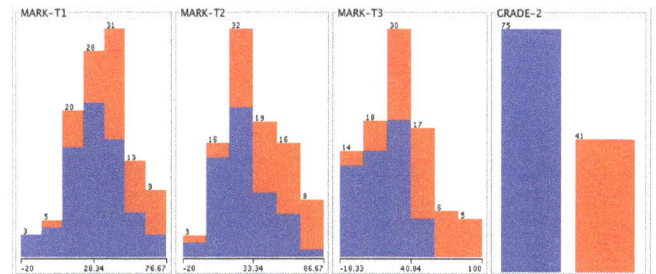

Fig. 1. Class distribution visualizing all marks (for tests T1, T2 and T3) in the first dataset (116 students). Blue color represents students that fail the evaluation (or absent themselves) and red color represents students that pass the evaluation. This chart is plotted by the Weka framework.

[12]: a model easily understandable by humans with some other additional advantages (learning with numerical or nominal data, robustness, verifiable reliability, etc.). Concretely we have selected the J48 algorithm (C4.5 [13] implementation coded in Weka), using it with its default configuration. When plotting the decision trees (Fig. 4, 5, and 6), the numbers present in the nodes (<first> / <second>) represent the number of examples that satisfy the branch (<first>) and the number of examples that, in addition, are incorrectly classified (<second> that it is not present when there is no errors).

B. Methods

Once we have described the datasets and the framework we have used, we can detail which methodology we have followed.

Firstly we have preprocessed the data in order to clean and prepare them. Data extracted from SIETTE are very rich and diverse, but nowadays, they cannot be directly exported to the kind of dataset supported by Weka (ARFF files). Some transformation steps were needed: discretization of numerical grade to ECTS grading scale, calculation of new calculated attributes, identification of missing values, etc.

The datasets used in this study have been progressively transformed to do more detailed mining process. Although the details will be presented in next section, we can advance that we have used 3 datasets derived from the original one.

The first dataset, with 116 examples (students), is described by 3 attributes (marks achieved in every test) and a binary nominal class (passing the subject or failing it – including absent students –). In Fig. 1 it is shown the class distribution for three different marks.

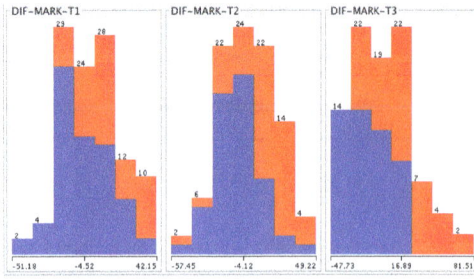

Fig. 2. Class distribution visualizing marks and differences with the average marks (for tests T1, T2 and T3) in the second dataset (116 students). Blue color represents students that fail the evaluation (or absent themselves) and red color represents students that pass the evaluation. This chart is plotted by the Weka framework.

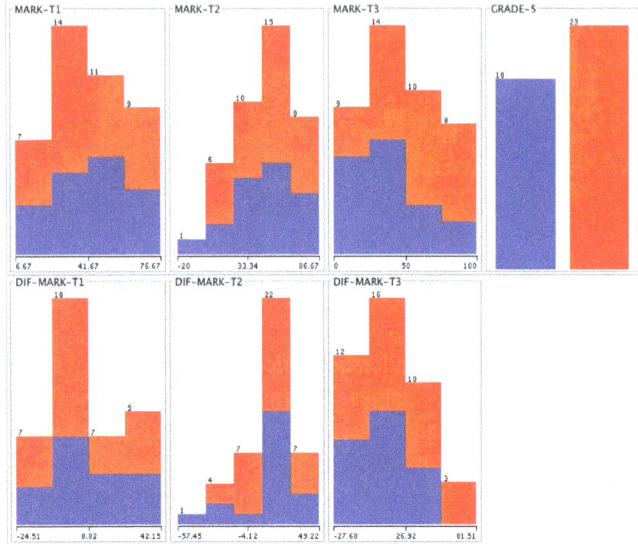

Fig. 3. Class distribution visualizing marks and differences with the average marks (for tests T1, T2 and T3) in the third dataset (41 students). Blue color represents students that have grades D or E and red color represents students that have grades A, B or C. This chart is plotted by the Weka framework.

In the next step we calculated the differences between the mark itself and the average valued achieved in that test by students during their course. Therefore, we incorporated 3 new attributes to the dataset. In Fig. 2 it is shown the class distribution for such new attributes.

Finally, once we have detected patterns to separate students that pass the evaluation and those students that do not pass it, we were interested in inducing some models that could find some pattern to differentiate between best students (with A, B or C grades) and the rest of students that pass the evaluation (D or E grades). In this dataset we only had 41 students so the induction algorithm had some problems with so few examples. To solve it we resample the dataset [14] making it five times bigger (205 examples) and configured J48 to examine a bigger number of examples before expanding (minimum of 20 examples) in order to avoid overfitting and reduce the complexity of the model [12]. In Fig. 3 it is shown the class distribution for this last dataset.

III. RESULTS

In this section we present the results that we have collected

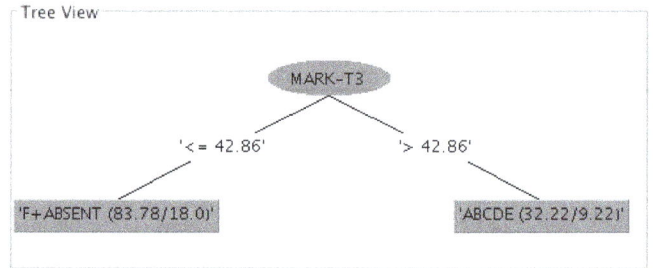

Fig. 4. Decision tree induced by J48 using the first dataset. Attributes are the marks for T1, T2 and T3; the binary class separate between students that pass (ABCDE) or not pass (F+ABSENT) the global evaluation process.

after applying the mining process to the data previously described. As we have explained, we have used a TDIDT algorithm (J48 implementation of C4.5), so the induced models are decision trees, what make possible an easy interpretation of the patterns. In addition, we can rely on the results, because validation processes show high confidence levels. The validation processes we have conducted are 10-fold cross validations.

For the first dataset, that which separates students in a binary class (pass or not pass the evaluation) and only include the marks achieved for every test (T1, T2 and T3), the pattern is easy to understand (even no TDIDT algorithm would be necessary because the class distribution in Fig. 1 shows a similar information). The most important attribute to determine the difference between two student profiles is the mark achieved for the last test (T3), the most close to the final exam. The decision tree, shown in Fig. 4, is not surprising, but reflects the ability of machine learning algorithms to find patterns. Furthermore, the validation shows 80% accuracy, quite reliable considering the number of examples and the class unbalance.

Analyzing the second dataset, extended with new attributes that summarize the differences between the own mark and the average value, some additional knowledge is extracted. Decision tree (Fig. 5) reveals that once we know the mark for T3 (root node), we can detect some other differences. In this case, the new added attributes reveal as important elements to determine the final achievement of the students. Particularly students that are below 42.86 points in T3, need to do best that the average in T1 and T2 to pass. So the requirements are not

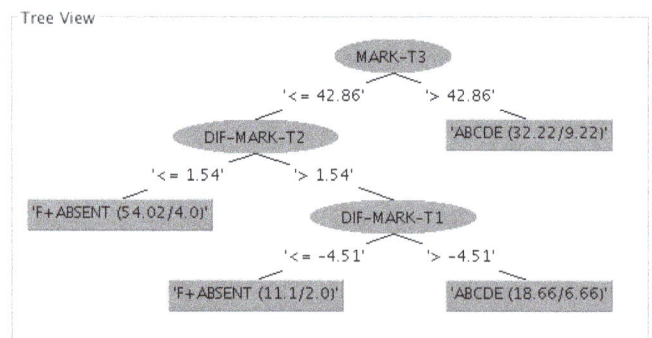

Fig. 5. Decision tree induced by J48 using the second dataset. Attributes are the marks (for T1, T2 and T3) and the difference with the average value; the binary class separate between students that pass (ABCDE) or not pass (F+ABSENT) the global evaluation process.

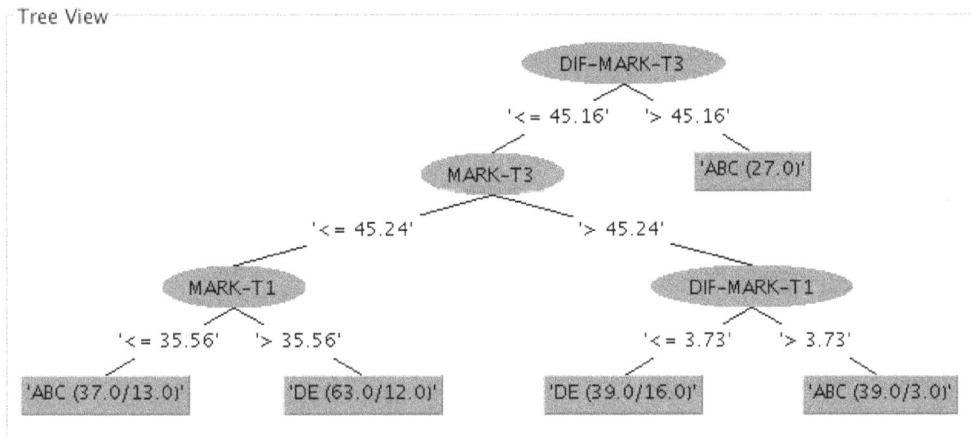

Fig. 6. Decision tree induced by J48 using the third dataset. Attributes are the marks (for T1, T2 and T3) and the difference with the average value; the binary class separate between students that achieve best grades (ABC) and the other ones (DE).

so restrictive for them, note that students do not need to pass the test (50 points out of 100), they only need to do best that the average value (close to 35 out of 100, see Table I)

Once again, the validation presents a quite reliable model (even higher than previous model) because we have 87% accuracy. This makes sense because we have added new attributes that help to better differentiate between student profiles.

Finally, once we have identified some criteria that determine differences between students that pass or not pass the final evaluation, we focus in those that pass the evaluation and how good their results are. Concretely we want to know if there is some element that reveals how they differ. In Fig. 6 we show the decision tree induced by J48 which reliability is relatively high (close to 80% accuracy).

Once again, the last test (T3) seems the most decisive element. It is logical, because this test includes and extends the concepts and abilities needed for the second test (T2). But this time the model differs substantially from previous ones because the actually important attribute is not the mark itself, but the difference with the average value. For every student (there is no exception, see most right-side branch in the decision tree) which T3's mark is beyond the average value in more than 45.16 points (out of 100), the final grade is better than D (A, B or C grade). Note that this difference is even greater than the standard deviation (27.84).

For those students that do not surpass the average value in such quantity, we find both kinds of students. In this case, differences between them are less clear and they could be even misunderstanding at a first moment. As it can be seen in the decision tree, first test information (T1) is selected to expand the tree in the deepest levels. It seems strange that students with lower marks ($\leq$ 35.56) get highest grades in the final evaluation, but we found some explanations that diminish the importance of such strangeness. On one hand, we can see that such asseveration is not so strong, because not all the examples are correctly classified (see <second> number in leaves), so some level of noise is present in that attribute. On the other hand, if we know that first test (T1) is conducted at the

beginning of the semester and its relation to final exam is very poor, we can think that dependencies are arguable; even more, we can suppose that good students with a "poor" mark in the first test can detect the necessity of strengthen the efforts because they did an incorrect initial calibration about the difficulty of the subject.

IV. CONCLUSION

In this paper we have studied, by using data mining techniques, the possibility that learning processes in the academic context could incorporate new and relevant knowledge that enables improvements in such processes.

In the conducted analysis we have detected that there are relations between the continual assessment carried out during the semester and the final evaluation. These relations, correctly used, can lead the adaptation of existing strategies or to boost the integration of new methods in subjects for future courses.

To a large extent, such improvements depend on having enough data about the evolution of the evaluations, on analyzing them continuously, on detecting anomalous behaviors; and on developing preventive and corrective actions (new exercises, individual tutorial actions, etc.). At this moment, Web-based Educational Systems offer tools to obtain and process that data, so its usage is highly recommended.

In addition, due to the flexibility of these systems, they can be adapted and extended. New functionalities can be added, and two different developments can be incorporated to progress in the previously mentioned improvements. As a first point, Web-based Educational Systems can collect more data, those that have shown their usefulness for data mining analysis (even calculating new fields). As a second feature, they could incorporate the mining process in the core of the system in order to offer a dual advantage: helping the teacher with the analysis tasks (assessment task) and helping the students by guiding their learning process (adapting task).

This study reveals many future research lines in different dimensions. There exists a wide diversity of techniques in the data mining field, so selecting other paradigms could improve the knowledge acquired (association rules, decision rules, etc.). If we are not so interested in the understandable knowledge (assessment task) and we prefer to provide the system with better guiding characteristics (adapting task), we have a perspective even broader because we could use many other strategies not so easily human-readable but very accurate (ensembles, neural networks, etc.).

Another promising area is the automatic or semi-automatic tune up of the Web-based Educational Systems. It is interesting to modify the educational system to respond to specific necessities of students [15]. This adaptation could even be implemented in real time, responding during the interaction with the student.

In this sense there are emerging new areas in machine learning and data mining related with data streams [16], very large (even non-ended) datasets that grow increasingly. Its usage fits very well with the dynamic of Web-based Educational Systems that are open constantly and can interact with students (and receive data) at every moment. Therefore, incorporating incremental algorithms [17] that can learn in this context would be positive. Additionally, as the student profile is not static, providing mechanisms to detect concept drift [18, 19] would contribute to create much more adaptable systems.

REFERENCES

[1] P. Brusilovsky, E. Schwarz and G. Weber, "ELM-ART: An intelligent tutoring system on World Wide Web", *Lecture Notes in Computer Science*, vol. 1086, pp. 261-269. 1996.

[2] C. Romero, S. Ventura, A. Zafra and P. de Bra, "Applying Web usage mining for personalizing hyperlinks in Web-based adaptive educational systems", *Computers & Education*, vol. 53, pp. 828–840. 2009.

[3] E. Melis, E. Andrès, J. Büdenbender, A. Frishauf, G. Goguadse, P. Libbrecht, M. Pollet and C. Ullrich, "ActiveMath: A generic and adaptive web-based learning environment", *International Journal of Artificial Intelligence in Education*, vol. 12, pp. 385-407. 2001.

[4] P. Brusilovsky and C. Peylo, "Adaptive and Intelligent Web-based Educational Systems", International Journal of Artificial Intelligence in Education, vol. 13, pp. 156-169. 2003.

[5] U. Fayyad, G. Piatetsky-Shapiro, P. Smyth, "From Data Mining to Knowledge Discovery in Databases", AI Magazine, vol. 17, pp. 37-54. 1996.

[6] C. Romero and S. Ventura, "Educational data mining: a review of the state of the art", *IEEE Transactions on Systems, Man, and Cybernetics-Part C*, vol. 40, pp. 601-618. 2010.

[7] E. Guzmán, R. Conejo, and J. Gálvez, "A Data-Driven Technique for Misconception Elicitation", *Lecture Notes in Computer Science*, vol. 6075, pp. 243-254, 2010.

[8] C. Romero, S. Ventura, and P. De Bra, "Knowledge discovery with genetic programming for providing feedback to courseware author", *User Model. User-Adapted Interaction*, vol. 14, pp. 425-464. 2004.

[9] R. Conejo, E. Guzmán, E. Millán, M. Trella, J. L. Pérez-De-La-Cruz and A. Ríos, "SIETTE: A Web-Based Tool for Adaptive Testing", *International Journal of Artificial Intelligence in Education*, vol. 14, pp. 29-61. 2004.

[10] M. Hall, E. Frank, G. Holmes, B. Pfahringer, P. Reutemann, I. H. Witten, "The WEKA Data Mining Software: An Update", *SIGKDD Explorations*, vol. 11, pp. 10-18. 2009.

[11] European Communities, "*ECTS Users' Guide*", Luxembourg: Office for Official Publications of the European Communities, 2009.

[12] T. M. Mitchell, *Machine Learning*. New York, McGraw-Hill, 1997, ch. 3.

[13] J. R. Quinlan, *C4.5: Programs for Machine Learning*. Morgan Kaufmann Publishers Inc., San Francisco, CA, USA. 1993.

[14] P. H. Lee, "Resampling Methods Improve the Predictive Power of Modeling in Class-Imbalanced Datasets", *International Journal of Environmental Research and Public Health*, vol. 11, pp. 9776–9789. 2014

[15] A. Kozierkiewicz-Hetmańska and N. Nguyen, "A method for learning scenario determination and modification in intelligent tutoring systems", *International Journal of Applied Mathematics and Computer Science*, vol. 21, pp. 69-82. 2011.

[16] J. Gama, *Knowledge Discovery from Data Streams*. Chapman & Hall/CRC. 2010.

[17] J. del Campo-Ávila, G. Ramos-Jiménez, J. Gama and R. Morales-Bueno, "Improving the performance of an incremental algorithm driven by error margins", *Intelligent Data Analysis*, vol. 12, pp. 305-318. 2008.

[18] J. Gama, I. Žliobaitė, A. Bifet, M. Pechenizkiy and A. Bouchachia, "A survey on concept drift adaptation", *ACM Computing Surveys*, vol. 46 (4), Article 44, 37 pages. 2014.

[19] I. Frías-Blanco, J. del Campo-Ávila, G. Ramos-Jiménez, R. Morales-Bueno, A. Ortíz-Díaz and Y. Caballero-Mota, "Online and non-parametric drift detection methods based on Hoeffding's bounds", *IEEE Transactions on Knowledge & Data Engineering*, to be published.

17

Assessing Road Traffic Expression

Fábio Silva, Cesar Analide, Paulo Novais,

Department of Informatics, University of Minho

Abstract — **Road traffic is a problem which is increasing in cities with large population. Unrelated to this fact the number of portable and wearable devices has also been increasing throughout the population of most countries. With this advent, the capacity to monitor and register data about people habits and locations as well as more complex data such as intensity and strength of movements has created an opportunity to contribute to the general wealth and comfort within these environments. Ambient Intelligence and Intelligent Decision Making processes can benefit from the knowledge gathered by these devices to improve decisions on everyday tasks such as deciding navigation routes by car, bicycle or other means of transportation and avoiding route perils. The concept of computational sustainability may also be applied to this problem. Current applications in this area demonstrate the usefulness of real time system that inform the user of certain conditions in the surrounding area. On the other hand, the approach presented in this work aims to describe models and approaches to automatically identify current states of traffic inside cities and use methods from computer science to improve overall comfort and the sustainability of road traffic both with the user and the environment in mind. Such objective is delivered by analyzing real time contributions from those mobile ubiquitous devices to identifying problematic situations and areas under a defined criteria that have significant influence towards a sustainable use of the road transport infrastructure.**

Keywords — **Traffic Expression, Smart Cities, Computational Sustainability**

I. INTRODUCTION

CURRENT trends such as smart cities and the internet of things has focused attention towards the quality of living and well-being inside big cities . It is also believed that most people will be living inside cities until 2050. If true, such statement would predict the increase of road traffic in cities that were not neither originally designed nor prepared to handle such influxes of traffic. Ambient Intelligence (AmI) is a multi-disciplinary subject that is equipped with procedures that may help solving such problems taking advantage of fields such sensing systems, pervasive devices, context awareness and recognition, communications and machine learning. It is currently applied in a number of applications and concepts in fields like home, office, transport, tourism, recommender and safety systems, among others [1] .

Road traffic analysis is an expensive and time consuming task which traditionally involves direct evaluation and field studies to assess and evaluate the impact of the flux of traffic in certain cities. An alternative to this is simulation experiments provide possible scenarios under which some assessment can be made. However, the downside of simulation lies in the use simplified models that are thought to mimic reality when in fact they may differ to some degree. Ubiquitous sensorization may be used to assess current traffic conditions, avoiding the use of costly field studies. Example of ubiquitous sensitization can already be found in certain areas such as traffic cameras and smart pressure detectors to assess traffic flow in specific points. This sensing is limited to the area it is implemented and does not provide information outside its operating range [2], [3]. More complex studies can be made with portable ubiquitous devices that follow drivers either because there a sensing device in the vehicle or the person driving carries a portable sensing device able to capture data related to driving. The nature of mobile ubiquitous devices also enable the possibility of direct analysis of driver behavior and community habits (points of congestion, high speed hazardous corners, aggressive sites) assessed trough the statistical treatment of driving records and offer safer alternatives for navigation with such information. These models have a direct impact diagnosing the current state of traffic and traffic behaviors to each route that may be used in modern GPS navigation systems, as an additional parameters.

Other approaches for the use of ubiquitous sensing devices involve real-time safety assessment, in [4] and [5] where a set of indicators is used to assess driving safety. Such indicators take into consideration the time of reaction, vehicle breaking time and whether or not there is a collision course. Yet, the analysis is still limited to the visible surrounding area and activities such as identification of other vehicles within the nearby space with the help of video interfaces disregarding sources of information outside that scope. In transport applications inside an area also known as Smart Cars, the AmI system must be aware not only of the car itself and its surroundings, but also of the driver's physical and physiological conditions and of the best way to deal with them [6]. The driver's behavior is important with several authors proposing machine learning and dynamic models to recognize different behaviors in drivers [7]. There are also examples of applications integrating AmI and ubiquitous principles in driving and traffic analysis. In [8], it is described a monitoring and driving behavior analysis system for emerging hybrid vehicles. The system is fully automated, non-intrusive with multi-modal sensing, based on smartphones. The application runs while driving and it will present personalized quantitative information of the driver's specific driving behavior. The quality of the devices used to perform such monitoring have a direct relationship to the quality of the measurement, thus, in this case, it is the main source of measurement error which needs to be controlled and contained to known error values order to make this study effective to production use. Other advantages include the possibility to increase information quality and create new routing styles in existing navigation systems taking into consideration aspects such as driver's driving style or accident or hazardous events rate during the routing planning phase.

Other approaches to analysis of driving behavior can be found in [9] a mobile application assesses driving behavior, based on the identification of critical driving events, giving feedback to the driver. The I-VAITS project [6] is yet another example that pretends to assist the driver appropriately and unobtrusively, analyzing real-time data from the environment, from the car and from the driver itself, by the way the driver uses the different elements of the car, their movements or image processing of their face expressions. In [10], in the context of a car safety support system, an ambient agent-based model for a car driver behavior assessment is presented. The system uses sensors to periodically obtain information about the driver's steering operation and the focus of the driver's gaze. In the case of abnormal steering operation and unfocused gaze, the system launches proceedings in order to slow down, stop the car and lock the ignition.

An alternative approach to the use ubiquitous sensing is to gather information about the condition of the environment the driver is in, mapping it to further use. In the Nericell system [11], from Microsoft Research, monitors road and traffic conditions using the driver's smartphone and corresponding incorporated sensors, but it can also detect honking levels, road condition and potholes as an example.

In what refers to devices used, there are today a wide range of options that can be used. The most effective should be portable devices that are always present and can perform complex tasks while not requiring user's direct attention. In such list, there are devices like smartphones, smartwatches, and intelligent wristbands. Those offer the advantage of accompanying user from one situation to another, however there are devices that can be used that are more specialized such as the internal computer of a car. In this last case the object itself becomes part of the car which might increase its production cost while on the other hand multi-purpose portable devices might suffice to the work described.

The work described in this paper tries to enhance ubiquitous sensing for driving applications with the objective to support the concept known as sustainable driving. It requires the gathering of information about traffic condition but also, consciousness about sustainability dimensions such as environment, economic and social. With this in mind optimization should consider more than just economic aspects of driving, but also consider fuel emissions and social aspects

such as driver's status, attention and driving style. Such work should complement existing other works and act as a platform for smart city traffic assessments. Moreover, the information generated by such system may be useful to third party systems which may use the knowledge base in their management applications and management systems.

II. COMPUTATIONAL SUSTAINABILITY

A. Computational Problem

The term computational sustainability is used by researcher such as Carla Gomes [12] to define the research field where sustainability problems are addressed by computer science programs and models in order to balance the dimensions of sustainability. It is accepted that the world ecosystem is a complex sustainability problem that is affected by human and non-human actions. In order to tackle these problems complex management systems should be put in practice in order to predict a number of attributes related to the sustainable problem at hand. Nevertheless, the pairing between computer science and the study of sustainability is as old as the awareness of sustainability and the availability computing systems. It is a fact that, as computational power capacity increased over time so did the complexity and length of the models used to study sustainability. The advent and general availability of modern techniques from artificial intelligence and machine learning allowed better approaches to the study of sustainability in a wide range of domains such as smart cities and transport systems.

Classical computational sustainability problems are not only found in smart grids, pollution, and distribution of energy but also city traffic. Considering the definition of sustainability and the topic of traffic expression, the use of computational methods to monitor and assess and optimize the transport efficiency are already used in systems today [2], [13], [14]. Nevertheless, the efficiency problem need to consider all dimensions of sustainability in order to become complete. The systems need to concern the optimization of not only traffic flow, economy and emission but also emissions, safety, and driver awareness.

In order to proceed to the collection of data and information required for the assessment of transport and traffic sustainability there are a number of topics under computer science that may be used. Perhaps, the most obvious would be the traditional methods of information acquisition through the sensorization of the environment and users, ambient intelligence, ubiquitous computing and information and data fusion. Less obvious techniques, concern the dynamic modelling of the environment and simulation of real world states when subjected to the conditions under study. The computational problem is therefore created by the means used to acquire this information and the resolution of the problem under the computational sustainability which include resource constraint optimizations, the satisfaction of dynamic models and preservation of statistical behaviors and actions.

B. Sustainable Driving

Traffic assessment is directly related to trending topics such as ubiquitous and pervasive methods that allow the balancing of economic, environmental and social factors needed for sustainable development. A new emerging and interdisciplinary area, known as Computational Sustainability, attempts to solve problems which are essentially related to decision and optimization problems in correlation to welfare and well-being. Due to its importance, some researchers have discussed and proposed quantification methods, and modelling process for sustainability [15], [16].

Often, decision and assessment are based on measurements and information about historical records. Indicator design provides an explanation on why such decisions are being made and it often uses information fusion to create and update its values. From a technological point of view, indicator analysis uses different and sometimes nonstandard data which sounds feasible by technological data gathering software that collect, store and combine data records from different sources. In the case of transportation systems, the assessment of the impact of a given driving pattern is made over sustainability indicators, like fuel consumption, greenhouse gas emissions, dangerous behavior or driving stress in each driver's profile.

Applications and systems that deal with this information acquisition and reasoning are already present in the literature. A system to estimate a driver profile using smartphone sensors, able to detect risky driving patterns, is proposed in [17]. It was verified whether the driver behavior is safe or unsafe, using Bayesian classification. It is claimed that the system will lead to fuel efficient and better driving habits. In [18], and in addition to car sensory data, physiological data was continuously collected and analyzed (heart rate, skin conductance, and respiration) to evaluate a driver's relative stress. The CarMa, Car Mobile Assistant, is a smartphone-based system that provides high-level abstractions for sensing and tuning car parameters, where by developers can easily write smartphone applications. The personalized tuning can result in over 10% gains in fuel efficiency [19]. The MIROAD system, Mobile-Sensor-Platform for Intelligent Recognition Of Aggressive Driving [20], is a mobile system capable of detecting and recognizing driving events and driving patterns, intending to increase awareness and to promote safety driving, and, thus, possibly achieving a reduction in the social and economic costs of car crashes.

In [21], an android application is depicted which makes use of internal vehicle sensors to assess driving efficient patterns. With information about throttle, breaking and consumption the application is able to provide driving hints in real time according to a set of predefined rule matrixes. In this case the application is focused on fuel efficiency. A more compressive study for the use of driving and traffic data can be found in [22]. This analysis considers the availability of data internal vehicle sensors, traffic data through internet services and historic driving patterns records to help the creation of more efficient navigation plans.

From the systems reviewed, there is clear focus on the lower level problem towards sustainable driving. The interest for the consideration of the sustainable problem across all of its dimensions is not the primary target of these systems but rather themes like safety, efficiency and driving profile through event detection. These applications do however fit in the category of computational sustainability as computational tool that may be used on a subset of the driving sustainability problem.

Our interpretation of the application of computational sustainability in sustainable driving is represented in figure 1, where sustainable driving is obtained through 3 types of problems that can be applied to each dimension of the classical definition of sustainability separately or in conjunction. Those 3 problems consider constraints optimization and reasoning problems, acquiring and storing information through sensorization statistics and machine learning procedure and building dynamic models that can express the state of the environment and its participants so that the impact of decisions may be assessed.

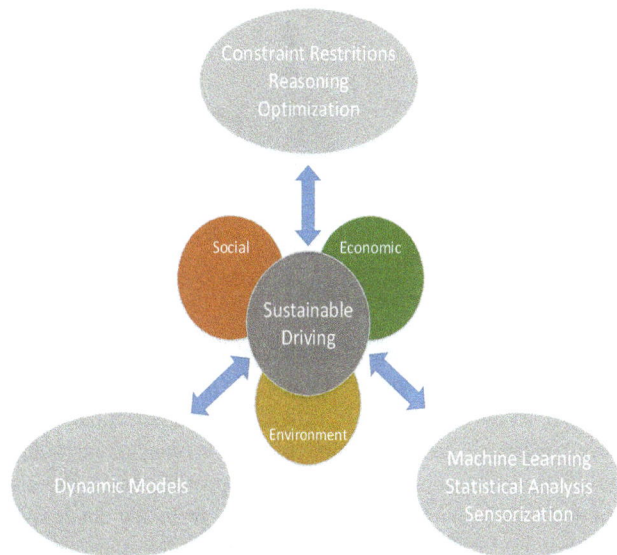

Fig. 1. Sustainable Driving Approach

In this work, the theme of sustainable driving will be addressed using a ubiquitous system for data acquisition integrated in a sustainability framework for the building of dynamic models that express the behavior of traffic and its conditions.

III. DRIVING EVALUATION

A. PHESS Driving System

The People Help Energy Savings and Sustainability (PHESS) project is being developed to help drive awareness towards the need for sustainable and energy efficient behaviors [23]. The framework is based on distributed system of multi-devices that generate data towards the creation and maintenance of indicators in the platform [24]. In order to drive awareness, specialized modules were developed that target user attention, mood and engagement. Through a set of usage real world scenarios the platform is being demonstrated across different applications [25].

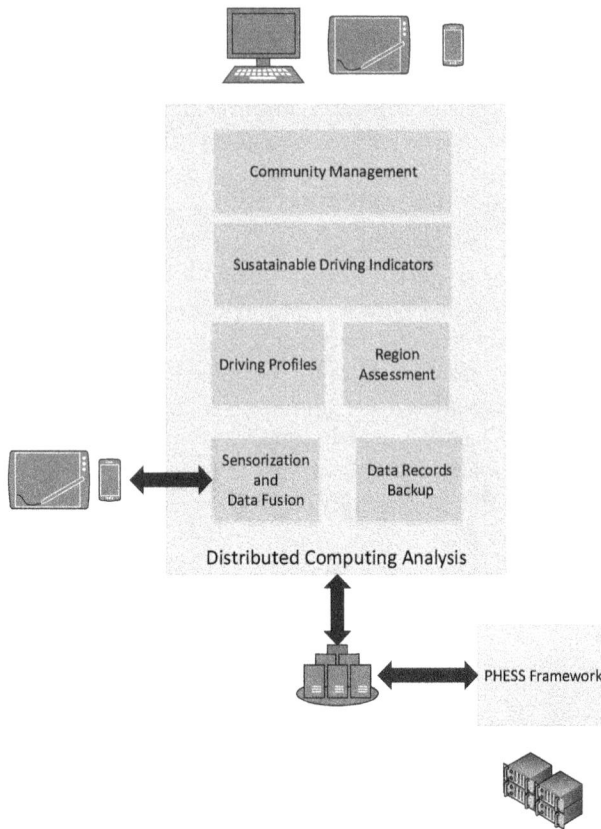

Fig. 2. PHESS Driving Architecture

In regards to the work presented in this paper, the driving scenario is being targeted. Figure 2, illustrates the system designed to monitor, assess efficiency and sustainability in driving actions. As a building blocks to assess sustainable driving the system uses both driving profiles from the data gathered while driving, event detection from the analysis of such data and indicators defined in the PHESS platform. The indicator definition is made with application both in the driving profile definition and the analysis of regions in cities. That procedure reduces complexity while it does not lose much of the information in the aggregation of data records and information and becomes more meaningful.

B. Indicator Development

In order to produce information about traffic flow and route safety it is necessary to gather information about relevant information about driving patterns. The focus of our analysis was derived from indicators accepted in related studies in the literature review. Towards this effect it was considered the following indicators:

- Average velocity;
- Average fuel consumption;
- Intensity of acceleration and breaking;
- Number of breaking and accelerating events per time unit;
- Standard deviation of velocity, intensity of breaking

and acceleration and number of breaking and accelerations;
- Number of turn events based on curvature detection;
- Intensity of force exerted in the vehicle during turns;

All of the indicators defined and built upon ubiquitous mobile sensorization, thus limited to their sensing abilities. Even so, some of the indicator defined are obtained through simple statistical procedures over the recorded data like average speed, number of breaks and standard deviations. The intensity of acceleration and breaking events is a more challenging task. Due to the usage of mobile smartphones, sensor access is not easily controllable. Efficiency measures make data reading uneven in time meaning sometimes there is oversampling where others there is under sampling. In order to mitigate such problem the assessment is made using the linear slope from the line connecting an initial and final velocity over a period of time as presented in equation 1. Such slope provides a mean to assess intensity that is independent of the size of the time interval.

$$Slope = \frac{v_f - v_i}{t_f - t_i} \tag{1}$$

Number of breaking and accelerating events are measured in time windows, referenced as per unit of time. Defining an event window is helpful because only accelerations and breaking inside such window are considered and can be analyzed and compared between time windows.

Average fuel consumption is obtained from the fusion of user input data and the distance travelled. As an initial setup the user is required to configure the smartphone with a number of initial variables such as vehicle model and vehicle average consumptions. The aggregation of different vehicles in a large scale analysis makes this indicator more relevant for analysis.

Curve and turn detection is a special event due to characteristic and driving difficulty. Due to car handling, driving inside curvatures can present a risky task specially if driven at too much speed or under high breaking or accelerating intensity. As a consequence, for this analysis, a special strategy is employed which monitor the degree of curvature trough smartphone sensors.

$$Direction = $$
$$\tanh(\sin(\delta_2 - \delta_1) * \cos(\varphi_2), \cos(\varphi_2) * \sin(\varphi_2)) \tag{2}$$
$$- \sin(\varphi_2) * \cos(\varphi_2) * \cos(-(\varphi_2 - \varphi_2))$$

Equation 2 demonstrates the formula used to track angle difference in the direction between two points. As the curvature becomes more intense the road curvature is identified as potentially more dangerous than others. The intensity of forces while driving inside curvatures is monitored using the intensity of the accelerometer vector, equation 3.

$$Intensity = \sqrt{Acc_x^2 + Acc_y^2 + Acc_z^2} \qquad (3)$$

The indicator analysis uses a three level classification scheme based on the statistical occurrence of the indicator value being accessed. For the classification definition, quartiles are used as a mean to identify outlier data in each indicator and classify it differently. Consequently the procedure adopted was to order the sample data records and classify data between the 80-95% quartile as yellow events, the data above the 95% quartile as red events and the rest as green events. This leads to the assumption that most drivers will have an adequate driving style for the most of their trips. In table 1 it is presented, the classification for each indicator represented in this paper.

C. Driving Profile

The usage of roads can be affected by driver's driving patterns. It is accepted that some drivers have a predisposition to drive more aggressively than others and there are significant deviations in their behaviors. Our approach uses this thought to gather the driving records from a community of users to classify different driving patterns to different people and link that data geographically in later analysis. The driving profile is based on the indicator defined in this papers plus attributes that respect directly to the person's driving profile. Thus the list of driving attributes considers the indicator plus the list of attributes:

- Time of day;
- Trip average duration;
- Standard deviation of trip duration;
- Average maximum and minimum velocity per trip;
- Number of cars driven.

With those measures, a complete profile can be designed and executed in applications that monitor current driver's performance. In [26], [27] other parameters were used to collect data from ordinary drivers in real traffic situations, such as wheel rotation, engine speed, ambient temperature, use of breaks and fuel consumption. In these studies, GPS data was also monitored, where each driving pattern was attributed to street type, street function, street width, traffic flow and codes for location in the city (central, semi-central, peripheral). It was concluded that the street type had the most influence on the driving pattern. The analysis of the 62 primary calculated parameters, resulted in 16 independent driving pattern factors, each describing a certain dimension of the driving pattern. When investigating the effect of the independent driving pattern factors on exhaust emissions, and on fuel consumption, it was found that there already studies with a common number of factors amongst the literature. Due to the decision to implement a pervasive system over mobile sensorization the work here described will account the attributes that are able to be collected by smartphone applications.

While these attributes characterize driving in a long term analysis, such strategy might miss spontaneous events that occur sporadically. An example of such is a sudden break with high intensity. In order to deal with these one-off events, other attributes are of relevance:

- Force exerted in the car;
- Slope of the line connecting initial to final velocity during breaking and accelerating events;
- Degree of the curvature of the road and force exacted in the car.

It is important to stress that these attributes are already accounted in the driving profile because they are also defined as indicators in the PHESS platform.

D. City Analysis

The usage of roads can be affected by driver's driving patterns. It is accepted that, if the majority of drivers have a predisposition to drive more aggressively in certain areas than others, then those areas are more dangerous. Our approach uses this thought to gather the driving records from a community of users and use them to calculate potential hazardous spots inside cities. Most evaluations are made using standard driving attributes, matured in the literature over a number of studies across different authors and projects. This kind of analysis is only possible with a dedicated user community that constantly updates and makes use of the platform supporting these models.

There is information that is dependent on external conditions of traffic and not related with driving itself. The platform developed will try to assess external condition using context estimation from the data gathered. The strategy employed uses indicator data linked with geographical data to define such context information. The indicator data is aggregated over a squares of geographical regions and their average value is computed. The granularity of the assessment is dependent on the size of the squared region. Nevertheless, such approach with an appropriate level of granularity is able to assess regions with high congestions rates or with high average speed, as an example. In this case, the velocity recorded by users is aggregated inside each square of terrain. The same analysis is available for other indicators in this systems and displayed in the same manner. Value added information produced in the system is published using a range of public web services. These web services provide public information about current traffic and driving conditions as well as, modelling analysis based on the historical data available in the platform.

Aside from driving study, other analysis can be made with the help of context conditions. Such conditions include weather, traffic congestion and time of day, for instance. Each example can have significant influence on the safety and on the assessment of attributes related to driving. Aggressiveness and dangerous behavior has different meanings in any of these conditions and while some concepts are broad enough to be used by all, others are situation specific meaning that what is dangerous in one situation might not be in another. Usually, driving patterns are defined and associated to the speed profile of the driver, but it can be expanded to other variables, as gear changing, and big changes on the acceleration [27]. Experiments with communities are often used to provide real time analysis of geographic conditions and events, with examples of such in the Waze platform [14]. However, they are the lacking features of historic analysis and historical supported suggestions.

The aim of this work is to focus on intangible and soft attributes which we define as attributes that are not directly observed by data records but rather computed with techniques from static analysis and machine learning processes. Such attributes should be used to find hidden patterns of road usage that might be missed in standard traffic flow simulations. Examples of such errors in simulation include driving aggravation due to unforeseen events even with normal traffic conditions.

IV. ANALYSIS AND DISCUSSION OF STUDY RESULTS

Although this work is not using the internal data from vehicle sensor system as in past research [26], the approach followed in this work uses smartphone data for ubiquitous and pervasive monitoring. Data gathering is made through sensors, which is pre-processed internally with data fusion methodologies to enrich data and provide richer information. The number of variables used to assess driving patterns is based on the indicators and driving profiles defined.

Using indicator assessment over normal car trips in the system it is possible to take note on each point detected by the GPS sensor the indicator classification. An example of such mechanism is present in figure 3.

Each view can be personalized taking into account all indicators or a subset of them according to their interest to the analysis. Using the indicator classification it is also possible to assess whether or not there are dangerous systematic behavior in each driver's driving profile. Such task may be accomplished by assessing each indicator present in the profile. Other approach would be to directly compare driver's inside the community by their driving profile attributes.

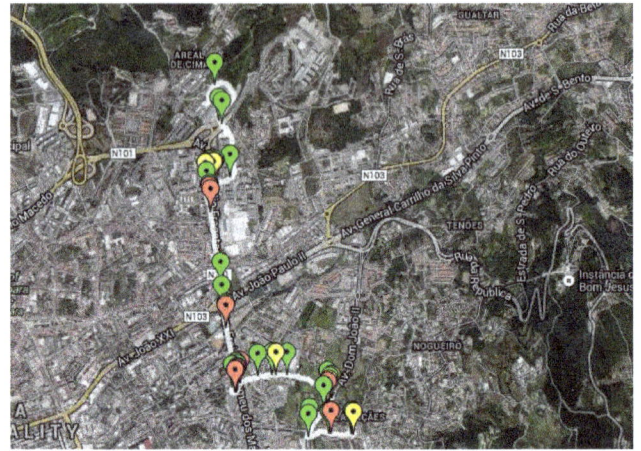

Fig. 3. Event Detection Based on Indictor Analysis

The model described in this article was tested as a complimentary module to a sustainability framework PHESS. Its aims are to produce and generate knowledge that can be used to perform decisions and suggestions that have a direct impact on sustainability and the sustainability of user's actions. More than a responsibility framework, it is intended to increase awareness to sustainable problems that arise from user's own actions and road usage by drivers.

Taking into consideration a test city with a community of 10 users, it is possible to assess the sample metrics and indicators defined. Using the strategy described in the section city analysis a demonstrative example of the region grid classification of indicator is made in figure 4. In order to analyze the classification and demonstrate that the scale has been appropriated to detect a small but significant set of yellow and red events. Such detection mechanisms can be improved with more technical data about dangerous events or even adjust the quartiles used for classification, nevertheless the proposed approach provides satisfactory results.

Each event is characterized in the map, and for the user it is possible to see the information relevant to that assessment. On the other hand, figure 4 does not provide event level explanation but rather a set of filters with each indicator defined that may alter the map zone classifications according to whether or not they are selected.

The map covered by the identification of low and high average speed squares is within the expected range but varies according the time of day, however the location of squares is preserved although with different averages. Our approach, identifies such metrics on daily basis but the identified spots are within 10% to 15% of the visible map.

As with the analysis on figure 3, the analysis present in figure 4 may only include a subset of indicators in its representation thus simplifying the analysis of the map.

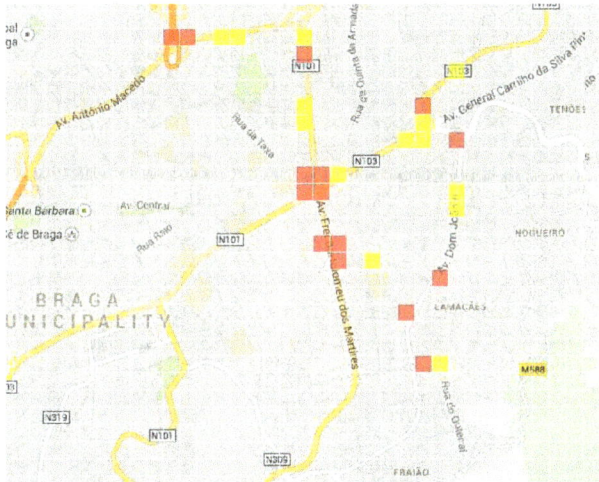

Fig. 4. Region Analysis

The results provided are based on web services which take information from the central distributed system to deliver information to graphical tools that display the information. In this case standard webpages were used.

While the results are satisfactory, future development should improve the social dimension of sustainable driving defining indicator directly associated with driver actions and emotions.

V. CONCLUSION

The use of pervasive devices already adopted by communities of users possess enough information and computing ability to build collaborative systems to tackle complex tasks. City traffic evaluation is one of such problems that are costly to audit and diagnose structural problem but can be simplified with crowd computing. Results are seem as satisfactory are reliable with the possibility to adjust according to specifics needs or needed improvement. The use of mobile sensors does constitute an additional effort to mitigate external influences such user involuntary movement, measurement and coverage errors. Nevertheless, the outputs generated in this platform were also found of relevance to the study of sustainability, where the intangible metrics and the structures employed to the indicator analysis pave the way to building sustainability assessing indicators able to join general purpose sustainability assessment frameworks such as the platform PHESS in discussion in this work.

In future iterations there are plan to update from grid analysis to road detection and road analysis becoming more accurate. Also, the validation of experiments on other cities are planned in order to prove both resilience and adaptation of the system. Integration of metrics found by this platform in common navigation systems are planned on the long term project, thus influencing routing options of people and acting as a true pervasive and ubiquitous object directing people away from dangerous situations into more comfortable and safe environments

VI. REFERENCES

[1] H. Nakashima, H. Aghajan, and J. C. Augusto, *Handbook of Ambient Intelligence and Smart Environments*. Springer, 2009.

[2] T. Bellemans, B. De Schutter, and B. De Moor, "Data acquisition, interfacing and pre-processing of highway traffic data *," vol. 19, pp. 0–2, 2000.

[3] G. Leduc, "Road Traffic Data: Collection Methods and Applications," 2008.

[4] A. Laureshyn, A. Svensson, and C. Hydén, "Evaluation of traffic safety, based on micro-level behavioural data: theoretical framework and first implementation.," *Accid. Anal. Prev.*, vol. 42, no. 6, pp. 1637–46, Nov. 2010.

[5] M. M. Minderhoud and P. H. Bovy, "Extended time-to-collision measures for road traffic safety assessment.," *Accid. Anal. Prev.*, vol. 33, no. 1, pp. 89–97, Jan. 2001.

[6] A. Rakotonirainy and R. Tay, "In-vehicle ambient intelligent transport systems (I-VAITS): towards an integrated research," in *Proceedings. The 7th International IEEE Conference on Intelligent Transportation Systems (IEEE Cat. No.04TH8749)*, 2004, pp. 648–651.

[7] J. Sun, Z. Wu, and G. Pan, "Context-aware smart car: from model to prototype," *J. Zhejiang Univ. Sci. A*, vol. 10, no. 7, pp. 1049–1059, Jul. 2009.

[8] K. Li, M. Lu, F. Lu, Q. Lv, L. Shang, and D. Maksimovic, "Personalized Driving Behavior Monitoring and Analysis for Emerging Hybrid Vehicles," in *Proceedings of the 10th International Conference on Pervasive Computing*, 2012, pp. 1–19.

[9] J. Paefgen, F. Kehr, Y. Zhai, and F. Michahelles, "Driving Behavior Analysis with Smartphones: Insights from a Controlled Field Study," in *Proceedings of the 11th International Conference on Mobile and Ubiquitous Multimedia*, 2012, pp. 36:1–36:8.

[10] T. Bosse, M. Hoogendoorn, M. A. C. A. Klein, and J. Treur, "A Component-Based Ambient Agent Model for Assessment of Driving Behaviour," in *Ubiquitous Intelligence and Computing*, vol. 5061, F. Sandnes, Y. Zhang, C. Rong, L. Yang, and J. Ma, Eds. Springer Berlin Heidelberg, 2008, pp. 229–243.

[11] P. Mohan, V. N. Padmanabhan, and R. Ramjee, "Nericell," in *Proceedings of the 6th ACM conference on Embedded network sensor systems - SenSys '08*, 2008, p. 323.

[12] C. Gomes, "Computational Sustainability.," *IDA*, 2011.

[13] H. Lee, W. Lee, and Y.-K. Lim, "The Effect of Eco-driving System Towards Sustainable Driving Behavior," in *CHI '10 Extended Abstracts on Human Factors in Computing Systems*, 2010, pp. 4255–4260.

[14] Waze Ltd, "Waze," 2014. [Online]. Available: https://www.waze.com/. [Accessed: 23-Sep-2014].

[15] V. Todorov and D. Marinova, "Modelling sustainability," *Math. Comput. Simul.*, vol. 81, no. 7, pp. 1397–1408, 2011.

[16] A. Kharrazi, S. Kraines, L. Hoang, and M. Yarime, "Advancing quantification methods of sustainability: A critical examination emergy, exergy, ecological footprint, and ecological information-based approaches," *Ecol. Indic.*, vol. 37, Part A, no. 0, pp. 81–89, 2014.

[17] H. Eren, S. Makinist, E. Akin, and A. Yilmaz, "Estimating driving behavior by a smartphone," in *2012 IEEE Intelligent Vehicles Symposium*, 2012, no. 254, pp. 234–239.

[18] J. A. Healey and R. W. Picard, "Detecting Stress During Real-World Driving Tasks Using Physiological Sensors," *IEEE Trans. Intell. Transp. Syst.*, vol. 6, no. 2, pp. 156–166, Jun. 2005.

[19] T. Flach, N. Mishra, L. Pedrosa, C. Riesz, and R. Govindan, "CarMA," in *Proceedings of the 9th ACM Conference on Embedded Networked Sensor Systems - SenSys '11*, 2011, p. 135.

[20] D. A. Johnson and M. M. Trivedi, "Driving style recognition using a smartphone as a sensor platform," in *2011 14th International IEEE Conference on Intelligent Transportation Systems (ITSC)*, 2011, pp. 1609–1615.

[21] R. Ara, R. De Castro, and R. E. Ara, "Driving Coach: a Smartphone Application to Evaluate Driving Efficient Patterns," vol. 1, no. 1, pp. 1005–1010, 2012.

[22] A. Fotouhi, R. Yusof, R. Rahmani, S. Mekhilef, and N. Shateri, "A review on the applications of driving data and traffic information for vehicles' energy conservation," *Renew. Sustain. Energy Rev.*, vol. 37, pp. 822–833, Sep. 2014.

[23] F. Silva, D. Cuevas, C. Analide, J. Neves, and J. Marques, "Sensorization and Intelligent Systems in Energetic Sustainable Environments," in *Intelligent Distributed Computing VI*, 2013, vol. 446, pp. 199–204.

[24] F. Silva, C. Analide, L. Rosa, G. Felgueiras, and C. Pimenta, "Social Networks Gamification for Sustainability Recommendation Systems," in *Distributed Computing and Artificial Intelligence*, Springer, 2013, pp. 307–315.

[25] F. Silva, C. Analide, L. Rosa, G. Felgueiras, and C. Pimenta, "Ambient Sensorization for the Furtherance of Sustainability," in *Ambient Intelligence-Software and Applications*, Springer, 2013, pp. 179–186.

[26] E. Ericsson, "Variability in exhaust emission and fuel consumption in urban driving," *URBAN Transp. Syst. Proc. ...*, no. 1980, pp. 1–16, 2000.

[27] E. Ericsson, "Independent driving pattern factors and their influence on fuel-use and exhaust emission factors," *Transp. Res. Part D Transp. Environ.*, vol. 6, no. 5, pp. 325–345, 2001.

18

A First Approach to the Implicit Measurement of Happiness in Latin America Through the Use of Social Networks

Francisco Mochón[1], Oscar Sanjuán[2]

[1]*Universidad Nacional de Educación a Distancia, Madrid, Spain*
[2]*Universidad Carlos III de Madrid, Spain*

Abstract - **This research paper can be classified as pertaining to the group of empirical studies that try to measure subjective well-being. The article presents as its greatest contributions the use of a subjective measurement of well-being based on social networks for the Latin American setting, as well as its comparative analysis with another traditional method.**

Keywords - **Happiness, subjective well-being, social networks, Twitter, Latin America, Latinobarómetro.**

I. INTRODUCTION

THIS research paper can be classified as pertaining to the group of empirical studies that for some years now have attempted to analyze subjective well-being in Latin America. Among them some of the most noteworthy are [15], [16], [17], [18], [28], [29], [30] y [5].

The novelty of this paper with respect to previous studies is that its objective is to verify to what extent the results of measuring the happiness of Latin Americans obtained following two radically different methods are consistent. One is based on the use of surveys from Latinobarómetro and the other on inferring the feelings of social network users from a semantic analysis of the words used in their communications and messages. A scientific method is followed in both cases.

The scientific study of happiness is not based on conjectures or presumptions but instead on research projects. Traditionally, researchers have analyzed factors that influence whether an individual defines herself as happy or satisfied [3] [4]. Psychology, sociology and economics have tried to explain the conditions that allow individuals to develop as happy persons [20], [13], [14] and [30].

Following [9][10], the notion of happiness generally used in economics identifies happiness and developing subjective well-being. In this sense, happiness or subjective well-being is no more than an assessment of life itself, regardless of pyschological judgments about momentary pleasure [2], [27]. In other words, happiness refers to how the individual evaluates the overall quality of her life [26], [7]. As such, the

happiness of individuals will depend entirely on an individual perception and it will be linked to concepts of quality of life and well-being. In any case, what matters is that that individual perception about the state of subjective well-being or happiness is measurable. This is the notion of happiness that we will use in the third epigraph of the paper, where we use data from Latinobarómetro to measure the happiness of Latin Americans. In the fourth epigraph, we take a completely different look at happiness, using information contained in messages sent over social networks—in particular, data from Twitter--to infer the feelings of individuals [8], [12].

The paper is organized as follows. In this section, we present a short introduction to the study. The following section gives a brief description of the different methods used to measure happiness. As already mentioned, Sections 3 and 4 present two alternative measurements of Latin Americans' happiness, one based on the information gathered from subjective surveys and the other inferred on the basis of information contained in social networks. The fifth section presents a comparative analysis of the results obtained following the two aforementioned methods. The sixth section presents the main findings and sketches out lines of future research that will be conducted to more deeply explore the subjects presented in this paper.

II. THE MEASUREMENT OF HAPPINESS: ALTERNATIVE METHODS

Happiness is measurable, and this is what enables us to speak of the science of happiness. In the new science of happiness, different methods have been used to measure happiness. Ed Diener and his collaborators presented a method to measure happiness based on the idea that individuals can consistently identify their level of satisfaction with life on a scale, and as such, what must be done is to ask people questions [7]. This way of measuring happiness is the one that justifies conducting surveys like the World Values Survey, and it is the most widely-used method [23].

Another method for measuring happiness is based on the sampling of experiences developed by the psychologist

Csikszentmihalyi and several researchers. This method consists of using locators (beepers) and afterwards using computers to contact individuals at random and ask them about their mood [24], [25].

A different approach is followed by a group of researchers led by Nobel Prize winner Daniel Kahneman. They created a method for measuring happiness based on following or reconstructing what people do at each moment of the day and asking them how they feel [19]. The main findings of this research specify that the three basic components of happiness are pleasure, commitment and meaning. Following this method and using messages on mobile telephones as an instrument of communication with those surveyed, Matthew Killingsworth identified happiness associated with a wide range of activities [22]. He points out that, until recently, researchers had to trust the assessments and appraisals that people made about their average emotional states over long periods of time. This inconvenience is avoided when following the method based on reconstructing what people do at different moments every day.

Recently, and amidst the impressive growth of social networks, there has emerged a new method for measuring happiness. This method consists of inferring the feelings of social network users on the basis of a semantic analysis of the words used in their communications and messages. Likewise, a study done by the Vermont Complex Systems Center uses information from Twitter to infer how happy or unhappy people in different states of the United States feel. Specifically, the researchers Dodds and Danforth have developed a method that, by incorporating the direct human evaluation of words, allows us to quantify levels of happiness on a continuous scale from a diverse collection of texts [8] [12], [21]. The method is transparent and able to quickly process texts from the Internet.

In the study carried out by Dodds and Danforth, on the basis of ten million "tweets," a code for determining to what extent each analyzed message can be catalogued as happy or sad was developed. The study focused on certain key words that were deemed to be indicative. Thus, "beauty" and "hope" are associated with happiness, while "hate" and "smoke" are associated with unhappiness. The researchers analyze the frequency with which the identifying words are used as good words and bad words in different states of the U.S.A. and qualify them as happy or unhappy. It is important to note that this study requires a highly complex task beforehand that allows us to obtain the terms to evaluate, that is, the words susceptible to be captured and measured. This list of words was obtained by directly asking English-speaking people about the words that evoke happiness for them. Once the list of words was obtained, it was then necessary to create a scale that reflected how one word was evaluated with respect to the following one. This scale was obtained through a similar method, asking people to order words according to the value in terms of happiness that each word had for each of them..

III. THE HAPPINESS OF LATIN AMERICANS ACCORDING TO LATINOBARÓMETRO

The measurement of happiness that this part of the study presents is in keeping with the literature that analyzes the answers of individuals to questions about subjective well-being in cross-section or panel surveys, and which is the most widely-used by researchers. The hypothesis on which these studies are based is that the subjective data provided by individuals can be treated ordinally in economic analyses so that greater subjective levels of well-being reflect greater levels of happiness [13]. In other words, it is argued that although everybody has their own ideas about happiness, individual happiness can be captured and analyzed.

Anyone can be asked how satisfied they feel with the life they lead, and behind the answer given in a survey, a conscious evaluation of their subjective well-being can be found. Supposedly, individuals are able to evaluate their subjective level of well-being with respect to certain circumstances. In addition, reliable studies indicate that the subjective well-being demonstrated by individuals is reasonably stable and sensitive to changes in circumstances. In fact, in research about happiness, individuals' answers to questions about their feelings are analyzed and consistent findings are obtained [6].

Specifically, in this section of the paper, there is a synthesis of a paper done in 2012 on life satisfaction in 18 Latin American countries [5]. The countries analyzed are Argentina, Bolivia, Brazil, Colombia, Costa Rica, Chile, Ecuador, El Salvador, Guatemala, Honduras, Mexico, Nicaragua, Panama, Paraguay, Peru, the Dominican Republic, Uruguay and Venezuela. The results obtained in the cited study are in general consistent with those already known for other countries, as well as with those obtained in different papers that refer to the region.

The data used come from annual personal surveys created by the Latinobarómetro Corporation for the period 2000-2009. The sample used includes 191,488 individuals and they are different every year. The distribution of the sample by countries over the period is presented in Graph 1, where the information about Brazil has been omitted, given that in the study based on social networks for this country, "tweets" were not analyzed because they were in a different language.

The key variable is the degree of satisfaction with individuals' current lives, as it is defined in the Latinobarómetro survey. The degree of a person's satisfaction with life falls into one of the following four categories: not at all satisfied, not satisfied much, quite satisfied and very satisfied. Graph 1 presents the percentage of individuals from the Latin American countries mentioned that indicate they were quite or very satisfied with life during the years 2000-2009. As can be seen, in eight of the 17 countries, more than 70% of the population was quite or very satisfied with their life at the time. Peru, Bolivia and Ecuador are the countries where people were least satisfied with life. In these countries, it can be seen that less than 54% of the people surveyed indicated that they were quite or very satisfied with life.

GRAPH 1

PERCENTAGE OF THE LATIN AMERICAN POPULATION THAT WAS
QUITE OR VERY SATISFIED OVER THE PERIOD 2000-2009

Peru	43.56
Bolivia	48.63
Ecuador	53.15
Nicaragua	61.35
El Salvador	65.02
Chile	65.03
Argentina	66.94
Paraguay	67.65
Honduras	68.15
The Dominican Republic	70.14
Uruguay	71.54
Mexico	72.07
Guatemala	73.97
Colombia	74.13
Panama	76.69
Venezuela	78.81
Costa Rica	81.32

Source: De Juan and Mochón (2012)

A more rigorous analysis of the happiness of Latin Americans is obtained by studying the satisfaction with life variable by countries. The results show that there are significant differences in the average level of satisfaction with life between the countries studied (Table 1). Of the group of countries analyzed, only eight show a coefficient of satisfaction with life higher than 2 (quite satisfied). These are Costa Rica (2.234), Venezuela (2.173), Panama (2.086), Colombia (2.068), the Dominican Republic (2.035), Guatemala (2.034), Honduras (2.022) and Mexico (2.010). Also, there are six countries that have an intermediate value, between 1.76 and 1.89. They are El Salvador (1.891), Uruguay (1.880), Paraguay (1.844), Nicaragua (1.835), Argentina (1.785) and Chile (1.764). The countries that show a lower coefficient are Ecuador (1.614), Bolivia (1.519) and Peru (1.484).

TABLE 1.
AVERAGE LEVEL OF SATISFACTION WITH LIFE

Countries	Satisfaction with life
Costa Rica	2.234
Venezuela	2.173
Panama	2.086
Colombia	2.068
Dominican Republic	2.035
Guatemala	2.034
Honduras	2.022
Mexico	2.01
El Salvador	1.891
Uruguay	1.88
Paraguay	1.844
Nicaragua	1.835
Argentina	1.786
Chile	1.764
Ecuador	1.614
Bolivia	1.519
Peru	1.484

Source: De Juán and Mochón (2012)

So, from the descriptive analysis carried out on the basis of the information provided by Latinobarómetro, it can be seen there are significant differences between countries in the level of satisfaction with life. These results indicate that the happiest individuals are those who live in Costa Rica, Venezuela, Panama and Colombia, while the least happy are those in Peru, Bolivia and Ecuador. It must be noted that these results are consistent with the results obtained in [5] using econometric techniques.

IV. THE HAPPINESS OF LATIN AMERICANS ACCORDING TO SOCIAL NETWORKS

The boom that social networks are currently experiencing is well-known, and their great reach justifies their use as a medium to measure opinion, interest in a subject or a person or even feelings and moods[31] [32] [11].

Especially relevant is the use of social networks in marketing and publicity, the measurement of audiences, opinion surveys, popularity and even as previews of election results. Resorting to social networks to obtain a barometer of opinion is especially common in the social network Twitter, where, since its creation, it has been possible to know the number of followers or the effect of a speceific term or tag [8]. Keep in mind that there are also tools that facilitate more rigorous analysis and establishing relationships, measuring impacts, etc.

To summarize, the reasons that can justify choosing Twitter as a tool for measuring interest, opinion or mood are as follows:

1. Availability of an API (Application Program Interface): the existence of a public API makes it possible to make consultations and recover information in a relatively simple way, through the creation of simple computer programs that facilitate recovery, storage and analysis using different techniques ranging from basic statistics to machine learning.

2. Simple content based on text: the most usual type of message on Twitter is the short text message, owing to its origin from when messages were sent and received via SMS. This characteristic requires the meaning of the messages to be direct, specific and simple in most cases, which helps in their analysis.

3. Instantaneity and transience: the instantaneity and simplicity of the messages on Twitter make it a good

mechanism for measuring what happens in almost real time or during a period of time. They are not reflexive, prepared publications but spontaneous, fast communications.

4. Profiling: many of the users on Twitter not only make comments but also have a public profile, which allows for their segmentation according to this data.

5. Geographical segmentation: within any mechanism for measuring opinion, a basic factor is knowing where we are measuring. On Twitter, this is possible through both the user profile and the location of a specific publication.

6. Global use: although Twitter is not the most used social network, it has many users and a very high level of participation [12].

The use of Twitter to measure subjective well-being as presented in this study is not completely new. As has been noted, there is a project called "hedometer" (http://hedonometer.org) that has taken a measurement of happiness (subjective well-being) in the United States of America [8]. This measurement is especially interesting as it demonstrates the possible use of social networks to measure happiness. In addition, it has other interesting characteristics, like being able to take the measurement in a large geographical area with a common language, and being a space where the use of social networks in general and Twitter in particular is very widespread.

Taking these characteristics as a framework of reference, a similar study has been undertaken in our case, in another relatively homogeneous geographical environment and in a common language. Specifically, the study was carried out for the Spanish-speaking countries of Latin America. Although a study with these characteristics can be valuable in itself, it was interesting to contrast the results with the results obtained when a traditional method of measuring happiness is used.

To produce the present research paper, the following considerations have been taken into account:

1. The recovery of "tweets" for a national geographical area is very complex and unreliable, since it can only be based on the data in the personal profile of each user, and this information is not usually contributed by the users. This is why we have decided to use the "tweets" recovered from the capitals of each country as a representative sample to analyze. This process is rather more simple than if we try to use the personal profile of each user, and more effective since Twitter allows consultations which indicate a geographical position and a sphere of interest.

2. The recovery of "tweets" has been done for a group of key words obtained, taking as a reference the group of key words that hedometer uses [8]. These key words are logically in English, which is why they have been translated. As we are dealing with key words, the idiomatic and semantic problems of translation can be managed. In any case, we have eliminated those that could present some problem. Obtaining this list (Table 2) has a certain value, since it was created on the basis of a thesaurus, considering the different words according to their meaning and impact as indicators of happiness based on the information provided by [8]. A thorough process of translation was applied to the original list, eliminating those words that make no sense in Spanish.

TABLE 2. LIST OF WORDS AND WEIGHTS

WORD	VALUE	WORD	VALUE	WORD	VALUE	WORD	VALUE	WORD	VALUE	WORD	VALUE	WORD	VALUE	WORD	VALUE	WORD	VALUE
carcajada	8,5	fin de semana	8	vacacion	7,92	ganador	7,78	bonus	7,68	mono	7,62	victorias	7,58	comodidad	7,5	leal	7,46
felicidad	8,44	celebrar	7,98	mariposas	7,92	delicia	7,78	brillante	7,68	entretenimiento	7,62	conseguido	7,56	felicitaciones	7,5	oportunidades	7,46
amor	8,42	comedia	7,98	libertad	7,9	belleza	7,76	diamantes	7,68	excitado	7,62	billón	7,56	magdalena	7,5	triunfo	7,46
feliz	8,3	bromas	7,98	flor	7,88	mariposa	7,76	dia libre	7,68	excitación	7,62	tartas	7,56	ganar	7,5	guau	7,46
reido	8,26	rico	7,98	grande	7,88	entretenimiento	7,76	suerte	7,68	broma	7,62	facilisimo	7,56	extraordinario	7,5	joyas	7,46
risa	8,22	victoria	7,98	luz solar	7,88	el más divertido	7,76	madre	7,68	millonario	7,62	flores	7,56	gloria	7,5	bosques	7,45
riendo	8,2	navidad	7,96	amorcito	7,88	honestidad	7,76	super	7,68	premio	7,62	regalos	7,56	gracioso	7,5	manzana	7,44
excelente	8,18	libre	7,96	dulcito	7,88	cielo	7,76	increible	7,66	consiguió	7,62	oro	7,56	luz de luna	7,5	sueños	7,44
risas	8,18	amistad	7,96	premio	7,86	sonrisas	7,76	angeles	7,66	exitosamente	7,62	mirra	7,56	optimista	7,5	fantasía	7,44
disfruto	8,16	diversion	7,96	chocolate	7,86	conseguido	7,76	disfrutar	7,66	ganadores	7,62	familias	7,54	en paz	7,5	comida	7,44
exitoso	8,16	vacaciones	7,96	jajajaja	7,86	maravilloso	7,76	amigo	7,66	brillas	7,6	guapo	7,54	romance	7,5	miel	7,44
ganar	8,12	amado	7,96	cielo	7,86	glorioso	7,74	amistoso	7,66	fenómeno	7,6	amantes	7,54	festivo	7,49	milagros	7,44
arcoiris	8,1	amados	7,96	paz	7,86	besos	7,74	madre	7,66	genio	7,6	afecto	7,53	atractivo	7,48	sexo	7,44
sonrió	8,1	amando	7,96	espléndido	7,86	promoción	7,74	beneficio	7,66	logro	7,58	caramelo	7,52	contento	7,48	cantar	7,44
gane	8,1	playa	7,94	exitoso	7,86	familia	7,72	mejor	7,66	tarta	7,58	tierno	7,52	abuelita	7,48	luz de las estrellas	7,44
placer	8,08	jajaja	7,94	disfrutando	7,84	regalo	7,72	mal día	7,64	brindemos	7,58	diamante	7,52	internet	7,48	agradecido	7,44
sonrie	8,08	besando	7,94	besado	7,84	humor	7,72	campeón	7,64	excitante	7,58	ganancias	7,52	agradable	7,48	gané	7,44
arcoiris	8,06	sunshine	7,94	atracción	7,82	romantico	7,72	abuela	7,64	bondad	7,58	interesante	7,52	ganancias	7,48	logro	7,42
ganando	8,04	hermoso	7,92	celebrado	7,8	magdalenas	7,7	jaja	7,64	abrazo	7,58	pac_ficamente	7,52	listo	7,48	adorado	7,42
celebración	8,02	delicioso	7,92	heroe	7,8	festival	7,7	beso	7,64	ingresos	7,58	piropo	7,52	navidad	7,48	afectivo	7,42
disfrutó	8,02	amigos	7,92	abrazos	7,8	jajajajaja	7,7	gatito	7,64	fiesta	7,58	rosas	7,52	bebés	7,46	afecto	7,42
saludable	8,02	divertido	7,92	positivo	7,8	honor	7,7	milagro	7,64	sonriendo	7,58	sábados	7,52	brindo	7,46	la vida es bella	7,42
música	8,02	sobresaliente	7,92	sol	7,8	relax	7,7	dulce	7,64	canción	7,58	fiel	7,51	coraje	7,46	esto es vida	7,42
celebrando	8	paraiso	7,92	cumpleaños	7,78	angel	7,68	bendiciones	7,62	éxito	7,58	cielos	7,51	entusiasmo	7,46	vivo	7,42
felicitaciones	8	dulcisimo	7,92	fantástico	7,78	mal día	7,68	brillo	7,62	sabor	7,58	apreciar	7,5	honesto	7,46		

Source: compiled by the authors from the list using in the "hedometer".

3. The use of Twitter to make the ranking of the different countries according to inferred happiness on the basis of the contents of "tweets" has the problem of showing a strong dependence on the intensity or frequency of Twitter use in each country. The creation of a coefficient was produced by taking into account the studies that are compiled in the source in Table 3 as a reference.

TABLE 3. LIST OF CAPITALS AND COEFFICIENTS OF TWITTER USE

City	Use Ratio
Caracas	0.276
Bogota	0.184
Montevideo	0.123
Buenos Aires	0.191
Mexico	0.1616
Santiago	0.156
Asunción	0.123
Guatemala	0.12
Lima	0.1
Quito	0.05
La Paz	0.05
Santo Domingo	0.05
Panama	0.05
San Jose	0.05
Managua	0.05
Tegucigalpa	0.05

Source: This coefficient is the result of the information referring to the percentage of Twitter use in different countries and its creation is based on what was done at: http://alt1040.com/2011/04/los-10-paises-mas-adictos-a-twitter; The Netherlands Ranks #1 Worldwide in Penetration for Twitter and LinkedIn http://bit.ly/1fh0Ql8; and Twitter Grows Stronger in Mexico - eMarketer http://po.st/1WR61k. For those places that did not have a reference value, we assigned 0.5 percent.

Once the plan for carrying out the study was established (how the data would be obtained and under what conditions), we proceeded to design an algorithm to extract the information. The extraction algorithm was executed on Twitter for the duration of the study in order to obtain a happiness ranking. The extraction and generation of the happiness ranking was done according to the following process:

1. For each country (City) on the list:
   a. For each word on the list:
   i. Recover the corresponding "tweets"
   b. They are added up
   c. The happiness factor of the word is applied
2. A total is obtained
3. The correction of Twitter use is applied
4. The list of countries is ordered according to the score obtained
5. The ranking is generated

This process was executed on Twitter for two months to obtain a sample size large enough to be able to obtain significant results. The number of "tweets" used was 100,000.

V. HAPPINESS IN LATIN AMERICA ACCORDING TO TWITTER

As has been mentioned, to be able to apply the algorithm described in the previous section, the first step was the creation of the list of key words to be used in the study. As already noted, since they are key words, most of them can be translated directly. In some cases, however, problems arise because the direct translation does not work well or because the translated term generates noise on making reference to words in radically different contexts. For these cases, we opted to follow one of the two alternatives below:

1. In those cases where, even if the direct translation is not valid, there is an equivalent word or expression, we treat this equivalence as valid.
2. When the direct translation is not valid and there is no equivalent word or expression in the same context, we eliminate that word from the list.

The list of words with their respective weights used in this study is compiled in Table 2. These key words are the ones that were used to recover the "tweets." According to the considerations in the previous section, in the capture of data the previously mentioned algorithm to generate the ranking was applied. Likewise, the correction coefficient based on Twitter use was applied; this information is shown in Table 3. This is how a ranking of feelings of happiness was obtained for Latin American countries according to Twitter data, as appears in Table 4.

TABLE 4. HAPPINESS RANKING ACCORDING TO TWITTER

City/Country	Score
Caracas (VENEZUELA)	582,461.34
Buenos Aires (ARGENTINA)	462,247.10
Bogota (COLOMBIA)	346,830.89
Mexico (MEXICO)	329,093.01
Santiago (CHILE)	244,561.10
Asunción (PARAGUAY)	179,175.01
Montevideo (URUGUAY)	83,808.53
Guatemala (GUATEMALA)	77,924.13
Lima (PERU)	62,399.62
Panama (PANAMA)	61,444.27
San Jose (COSTA RICA)	34,071.67
Santo Domingo (DOMINICAN REP.)	32,397.79
Quito (ECUADOR)	21,650.52
Tegucigalpa (HONDURAS)	14,668.87
Managua (NICARAGUA)	10,229.80
La Paz (BOLIVIA)	2,521.12

Source: compiled by the authors

If we compare these results to the ranking obtained on the basis of the surveys from Latinobarómetro (Table 1), we see that there are notable discrepancies, especially in some countries with relatively high rates of Twitter use (Argentina, Chile, Paraguay, Uruguay and Peru; see Table 3). It seems as though in those countries where the use of Twitter is greater, there is a strong upward bias, such that they appear in relatively high positions in the happiness ranking presented in Table 4. This might be because social networks have a viral, disseminating effect, so both positive and negative messages are spread, and as a result, the values are much more extreme than the simple indicative proportion of number of users. This could also be interpreted to signify that countries with a greater number of users not only have more users, but also more active users. On the contrary, some countries with relatively low coefficients of Twitter use like Costa Rica, Panama and the Dominican Republic, precisely because of the absence of the aforementioned viral effect, occupy relatively low positions in the ranking shown in Table 4, while in the ranking made on the basis of Latinobarómetro (Table 1), they are in high positions.

As part of the experiment and with the aim of finding a correction factor that encourages making future evaluations at different temporary moments and including other factors, we decided to calculate a weighting or adjustment factor that would allow us to equate the results obtained through the use of social networks with those derived from the Latinobarómetro surveys. One justification for calculating this weighting factor is to try to offer additional information that contributes to explaining the differences between using both methods to infer the happiness of Latin Americans. In considering this weighting factor, it is observed that the weighting necessary to adjust the result is greater in smaller countries with lower rates of Internet and social network use, which supports the previously formulated hypothesis for explaining the differences between the ranking of Tables 1 and 4.

In analyzing the content of Table 5, the case of Bolivia deserves to be highlighted. Although its position in the ranking with data from social networks (18th in Table 4) is not very different from its position in the Latinobarómetro ranking (15th in Table 1), on a quantitative level, it presents a very big lag compared to the other countries. Everything seems to indicate that once again we see a polarization of the results owing to the scant use of social networks in this country.

TABLE 5

ORD1	ORD2	City/Country	Score	Factor	%Factor	Obj. Objective	Weight.Fact.	Final Score
11	1	San Jose (Costa Rica)	34,071.67		0.0023867	0.239%	81.32 17.68	602,370.3704
1	2	Caracas (Venezuela)	582,461.34		0.0001353	0.014%	78.81 1.00	583,777.7778
10	3	Panama (Panama)	61,444.27		0.0012481	0.125%	76.69 9.25	568,074.0741
3	4	Bogota (Colombia)	346,830.89		0.0002137	0.021%	74.13 1.58	549,111.1111
8	5	Guatemala (Guatemala)	77,924.13		0.0009493	0.095%	73.97 7.03	547,925.9259
4	6	Mexico (Mexico)	329,093.01		0.0002190	0.022%	72.07 1.62	533,851.8519
7	7	Montevideo (Uruguay)	83,808.53		0.0008536	0.085%	71.54 6.32	529,925.9259
12	8	Santo Domingo (Dom. Rep.)	32,397.79		0.0021650	0.216%	70.14 16.04	519,555.5556
15	9	Tegucigalpa (Honduras)	14,668.87		0.0046459	0.465%	68.15 34.41	504,814.8148
6	10	Asunción (Paraguay)	179,175.01		0.0003776	0.038%	67.65 2.80	501,111.1111
2	11	Buenos Aires (Argentina)	462,247.10		0.0001448	0.014%	66.94 1.07	495,851.8519
5	12	Santiago (Chile)	244,561.10		0.0002659	0.027%	65.03 1.97	481,703.7037
16	13	Managua (Nicaragua)	10,229.80		0.0059972	0.600%	61.35 44.42	454,444.4444
14	14	Quito (Ecuador)	21,650.52		0.0024549	0.245%	53.15 18.18	393,703.7037
18	15	La Paz (Bolivia)	2,521.12		0.0192890	1,929%	48.63 142.88	360,222.2222
9	16	Lima (Peru)	62,399.62		0.0006981	0,070%	43.56 5.17	322,666.6667

ORD1: Order based on twitter data.
ORD2: Order after applying the Weight Factor.

VI. CONCLUSIONS AND FUTURE RESEARCH

In this article, a first approach to measuring happiness in Latin America through the use of social networks is presented. Specifically, the social network used is Twitter, although we do not rule out the possibility of undetaking future studies with Facebook or other social networks. We have used Twitter because of its characteristics (ease of use, availability and popularity, geographical data, etc.). We have developed a process that permits the extraction of data and generation of a new ranking quickly and easily, which allows us to easily repeat the experiment with additional conditions, parameters and searches.

We can extract the following points as our main conclusions:

- The measurement of happiness through the use of social networks seems viable, and it is tremendously simple compared to traditional methods (e.g., surveys).

- The measurement of happiness through social networks like Twitter involves considering several factors in order to obtain reliable results. The most evident factors are the use of Internet and the use of social networks.

- The method used in this work consists of inferring the feelings of social network users on the basis of a semantic analysis of the words used in their communications and messages.

- It is possible to calculate, via objective and empirical means, factors that allow us to correctly interpret data collected through the use of social networks.

- In time, as the use of Internet and social networks increases, the use of these tools will be more precise.

As lines of future research, we propose the possibility of:

- doing new studies which incorporate data gathered over longer time periods
- including only countries with similar socio-economic conditions
- refining the creation of that weighting factor which could be converted into a rating
- including not only positive terms but also negative ones in order to improve reliability
- doing other studies that, instead of key words, are based on iconographic elements like "smiley faces."

REFERENCES

[1] Ahn, N. and Mochón, F. (2010). La felicidad de los españoles: Factores explicativos. Revista de Economía Aplicada, 54(XVIII), 5-31.

[2] Ahn, N., Mochón, F. and De Juan, R. (2012). La felicidad de los jóvenes. Papers Revista de Sociología. 97/2, 407-430.

[3] Argyle, M. (1999). Causes and correlates of happiness, in Kahneman, Diener and Schwarz (eds.): Well-Being: The foundations of hedonic psychology. New York: Russell Sage Foundation, op. cit., 353-373.

[4] Argyle, M. (2001). The psychology of happiness. New York: Routledge.

[5] De Juan, R and Mochón, F. La felicidad de los latinoamericanos en el periodo 2000-2009. (2012). Working paper 0212. Departamento de Análisis Económico. Uned.

[6] Di Tella, R., MacCulloch, R. (2006). Some Uses of Happiness Data in Economics. Journal of Economic Perspectives, 20(1): 25-46.

[7] Diener, E, Emmons, R.A, Larsem, R.J. and Griffin, S.A..The Satisfaction With Life Scale. Journal of Personality Assessment, 1985, 49, 1, pp 71-75.

[8] Dodds, P. and Danforth, C. (2009). Measuring the Happiness of Large-Scale Written Expression: Songs, Blogs, and Presidents. Journal of Happiness Studies. doi:10.1007/s10902-009-9150-9.

[9] Easterlin, R.A. (1995). Will raising the incomes of all increase the happiness of all? Journal of Economic Behavior and Organization, 27, 35-47.

[10] Easterlin, R.A. (2001). Income and Happiness: Towards a unified theory. The Economic Journal, 111, 465-84.

[11] Fowler, J.H, Christakis, N.A.Steptoe. and Roux,D. (2009) A Dynamic Spread of Happiness in a Large Social Network: Longitudinal Analysis of the Framingham Heart Study Social Network British Medical Journal. Vol. 338, No. 7685 , pp. 23-27.

[12] Frank M., Mitchell L., Dodds P., and Danforth C. 2013. Happiness and the Patterns of Life: A Study of Geolocated Tweets. Scientific Reports 2013. Vol. 3, No: 2625, doi:10.1038/srep02625

[13] Frey, B.S. and Stutzer, A. (2002a). What Can Economists Learn from Happiness Research? Journal of Economic Literature, XL(June), 402-435.

[14] Frey, B.S. and Stutzer, A. (2002b). Happiness and Economics. Princeton: Princeton University Press.

[15] Fuentes, N. and Rojas, M. (2001). Economic Theory and Subjective Well-Being: Mexico. Social Indicators Research, 53(3), 289-314.

[16] Gerstenblüth, M., Melgar, N. and Rossi, M. (2010). Ingreso y desigualdad: ¿Cómo afectan a la felicidad en América Latina? (Working paper, 09/10). Montevideo: DECON.

[17] Graham., C. and Pettinato, S. (2001). Happiness, markets, and democracy: Latin America in comparative perspective. Journal of Happiness Studies, 2, 237-268.

[18] Graham, C. and Pettinato, S. (2002). Happiness and Hardship: Opportunity and Insecurity in New Market Economies. Washington, DC: The Brookings Institution.

[19] Kahneman D, Krueger A, David A. Schkade, Schwarz, N, Stone, A.. (2004). A Survey Method for Characterizing Daily Life Experience: The Day Reconstruction Method. Science 3 December. Vol. 306 no. 5702 pp. 1776-1780))

[20] Kahneman, D., Diener, E. and Schwarz, N. (1999). Foundations of hedonic psychology: scientific perspectives on enjoyment and suffering. New York: Russell Sage Foundation.

[21] Knibbs K.(2013). Can 10 million tweets measure happiness? New research says so. http://www.digitaltrends.com/social-media/can-10-million-tweets-measure-happiness/.))

[22] Killingsworth M. (2012). The future of happiness research. Harvard Business Review. January February. pp. 88-89))

[23] Krueger, A. B. and Schkade, D.A. (2007). The reliability of subjective well-being measures. (Working Paper, 13027). Cambridge: National Bureau of Economic Research.

[24] Larson, R. Csikszentmihalyi, M. and Graef, R. (1980). Mood variability and the psychosocial adjustment of adolescents. Journal of Youth and Adolescence. 9 (6). pp. 469-490.

[25] Larson, R. and Csikszentmihalyi, M.(1983). The experience sampling method. I. H. T. Rels (ed.), Naturalistic approaches to studying social interaction new directions for methodology of social and behavioral science. pp.41-56. San Francisco- Josey-Bas.

[26] Layard, R. (2005). Happiness: Lessons from a new science. New York: Penguin.

[27] Pena López, J.A. and Sánchez Santos, J.M. (2009). Economía y felicidad: Un análisis empírico de los determinantes del bienestar subjetivo de la población. Universidad de la Coruña, mimeo.

[28] Rojas, M. (2006) Life Satisfaction and Satisfaction in domains of Life: It is a simple relatioship? Journal of Happiness tudies, 7, 467-497,

[29] Rojas, M. (2007). Heterogeneity in the relationship between income and happiness: A conceptual-referent-theory explanation. Journal of Economic Psychology, 28, 1-14.

[30] Rojas, Mariano (2011) El Bienestar Subjetivo: Su Contribución a la Apreciación y la Consecución del Progreso y el Bienestar Humano". Realidad, Datos y Espacio: Revista Internacional de Estadística y Geografía, 2(1), 64-77.`

[31] Wilson, C, Boe, B, Sala, A, Puttaswamy, K. P.N. and Zhao, B.Y. (2009). User Interactions in Social Networks and their Implications. Proceedings of ACM SIGOPS/EuroSys European Conference on Computer Systems (EUROSYS 2009) Nuremberg, Germany, April.

[32] Wang. G, Konolige. T, Wilson. C, Wang. X, Zheng. H, and Zhao. B. Y. (2013). You are How You Click: Clickstream Analysis for Sybil Detection. Proceedings of Usenix Security. Washington DC, August., Software Agents, Modeling Software with BPM, DSL and MDA.

19

Social Networks as Learning Environments for Higher Education

J.A.Cortés[1], J.O.Lozano[1]

[1]Systems Engineering Program, Cooperative University of Colombia

Abstract — **Learning is considered as a social activity, a student does not learn only of the teacher and the textbook or only in the classroom, learn also from many other agents related to the media, peers and society in general. And since the explosion of the Internet, the information is within the reach of everyone, is there where the main area of opportunity in new technologies applied to education, as well as taking advantage of recent socialization trends that can be leveraged to improve not only informing of their daily practices, but rather as a tool that explore different branches of education research. One can foresee the future of higher education as a social learning environment, open and collaborative, where people construct knowledge in interaction with others, in a comprehensive manner. The mobility and ubiquity that provide mobile devices enable the connection from anywhere and at any time. In modern educational environments can be expected to facilitate mobile devices in the classroom expansion in digital environments, so that students and teachers can build the teaching-learning process collectively, this partial derivative results in the development of draft research approved by the CONADI in "Universidad Cooperativa de Colombia", "Social Networks: A teaching strategy in learning environments in higher education."**

Keywords — **Collaborative learning; Digital Environments; mobile devices; Social Networks.**

I. INTRODUCTION

EDUCATIONAL environments, are immersed in the processes of innovation, which are framed in a set of social and technological transformations. These are given by the changes in information and communication, this is why the social relations and a new conception of relations technology-society identified trends in society. Communication networks introduced a technological configuration that enhances learning more flexible and, at the same time, the existence of new learning scenarios in particular as regards the use of social media in education.

Castells [3] defines Technology as "the use of scientific knowledge to specify ways of doing things in a reproducible manner." This technology has caused a profound revolution in all fields including Education, especially characterized by the appearance of multimedia devices and a dramatic expansion of such as Telecommunications networks. The speed of information processing grows constantly and almost unlimited storage capacity.

Currently, there are terms such as e- Learning, e-Commerce, e -Business, etc., extending the terms m -Learning, m - Commerce related to mobile environments and finally comes to personal atmosphere defined as PLE (Personal Learning Enviroment); these are the different ways to characterize the population living with technology and are the further evolution of structures and components thereof whenever required. These generations and environments that have been incorporated into their lifestyles will be located in the third wave posed Toffler [19] as the " Information Society " supported by advances in information technology and telematics. With the advancement of telecommunications is expected to be greater participation of individuals in the production of information; production with concern Cartier [2] investigated and called the term media, whose object of study is the content traveling the net and how they can be interpreted in a more meaningful way to integrate various means of expression such as text, sound, images defined as static and dynamic Multimedia. These changes and concepts are reflected in the concept of service integration and favorable to smart growth "smart" communications devices that are being used throughout the Knowledge Society and Information Technologies.

A. Research Problem

Social networks are structures composed of groups of people, which are connected by one or more types of relationships, such as friendship, kinship, common interest or shared knowledge.

Today, virtual learning environments (VLE), provides a space for academic interaction mediated by information technology and telecommunications (ICT's), which offer many features, resources and tools for collaborative work, making it a good tool for development of formative research, as frequent interaction among members generates diversity of ideas, approaches and insights that lead to the achievement of a joint and meaningful learning.

Within virtual learning environments, are several services that enable you to perform the educational process by encouraging the learning of students or users. These development platforms have allowed adapt educational environments such as LMS (Learning Management System) or learning system manager. Currently exist various digital platforms which are used massively in educational environments due to its low cost ; the use of software platforms such as Blackboard or Moodle, to allow virtual support the academic, automating these processes together have enabled the emergence of new models of teaching and learning; these models have allowed each student to have personal computers available for exclusive use in any environment, in addition to the resources of the institution Hunt [9], as their own . These new scenarios in which students interact with information networks creating interactive generation where the use of " Netpods " and social networks , however not taken into account the institutions for their educational processes, is common that the students and executives continue to consume Blackberries, SmartPhones and Tablets in their daily work , while schools and other educational institutions remain in a primitive "pre- digital " state, due to disuse of distributed equipment Gagné [5].

Understanding technology as a support to improve educational processes, means that institutions regularly do a review of their learning environments (data centers, licenses, software, broadband, electronic library, laboratories, etc.). What it is to take stock: what is, what is obsolete, what needs to be renewed or updated? This knowledge, ultimately, will allow institutions to have a true picture of their technological capacity and act promptly without incurring higher costs.

This project aims to define models that have access to the virtual learning environments (VLE) through social networks, in order to extend the benefits of the students in the Cooperative University of Colombia, these settings allow you to manage learning through online courses generated by teachers, where students have greater access to courses and information, these computing resources are expected to generate new knowledge

The information society produces spaces of flows such as technology, places and people called by Castells real virtuality: time without time and without space [3]. These concepts described in "The global village of McLuhan," McLuhan [11] where the presence and incorporation of these technologies into educational models allow us to reduce the time and distance in communication processes. Valuable contributions on these concepts arise and emerge new features such as the cyber society, Joyanes [10] (Formation of social networks) cyberculture (Knowledge of the culture of the society in Red) and cyberspace (feeling in the same spaces in different places).

The development of virtual courses is known in various ways and with varying purposes such as Virtual Education, e - learning, e- trainning, among others, but still varied views about the issue persists; Institutions prefer to acquire technological platforms such as WebCT, Learning Space, Blackboard, Moodle among others, and the institutions are counted starting from scratch development to support the development of courses on NET, Driscoll [4]. It seems that the constant was no longer invest in developing a technology platform but rather lead them to those efforts that teachers begin their process of building materials and can locate trouble on the platform , initiating a "virtual dialogue" with Barabasi [1] , Driscoll [4] students .

In this context the question arises on which this study will answer:

What would be the Mediations supported in Connectivism and social networks, which could be appropriated as differentiators in today's learning environments in higher education?

II. METHODOLOGY(MATERIALS AND METHODS)

A. Hypothesis

The hypothesis proposed research is: Current developments and likewise investigations have relied on traditional pedagogical schools that have been oriented logography , however latest studies and research in virtual learning environments have concluded the iconography and connectivist environments are aspects that due to technological development , are impacting today's learning environments. We believe in this concept, it is necessary to investigate how new developments on the Internet, in particular how social networks are impacting the people and in particular in education, in order to define new educational models that reinforce learning in Higher Education.

Methodology

This research is descriptive qualitative ethnographic court, as it seeks to establish as new generations of students entering higher education using new technologies; where the proposed development is constructed with the use of tools and technologies used by students and described in the background such as forms of connectivity to new Internet services , web development service through LMS and application semantic Web and Social Networks as primary implementation tools for the implementation of effective and efficient services in academic consultations.

For Valles [20], qualitative research is one where the quality`s study of the activities, relationships, business, media, materials or instruments in a given situation or problem is studied. The emphasis is to document all information that is given daily in a given situation or scenario, observe and carry out full and continuous interviews, trying to get the minimum of detail being investigated.

The analyzed population were different semesters of the Faculty of engineering students, and area teachers, different instruments were used in gathering information, in the same way as it is a research of qualitative cutting, were used statistical evaluation tools such as Atlas TI (evaluation in qualitative analysis Software).

The sample was taken on about 35 students and 20 teachers from different semesters of the Faculty of Engineering was applied to a survey; 30 teachers from different educational institutions working in different educational levels underwent a group interview.

Phase 1: Information Gathering and dissemination activities

After selecting the appropriate research design and adequate sample, data on the variables involved in the research were collected, the data is classified and the variables involved in the process were determined, observations were recorded and the data were coded with order to have grounds to establish the instruments. It was very important to find support teachers who formed for this purpose a team of teachers, and helped both of these tasks in preparation , as well as propaganda of the activities in the student community which had incidence direct or indirect.

Techniques and instruments for data collection.

In this phase was designed each one of the instruments used for gathering information such as: the development of surveys, design group interviews and activities that had to do with the proper format of the behavior of individuals who participated in the process, just as processes were determined application of tools and means of collecting training. To carry out the study and information gathering used instruments listed below:

- Surveys
- Semi-structured individual and group interview

Methods and Procedures

In order to discover the concepts and relationships of data (interviews, observations) found, then organize and carry out the analysis to the following:

1. To identify and characterize the categories from sociology and psychology allow linking learning, learning styles and technologies used by students in the new learning environments, we use a theoretical review, supported on different texts authors consulted experts on the subject.
2. Selection and constitution of the group from a survey conducted in the Cooperative University , as a result it was decided to work with students from different semesters and teachers from both the University and faculty who work in Educational Institutions Secondary level facilities the study, which students completed their informed consent.
3. With the purpose of identifying emerging categories in particular learning styles and technologies of communication and used social networks, as well as the environments of learning in different educational institutions, applied surveys and group interviews.
4. The identification of categories of styles of learning and different learning environments, as well as technologies, was based on data from surveys and interviews, which were supported by arrays of relationship. The interpretation of the data was made at a later stage with Atlas TI.

Phase 2: Adaptation of Tools and Implementation
Theoretical review

The study of learning environments supported by technologies, requires that the methodology should be consistent with the theoretical framework, thus establishing the foundations of learning and learning styles and the use of technologies supported by different authors as Papert [14] and Siemens [16] selected by the researchers, we find that the patterns we detect and how we do it in the research on learning with the support of the mass media and connectivism.

Ethnographic method that sets a reflective process and allows approach and establish a trust relationship with the student group allowing inquiry through communication was considered. Authors such as Taylor & Bogdan[18], reflexive ethnography Hammersley & Atkinson[8], is very important for the group of researchers, understand and assess this methodology should be dynamic, and each instrument is considered as a prototype arrangement changes were reviewed that establish the objectives of the research.

The Survey

We consider the use of the survey to do a scan on the stakeholders of the University Cooperative Researchers case Bogotá. This investigation aimed to consider the profile of the groups, the relationship could be established with them to do the work and an exploration of the means used to access knowledge through learning. Was performed in 2 groups of 35 students each and two groups of 30 teachers. Knowledge and use of social networks, virtual learning environments and networks: The question and explores three areas of information questionnaire.

This instrument is applied in physical and the student fills out the form printed in the same format giving the option to fill in the digital format also provided and these were sent to a digital mailbox researcher. The information obtained was tabulated in Excel and recurring topics were entered in the answers.

Group interview

Valles [20] said that the group interview is to expose a group of people to a semi -structured interview guideline, not addressed to an individual but to a group where some free stimuli and sometimes structured to allow established as a structured and free response questions through this structure.

It is in the group interview the possibility of approaching the group of teachers, this strategy allows to inform aspects of research actors, their interests, as well as allowing an everyday space group is allowed to explore the opinions, beliefs, representations on the topics of study.

Two groups of 30 teachers of Informatics Master of Education, for a conversation which was set for an hour and a half where they felt about the concepts of teaching and learning, Knowledge, Intelligence and Virtual Learning Environments were selected. The dialogue was developing with driving interviewer, allowing diverse opinions that were recorded and then applied a survey where personal concepts of each of the participants were embodied.

Phase 3: Data analysis and presentation of results

At this stage there was an emphasis on understanding and interpretation of quantitative and qualitative methods of analysis, the proper interpretation of the data, coding the data, defining research categories and their relationships, generation diaries field, the use of statistical tools and the definition of an appropriate methodology for this research. This step was essential to use a number of techniques and statistical procedures for the collection of information by researchers, was the basis for the evaluation and validation of the effectiveness of strategies designed and necessary process feedback.

Phase 4: Defining Models in social learning environments

As a result of analysis, the results were validated in order to determine standard and suitable for use in environments learning strategies, which required the collaboration of teachers and policy programs, in order to validate the proposed adequate to different teaching practices and the use of technology in educational process

III. RESULTS

The results that are part of this investigation and to allow a methodological proposal to consolidate ICT mediated, social networking and multimedia tools in structuring a virtual learning environment, in accordance with the present trends. In this approach we analyze how the structure should be in virtual courses, detailing each of its parts, especially the theories of David Merrill[12] and Robert Gagné[5] precursors of instructional design; how should be learning in environments connectivist, what should be the new styles of teaching - learning and how educational models should be used in these new environments.

A. Connectivist Learning Environments.

The Horizon Report [6] has raised the technology adoption trends in learning environments for the next years, which are summarized as follows:

1. Knowledge is decentralized, the amount of resources that are available online that allow the production and distribution of content in multiple ways to facilitate the acquisition of knowledge both teachers and students.
2. Technology continues to affect our way of life in all environments, the digital divide is diversified with more products related to access and digital literacy skills, informational, media literacy, creating new environments inequality gaps
3. Technology is not only a medium that has been used to train students, but has become a means of communication and relationships, and a ubiquitous, transparent part of their lives. Social relations is one that has been felt more impact, especially in education. Communication between all stakeholders in education has become more open, multidisciplinary, and multisensory and becomes integrated gradually into all our activities.
4. Teachers and educational institutions are gradually making inroads into digital technologies. So, are increasingly beginning teachers to use in their educational practices different technological resources, from email to those provided in the web such as social networks.
5. How we think about learning environments is changing. In traditional education, learning environments are associated with physical spaces and presentiality. Today, however, the "spaces" where students learn are becoming more community and interdisciplinary time and are supported by technologies associated with communication and virtual collaboration. The time and space are transformed to combine the classroom with the virtual, blurring the boundaries between the two worlds, which are experienced by students as one.
6. Current technologies rely increasingly on cloud computing structures , supporting the information technology tends to be decentralized , deployment of cloud applications and services are changing not only the way you configure and use the software and data storage , but also how to conceptualize these functions. No matter where we store our work; what matters is that our information is accessible no matter where we are or what device they have chosen.

B. Teaching Styles.

The proposal of a new distance education model emphasizes mostly on employment of connectivism and ITC in this type of education. Today is orienting the educational process to the use of technology and new learning models. That according to Siemens [16] in his lecture "Connectivism: Creativity and innovation in a complex world", emphasized that education should aim to promote the development of creativity and innovation in students. This suggested that students should be involved in creating learning content constantly looking to learn creating something new. The current education system reduces the listed capacity, according to the expert, who noted that technology in the classroom allows students to be co-creator and an active participant in their learning.

Teaching styles are linked to the peculiar way that each educator to implement and lead teaching their students. The concept of teaching style or style of education focuses not only on learning, but also in the way of how the individual undertakes, aims or combine various educational experiences. Therefore, the teaching style must be a social environment.

C. Connectivism as way Learning Network

If knowledge is changing so quickly and if such important challenges such as climate change or global warming are tackled , if the technology is changing every day , ¿how can we prepare our students today to take on the challenges of tomorrow? . Today we are moving from a model where education systems and the types of courses we teach are created in advance in the institutions and the student comes to the classroom after we have already created textbooks, materials and resources.

One of the things we have to do, is stop treating intelligence as something that exists inside a person 's head, but rather realize that intelligence exists as a result of contextual knowledge of our environment and relations , social media and technology for our students and ourselves are participants and which are collaborative members. The very structure of learning creates connections in neural networks can be found in the form of linking ideas and ways in which we connect with people and information sources. Our expertise lies in the connections we form, either with others or with information sources such as databases and information systems available on the Web. This appears as a connectivism learning theory for the digital age

In his article "Connectivism: A learning theory for digital age", Siemens [17] summarizes his theory on the following principles: among which can be highlighted which have to do with learning and its relationship with the networks, the taking of decisions and connections among ideas and concepts.

D. Designing Courses
Structure of virtual courses

To design the Courseware or supported in Virtual environments, now called VLO (Virtual Learning Objects) courses we rely on theories of Instructional Design. The teaching and learning processes are possible because, through an appropriate instructional design, media are used to facilitate students in rich learning situations. Technology is the means, not the end. Unable to assess the usefulness of a specific technology without verifying instructional design.

For proper operation of a virtual course should include the following parameters:

- Building a theoretical model of instructional design that supports the development of the course
- Establish activities that enable online interaction
- Designing the navigation tools
- Structure course information such that the user quickly place
- Design a suitable user interface

The development of the course was considered the following: design course, design of instruction, characterization of the instructor, characterization of the student, the platform components.

E. Design Platform

To build the platform the XP methodology was used, this is an agile methodology focused on enhancing interpersonal relationships as a key to success in software development, promoting teamwork, worrying about learning developers, and fostering a good working environment. XP is based on continuous feedback between the customer and the development team, communication between all participants in the solutions implemented simplicity and courage to face the changes. XP is defined as particularly suitable for projects with very vague and changing requirements, and where there is a high technical risk.

Step 1. Research or planning

Gather all the information regarding the project requirements in order to get to know the specifications of the problem. For this stage theoretical information of each of the key elements of the research, which correspond to Student Virtual Learning Environment and Social Media, was collected. This information was used to design the best way to know that teachers present concepts of the Cooperative University of Colombia on the subject, so propose a solution to the problem recorded in chapter 1 and allows interaction thereof.

Step 2. Analysis

Validate the information gathered above to specify the problem to be solved. Once the survey asks the following analysis and conclusion in general that allowed validating the trend in social networks and knowledge of virtual learning environments were generated.

Step 3. Design and Build

The conceptualization of the information gathered in the investigation phase was conducted.

- Design of the images used in the software.

- GUI design.
- Development of prototype software programming on Moodle.
- List of use cases.

Step 4. Simulation

At this stage all the necessary components were integrated to operate the project. Were designed and set up the courses, the interface was done with social networks and the tests were performed in order to test its operation. Simulation software in the emulator which is located in www.jairolozano.com/uccvirtual.

Fig1. Configuration and integration of Moodle with social networks. Source: Authors

IV. CONCLUSION

This paper presents the results of research by determining an appropriate structure to develop and integrate virtual learning environments in network. With advances in technology and its foray into the information society and knowledge, it is undeniable that online education is expanding its worldwide coverage , so you cannot be oblivious to its structure and consolidation that allow not only ownership its architecture, but of teaching and learning strategies that flow from them . Productivity Internet, where you see a preview is from 2010, from the semantic web concepts to the detriment of conceptualizations of web 2.0 for this purpose the following activities were performed:

- A proposal for a course mediated by the use of multimedia technologies, Software and Connectivity Social Networking was designed to improve learning environments in higher education
- Some appropriate strategies for using social media in learning environments were defined
- A learning environment where multimedia technologies and social networks such as mediation in the teaching-learning process used was structured.
- Measuring instruments defined methodological spaces in social networks supported in connectivism, which will enable teachers and students to improve the quality of teaching and learning processes.

- Of the above, and as a result of the interviews, are evident new roles, methods, trends and architectures in different Virtual Learning Environments (VLE), these are:

Trends in digital learning environments.

- Teacher training for using digital media in teaching and learning remains a challenge. Know and understand the educational potential of these technologies promote their use in the classroom. The training of teachers from a holistic perspective that incorporates the use of technology resources as an inseparable part of the practice of teaching and learning is the first condition for significant incorporation of digital media in all educational levels.

- Comprehensive change management in higher education must be understood from a systemic and transformative approach that contributes to economic growth, human development and social cohesion. While educational policies cannot be imposed, it is the responsibility of those who have been chosen for this consider, reflect and make decisions to promote the necessary changes; otherwise, we risk that they never occur. This includes a change of role in forcing educational institutions to avoid reflections that everything remains the same, allowing shoot tangible and sustained changes. A redefinition of the educational model that includes new ways to generate, manage and transmit knowledge is required.

- Digital literacy must become an essential skill of the teaching profession. Although there is general agreement on its importance, training in techniques and skills related to the digital realm remains an exception in teacher education programs. The based tools and platforms skills and standards have proven to be something ephemeral, given that digital literacy has less to do with tools like the thought: digital skills have multiple faces (technology, information, multimedia content, and digital identity) and require a comprehensive way to be faced.

- The training of students in the use of new media and audiovisual communication languages is critical. Students need new knowledge and skills in the field of writing and communication, other than those that were needed a few years ago only. Increasingly it is necessary to possess technological expertise to collaborate globally and be able to understand the content and design of new media. For this reason, must be integrated into the curriculum new literacies, and their evaluation, which requires understanding, in its entirety, the meaning and scope of these new skills and competencies.

- Using technology to appropriate treatment of information and knowledge building is still too rare. A key challenge is to not only reflect on the use of emerging technologies by themselves, but put them in the dialectic of information processing to solve complex problems of society, being one of the challenges of higher education. It is not only incorporate technologies or not, but to put forward the needs of student understanding and thinking new ways of working with complex reality we face , to be able to build knowledge about the same .

- Adapting teaching practices to the requirements of the digital society and knowledge is required. Technologies place the student as the protagonist and author in different spaces, but their role is still predominantly receptor in the contexts of formal education. The underlying this phenomenon is that it cannot reduce the proliferation of the use of technology, since many other sociocultural aspects are driving change in existing education and labor practices. The low velocity in the appropriation of technology by the education sector may be due, among other causes, to which teachers are trained as users and not as leaders in the design and implementation of the use of technology for educational purposes. These trends and challenges have a profound effect on the way we experience with emerging technologies and how they implement and use in the educational world.

ACKNOWLEDGEMENTS

The research is part of the final report of the research project supported by the Cooperative University of Colombia. JA Cortés, Cooperative University of Colombia, Bogotá, Colombia

REFERENCES

[1] A. L. Barabási. Linked: The New Science of Networks, Cambridge, MA, Perseus Publishing. 2002.

[2] M. Cartier. La médiatique. Editions du Laboratoire de Télématique. Université du Québec à Montréal. Montréal, Canada. 1980

[3] M. Castells (pp 31-419).La era de la información. La Sociedad Red. Tercera Edición. Vol. 1 España: Alianza Editorial. 2005.

[4] M. Driscoll. Psychology of Learning for Instruction. Needham Heights, MA, Allyn & Bacon. 2000.

[5] R. M. Gagné. Las Condiciones del Aprendizaje. Madrid: Ed, Interamericana. 1979.

[6] I. García, I. Peña-López,; L.Johnson, , R.Smith, , A. Levine, , & K. Haywood. Informe Horizon: Edición Iberoamericana 2010. Austin, Texas: The New Media Consortium. 2010

[7] J.C. Gleick. The Making of a New Science. New York, NY, Penguin Books, 1987.

[8] M. Hammersley, & P. Atkinson. Etnografía. Métodos de investigación.2001

[9] T. Hunt. Desarrollar la capacidad de aprender. La respuesta a los desafíos de la era de la información. Editorial Urano. Barcelona. 1997

[10] L. Joyannes. Cibersociedad. México: Ed. Mc Graw Hill. 1997

[11] M. McLuhan. La Aldea Global. Madrid: Ediciones GEDISA. 1995

[12] D. Merril. Educational Technology, New York: Li & Jones, 1991.

[13] N. Negroponte. Ser Digital. Ed. Atlántida. 1995.

[14] S. Papert. La informática en el aula: Agentes de Cambio.By Seymour Papert This article appeared in The Washington Post Education Review Sunday, October 27, 1996 The Washington Post Revisión de la Educación Domingo, 27 de octubre 1996

[15] H. Rheingold. La Comunidad Virtual. Ed. Addison Wesley. 1993.

[16] G. Siemenes. Conectivismo: Creatividad e innovación en un mundo complejo Universitat de València (España), coordinado por Beatríz Gallardo, George Siemens, Dolors Capdet y Paz Villar. 2011.

[17] G. Siemens. Conectivismo: Una teoría de aprendizaje para la era digital. 2004.

[18] S.J.Taylor & R. Bogdan. Introducción a los métodos cualitativos de investigación social. 1996.

[19] A. Toffler. La Tercer Ola. España: Plaza & Janes.1980.

[20] M. Valles. Técnicas Cualitativas de investigación de Social. 1996.

A Case-based Reasoning Approach to Validate Grammatical Gender and Number Agreement in Spanish language

Jorge Bacca, Silvia Baldiris, Ramon Fabregat, Cecilia Avila,
University of Girona, Spain

Abstract — **Across Latin America 420 indigenous languages are spoken. Spanish is considered a second language in indigenous communities and is progressively introduced in education. However, most of the tools to support teaching processes of a second language have been developed for the most common languages such as English, French, German, Italian, etc. As a result, only a small amount of learning objects and authoring tools have been developed for indigenous people considering the specific needs of their population. This paper introduces Multilingual–Tiny as a web authoring tool to support the virtual experience of indigenous students and teachers when they are creating learning objects in indigenous languages or in Spanish language, in particular, when they have to deal with the grammatical structures of Spanish. Multilingual–Tiny has a module based on the Case-based Reasoning technique to provide recommendations in real time when teachers and students write texts in Spanish. An experiment was performed in order to compare some local similarity functions to retrieve cases from the case library taking into account the grammatical structures. As a result we found the similarity function with the best performance.**

Keywords — **Authoring tool, Second language acquisition, indigenous people, Case-based reasoning, local similarity functions.**

I. Introduction

IN bilingual virtual training programs for teachers that have an indigenous language as mother tongue [1] [2], there are some difficulties when teachers design and create learning objects to teach Spanish as a second language for indigenous population. Some of those difficulties were reported in [3] and are mainly related to the process of writing texts, in particular the use of grammatical gender and number in the Spanish language. The main cause of this situation is that some indigenous do not have masculine or feminine distinction, or there are particular ways to express grammatical number that differs significantly from Spanish language.

In consequence, teachers have to be aware of some rules in order to properly apply the grammatical rules of Spanish and take care of teaching them correctly to their students. Nevertheless, in some cases, indigenous teachers of Spanish language use some didactic strategies, as reading from

textbooks and the language class [4], designed to teach indigenous languages but they apply them to teach Spanish language. This situation can create some problems in students, because they do not reach a good Spanish level, so it will affect the learning process of other subjects in the future. As a result of these issues, some learning objects that are written by indigenous teachers in Spanish may contain some grammatical errors in the texts.

As a solution, in this paper we introduce Multilingual-Tiny, a web authoring tool based on the TinyMCE [5] web content editor which consist of a complete set of plug-ins and online services for teachers to support them in the learning objects design and development. Multilingual-Tiny also has a module that applies Case-based reasoning (CBR), in order to provide recommendations (based on the grammatical structure of sentences) and taking into account the previous experience of skilled teachers from writing Spanish texts and well-formed texts obtained from the Internet. All of this process support teachers of Spanish language when they are creating their learning objects, mainly when they are writing texts in Spanish language.

This document is organized as follows: In section 2, some concerns about teaching Spanish as a second language are presented. In section 3 the architecture design of Multilingual-Tiny is described, including the applied CBR cycle. Section 4 describes an illustrative scenario which present the complete process performed by Multilingual-Tiny and also how the CBR technique was applied. Section 5 describes the followed validations process as well as the obtained results. Finally conclusions are presented in section 6.

II. Teaching Spanish as a Second Language

Teaching Spanish as a second language to indigenous communities is not a trivial task. It supposes a challenge to governments and universities in which is important to promote effective Bilingual Intercultural Programs (BIE) and at the same time, training teachers effectively in order to introduce Spanish in a coordinated bilingualism method [4], in which both, mother tongue or L1 and second language or L2 are developed at the same time. In this context, the mother tongue (which is an indigenous language), is acquired by a natural process [6]. The second language – L2, in this case, the Spanish language, is taught for facilitating indigenous people communication with Spanish speakers and also to receive instruction in some knowledge areas which are taught in Spanish.

Despite the efforts and advances obtained by applying the Bilingual Intercultural Programs in some countries such as Mexico and Peru, teachers of Spanish language may face some difficulties when they have to teach indigenous people how to read and write in Spanish and in the indigenous language [7] at the same time. Some of those difficulties are due to the fact that teachers of Spanish have an indigenous language as mother tongue and they learnt Spanish in a non-systematic way. The consequence is that those teachers use the same strategies for teaching both languages, so it could be counterproductive in student's learning process [8].

When teaching Spanish, teachers usually can follow two complementary strategies: reading from textbooks and the language class [5]. The former is a strategy in which teachers introduce and explains the topic in the indigenous language and then students read the book in Spanish language so that students identify vocabulary and pronunciation. Finally, the teacher explains vocabulary or concepts that students may have lost in the reading. The latter strategy is the language class, in which teachers of Spanish compare the indigenous language with the Spanish language in terms of grammar, vocabulary and structure in order to promote reflection and develop the meta-linguistic awareness [5].

In this context, in teacher's training, when universities are preparing indigenous students that will be future teachers of Spanish language for teaching in their indigenous communities, students have to develop competencies and skills in order to effectively apply the teaching strategies mentioned above and other didactic and pedagogic methods. Multilingual-Tiny, the web authoring tool developed, takes a relevant role in this task; giving recommendations to teachers to avoid grammatical errors. As a result, teachers can create quality educational content to teach Spanish and create learning objects in their mother tongue.

III. MULTILINGUAL-TINY APPROACH

A. Overview Architecture

Multilingual-Tiny is a web authoring tool developed in order to support indigenous students which will be future teachers of Spanish language in indigenous communities and teachers of this population, when they are creating learning objects, in particular, when they have to deal with some grammatical structures of sentences in Spanish. Multilingual-Tiny consist of plug-ins and online services to provide a virtual environment to design and develop learning objects in Spanish and indigenous languages and has a module based on the case-based reasoning technique, to provide recommendations in order to avoid grammatical errors and develop quality educational content.

The architecture of Multilingual-Tiny is depicted in Fig. 1. The architecture has 4 layers, from top to the bottom: The users layer, represent indigenous teachers and students that interact with Multilingual-tiny. The interface layer includes the authoring tool and shows the recommendations that come from the CBR module. The services layer provides a group of services for text processing and includes the CBR based module to provide the recommendations. Finally the data access layer includes services for data storing, such as the case library.

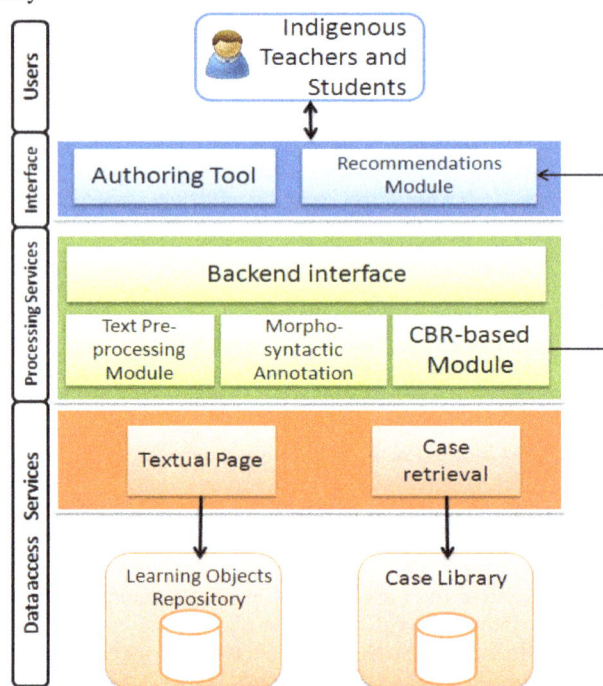

Fig. 1. Multilingual-Tiny Architecture.

B. Layer Description

The following paragraphs provide a detailed description about each of the layers mentioned before.

1) Users Layer

This layer represents the users that interact with Multilingual-Tiny, for instance, indigenous students that will be future teachers of Spanish language in indigenous communities and indigenous teachers. These users interact with the interface layer to use the service in order to create the learning objects.

2) Interface Layer and Authoring tool

Interface layer includes the authoring tool and the recommendations. The authoring tool is based on the TinyMCE [5] web content editor, which is an open source JavaScript based web editor that provides a group of services in order to create web pages without worrying about HTML code, because HTML is generated by it. The authoring tool can be integrated in the ATutor [9] e-learning platform or in other platforms.

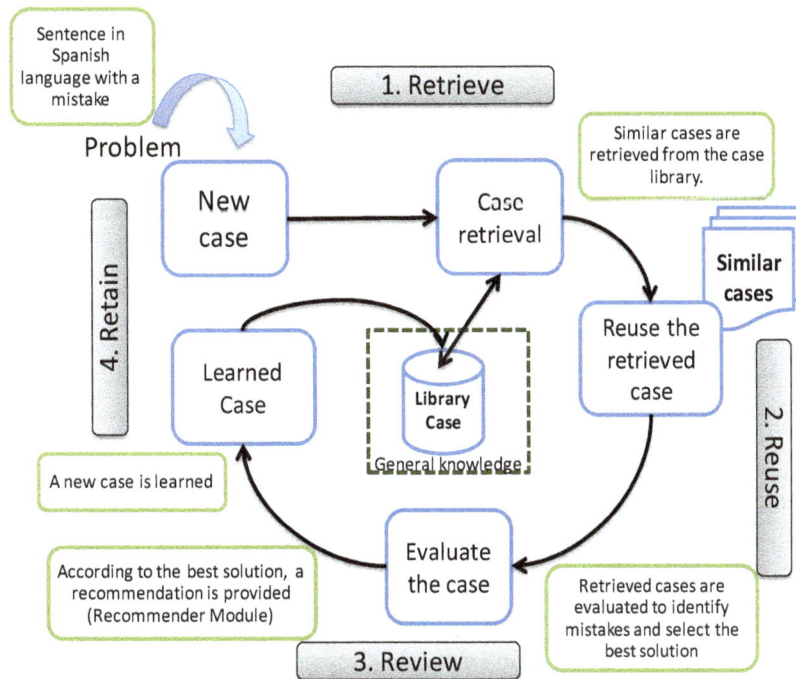

Fig. 2. CBR Cycle applied to generate grammatical recommendations for indigenous population.

As a result teachers can easily create web pages which will be part of a course in the ATutor e-learning platform as learning objects.

The authoring tool establishes communication with the Processing services layer when a learning object is being created. The text written in Spanish by indigenous teachers or students in the authoring tool is then sent to the Processing services layer to be analyzed.

The recommender module in the interface layer shows the recommendations that come from the CBR based module. These recommendations include suggestions on how to correct grammatical errors. The recommendation process is described in detail in next sections.

3) Processing Services Layer

This layer includes the services for text processing, the morpho-syntactic annotation module and the CBR based module. Those services are combined in order to provide recommendations to teachers when they are creating the learning objects to teach Spanish. The input of this layer is the text of the learning object that is being created in the authoring tool. The components of this layer are:

1) **Text Pre-processing Module:** The text pre-processing module is based on the open source FreeLing [10] library for Natural Language Processing. The input of this module is a text which has been written by the teacher as part of a learning object. This text is automatically split into sentences and the resultant sentences are split into words. This process is based on dictionaries and rules of the FreeLing library. The result of this process will be the input of the morpho-syntactic annotation module.

2) **Morpho-syntactic Annotation Module:** This module provides the morpho-syntactic annotation, which is a process of assigning tags for every word in the text, depending on the grammatical category. This process is based on the PoS (Part of Speech) tagging of FreeLing library. The input of this module is the output of the pre-processing module (which is a group of words). The PoS tagging is based on the EAGLES [11] recommendations. EAGLES define a group of standard tags for every grammatical category. As a result, each word of the text is assigned a tag depending on the context and grammatical structure of each sentence. The outputs of this module are groups of part-of-speech tags which represent a sentence. These tags will be an important component of the case representation in the case based reasoning module.

3) **Cased-based Reasoning Module:** This module is based on the Case-based reasoning technique. It takes the output of the morpho-syntactic annotation module, and executes the CBR cycle. As a result it provides the recommendations to indigenous teachers and students in order to correct grammatical errors when they write texts in Spanish language during learning objects creation. The module was built with jCOLIBRI framework [12], each case from the case library consists of a group of tags (part of speech tags) which represent a well-formed sentence. The CBR cycle, which includes 4 steps (Retrieve, Reuse, Review and Retain), is applied to grammatical sentence analysis in Spanish language and the process is depicted in fig. 2.

The main steps of this process are:

1) The Retrieve step: In this step a new case that comes from the morpho-syntactic annotation module, which is a new sentence, is compared with the cases stored in the case library by means of the similarity algorithm. As a result the most similar cases are retrieved. Both components are used:
    o Case library: Composed by a group of cases which are well-formed sentences in Spanish language obtained from a wide variety of texts from Internet. The case library is updated and new cases are stored when teachers add a new sentence structure. The case library is part of the Data Access layer which establishes communication with the services layer in order to store and retrieve cases.
    o Similarity Algorithm and Retrieve Component: Based on the JCollibri framework, the nearest neighborhood algorithm K-NN [13] is applied in order to retrieve the most similar cases when a new sentence is being analyzed. This process uses a global similarity function and a local similarity function for each attribute from the case.

2) The Reuse Step: In this step the K most similar cases obtained, by computing similarity, as described above are selected and the CBR Module organizes the cases according with the weights defined by the Morpho-syntactic Annotation Module.

3) In the next step, Review, the cases are evaluated in order to identify if the sentence is correct or if the sentence has a grammatical error. Besides, the case could be adapted or transformed to provide a recommendation about how to properly write the sentence. Further details about the overall process are depicted in section 4.

4) In the next step, which is called Retain, a new case obtained from the adaptation of the retrieved case is converted into a new case. Which is part of the recommendations provided by the recommender module and on the other hand it is stored in the case library as a new case. As a result from the process, grammatical errors in terms of using gender and number could be identified and a recommendation on how to correct it is provided to students.

4) Recommender Module

This module takes the cases retrieved from the case library as an input for providing recommendations to teachers or students on how to correct the sentence if a grammatical error in gender and number is identified. These recommendations take into account the indigenous language of teachers and students in order to explain why the sentence was incorrect from the indigenous language grammatical perspective.

IV. AN ILLUSTRATIVE SCENARIO

As well known, the CBR cycle includes 4 steps (Retrieve, Reuse, Review and Retain) as shown in Fig. 2. In this section a step-by-step illustrative case based on CBR cycle is applied in order to show how the grammatical sentence analysis in Spanish language is performed in Multilingual-Tiny to provide recommendations to students and teachers.

Step 1 - Writing the text:

Indigenous students which are preparing to be future teachers of Spanish language write a text in the web content editor when they are creating learning objects. In this step it is probably that students make mistakes in terms of grammatical issues when they write a text in Spanish but they are frequently thinking in their mother tongue which is an indigenous language. For instance:
— *Me gustan el gatos blancos* (sentence with a mistake in Spanish).
— *I like white cats* (English translation only for illustrative purposes).

The above sentence in Spanish has a mistake in the definite article ("el") because it is in singular form but it must be in plural form ("los").

Step 2 – Text Pre-Processing (Morpho-syntactic annotation of the initial text):

In this step the system takes the initial text and applies the morpho-syntactic part-of-speech annotation of the text according to EAGLES recommendations [11]. Taking the example mentioned above the morpho-syntactic annotation is depicted in table 1. It is important to remark that in table 1 for English language the sentence seems to be grammatically correct, but in Spanish language there is a mistake when using the definite article ("el") (which in English is "the") in singular form with a noun "gatos" (in English "cats") in plural form.

Step 3 – Case retrieval

Based on the morpho-syntactic annotation from step 2, in which each word has a specific tag (as depicted in table 1), a new case is created; this case is composed by the group of EAGLES tags. The new case could be: Case[PP1CS000, VMIP1P0, DA0MS0, NCMP000, AQAMP0]. This case is equivalent to the sentence: "*Me gustan el gatos blancos*" (in English: I like white cats). The new case is compared by means of the nearest-neighbor algorithm [13] with cases previously stored in the case library. The most similar cases are retrieved. For instance, if the following case is retrieved: Case=[PP1CS000,VMIP1P0,DA0MP0,NCMP000,AQAMP0], with a computed similarity of 96% from the global similarity function. It is important to remark that cases stored in the case library have been obtained from texts without grammatical errors.

TABLE I
MORPHO-SYNTACTIC ANNOTATION OF THE EXAMPLE.

Word – Token		Part-of-Speech tagging (EAGLES)	Meaning of the tag assigned to each word
Words in Spanish	Translation to English		
Me	I	PP1CS000	Personal pronoun, first person, common gender in singular form.
Gustan	Like	VMIP1P0	Main verb, indicative, present form, first person and plural.
El	the	DA0MS0	Define article, masculine, in singular form.
Gatos	cats	NCMP000	Common noun, masculine, in plural form.
Blancos	white	AQAMP0	Qualified adjective, masculine in plural.

Step 4 – Comparison of cases and recommendations

In this step a comparison between the new case and the most similar case retrieved is performed in order to find differences in terms of the sentence grammatical structure. By means of this comparison and the analysis performed is possible to identify for example if there are mistakes of grammatical gender or grammatical number which are common when indigenous people is learning Spanish. For instance the comparison of the example proposed ("Me gustan el gatos blancos" in English "I like white cats") with the case retrieved from the cases library is depicted in Fig. 3.

As a result of the comparison in this example, the system identifies a difference in the third element of the new case (DA0MS0) and the corresponding element of the re-trieved case (DA0MP0). Those tags are described as follows:
• DA0MS0 = Definite article (DA), Neutral (0), Masculine (M), in singular form (S), is not a possessive article (0).
• DA0MP0 = Definite article (DA), Neutral (0), Masculine (M), in plural form (P), is not a possessive article (0).

The difference was identified around the use of the grammatical number: In the new case the article is in singular form, but in the retrieved case (which has been extracted from a text correctly spelled) the article is in plural form. When the mistake has been identified, a recommendation is provided in order to correct the sentence; this recommendation takes information from the case retrieved in the CBR cycle in order to suggest the correct form that the sentence should have. As a result indigenous students and teachers can also learn by interacting with the authoring tool. Fig. 4 shows the graphical

user interface of the CBR module. In this case the interface shows the sentence that will be analyzed to identify possible mistakes in grammatical number and gender agreement.

NewCase = [PP1CS000, VMIP1P0, **DA0MS0**, NCMP000, AQAMP0]

Difference Identified

RetrievedCase = [PP1CS000, VMIP1P0, **DA0MP0**, NCMP000, AQAMP0]

Fig. 3. Comparing the example proposed with a retrieved case from case library.

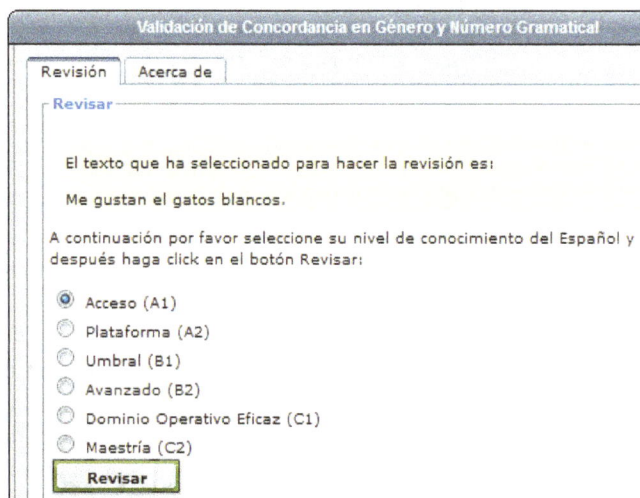

Fig. 4. CBR Module graphical interface in TinyMCE.

V. EVALUATION

A. Description

The purpose of the evaluation process in to validate the main of our approach which is to support indigenous teachers and students when they write a text in a web content editor for creating learning objects. As mentioned before, the support we offer to indigenous teachers and students refers to automatically generate recommendation in order to avoid grammatical errors and develop quality educational content.

In particular, we validate the case retrieval process, because this is the process that ensures that the offered recommendation is the best one that the user could receive.

We applied the K-NN algorithm [13] in order to retrieve the most similar cases from the case library to check the grammatical number and gender agreement.

The K-NN algorithm in the jCOLIBRI framework uses a local similarity function and a global similarity function. The former is used to compute the similarity in every attribute of the cases; the latter is used to compute de global similarity considering the results of the local similarities from all the attributes of the case. We design an experiment to compare and choose the best local similarity functions that allows retrieving the most similar sentence to check the grammatical number and gender agreement. In this section we describe the methodology and the main results of the comparison.

TABLE II
TEST PERFORMED IN THE EXPERIMENT WITH VALIDATION METHOD AND VOTING METHOD APPLIED

Similarity Function	Validation method					
	Leave One Out			N-Fold Random Crossvalidation		
	Weighted Voting Method	Majority Voting Method	Unanimous Voting Method	Weighted Voting Method	Majority Voting Method	Unanimous Voting Method
Levensh.	Test 1	Test 2	Test 3	Test 4	Test 5	Test 6
Overlap	Test 7	Test 8	Test 9	Test 10	Test 11	Test 12
Smith-Waterm.	Test 13	Test 14	Test 15	Test 16	Test 17	Test 18
Jaccard	Test 19	Test 20	Test 21	Test 22	Test 23	Test 24
Dice	Test 25	Test 26	Test 27	Test 28	Test 29	Test 30
TokensC	Test 31	Test 32	Test 33	Test 34	Test 35	Test 36

B. Methodology

In each case stored in the case library, the attribute with the highest weight is the morpho-syntactic annotation, which is basically a group of tags where each tag has been assigned to each word in the sentence according to the context and the grammatical structure. Since this group of tags is represented by means of a string data type, the local similarity function applied to this attribute should be able to compute the similarity between strings. There are many similarity functions for strings in literature some of them are described in [14], [15]. There are similarity functions based on fuzzy sets [16], and set-based string similarity [17] and [18].

For this experiment we chose four similarity functions commonly used in textual case-based reasoning. In addition we improved two similarity functions to consider the word order of the sentences during the analysis and deal with disambiguation by means of the FreeLing library. These are some important drawbacks described in [19] to be tackled in information retrieval and textual CBR. As a result 6 similarity functions were applied in the experiment. These are listed in table 3.

The validation methods used in this experiment were:
- Leave One Out
- N-Fold Random Cross-validation with 10 folds.

The voting methods selected for the K-NN algorithm were:
- Weighted Voting Method
- Majority Voting Method
- Unanimous Voting Method

The experiment was performed by means of 36 tests that combine the validation methods and the voting methods. Table 2 summarizes the group of tests in the experiment. For each test we obtain the following data:

- Precision
- Recall
- True Negatives
- False Positives
- False Negatives
- True Positives

TABLE III
SIMILARITY FUNCTIONS APPLIED IN THE EXPERIMENT

Simmilarity Function	Description	Improvement
Levenshtein	Also known as the edit distance	-
OverlapOrdered	Overlap Coefficient	Improved to consider the order of the tokens in each sentence analyzed.
Smith-Waterman	Based on the algorithm of dynamic programming for local alignment.	-
Jaccard	Jaccard coefficient for strings of characters	-
Dice	Dice coefficient for strings of characters	-
Tokens Contained Weighted First	Based on the TokensContained similarity function of jCOLIBRI framework.	Improved to consider the order of the tokens in each sentence analyzed.

C. Results

The following paragraphs summarize the main results we obtained in the experiment.

The results of tests 1,7,13,19,25 and 31 are show in fig. 5. In this case the validation method was Leave One Out and the voting method was Weighted Voting Method. On the other hand Fig. 6 shows the results of tests 4, 10, 16, 22, 28 and 34 where we used the same voting method but using the N-Fold random cross-validation method.

The F-Measure graphic in fig. 5 shows that the OverlapOrdered and TokensContained functions outperform the other functions compared, and the F-measure graphic in fig. 6 shows that Smith-Waterman, OverlapOrdered and TokensContained are functions with a better performance than the others.

F-Measure with β=1

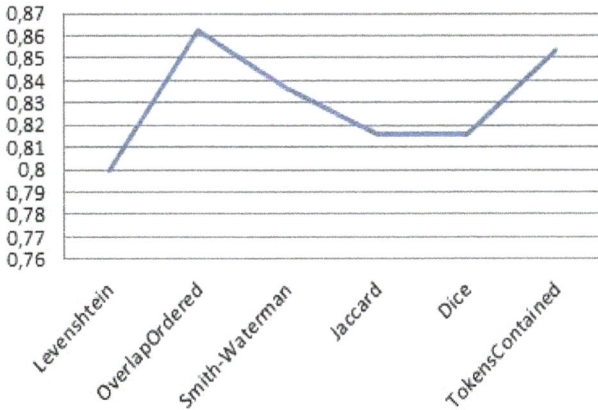

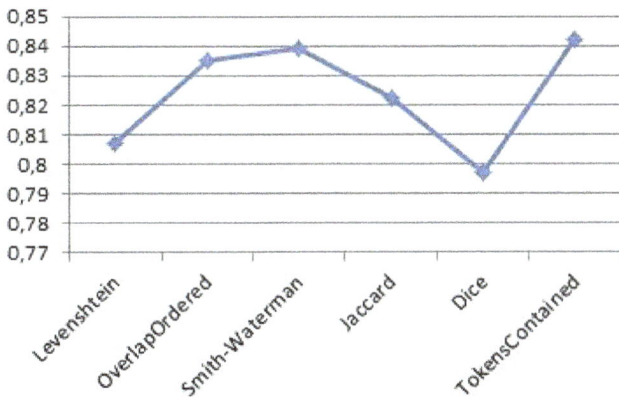

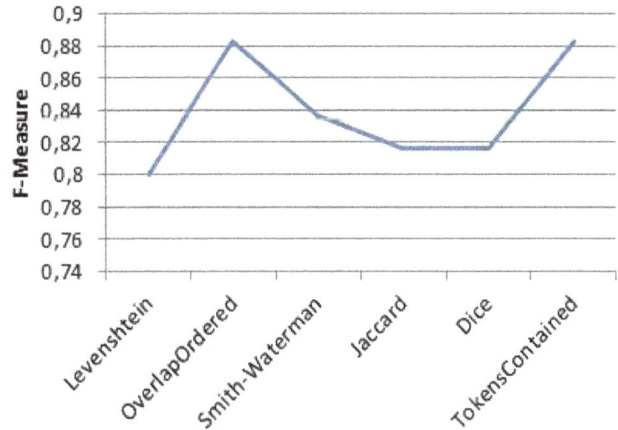

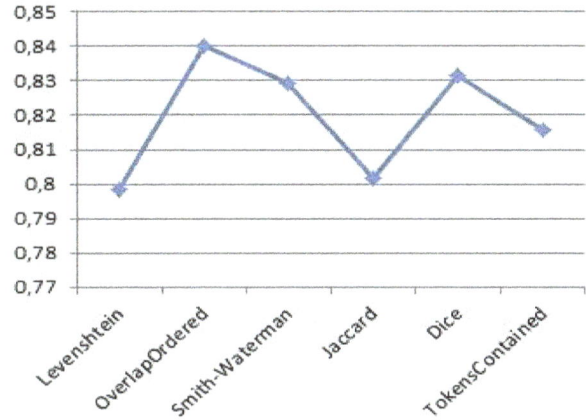

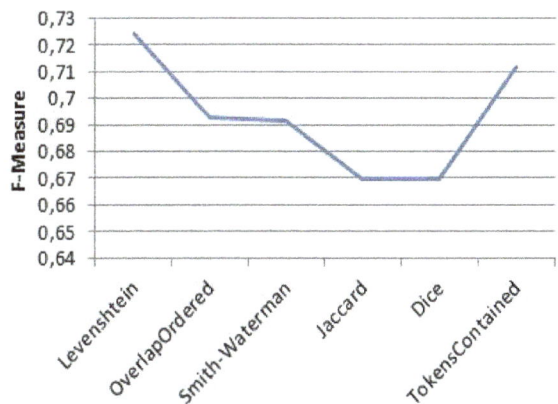

Fig. 5. Results of tests 1, 7, 13, 19, 25 and 31.

F-Measure with β=1

Fig. 6. Results of tests 4, 10, 16, 22, 28 and 34.

F-Measure with β=1

Fig. 7. Results of tests 2, 8, 14, 20, 26 and 32.

F-Measure with β=1

Fig. 8. Results of tests 5, 11, 17, 23, 29 and 35.

F-Measure with β=1

Fig. 9. Results of tests 3, 9, 15, 21, 27 and 33.

The results of tests 2, 8, 14, 20, 26 and 32 are shown in fig. 7 using the Leave One Out validation method as mentioned before, but in this case using the Majority Voting Method for the K-NN classifier.

In contrast fig. 8 shows the results of tests 5, 11, 17, 23, 29 and 35 using the same voting method but using the N-Fold random cross-validation method. In this case the results also shown that OverlapOrdered offers better performance than the other functions evaluated and TokensContained has a good performance when the voting method is the Majority Voting Method.

F-Measure with β=1

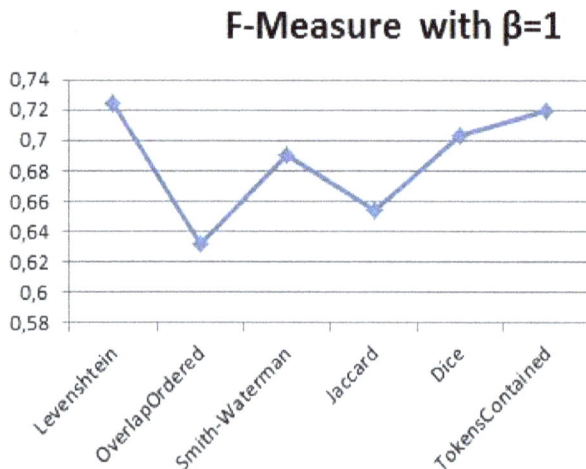

Fig. 10. Results of tests 6, 12, 18, 24, 30 and 36.

Finally the results of tests 3, 9, 15, 21, 27 and 33 are shown in fig. 9, in this case we use the same validation method (Leave one out), but we use the Unanimous Voting Method in the K-NN classifier.

In contrast, fig. 10 shows the results of tests 6, 12, 18, 24, 30 and 36 using the same voting method but the N-Fold random cross-validation method. In this case the Levenshtein distance has a better performance than OverlapOrdered in both methods of validation but Tokens Contained has almost the same performance than Levenshtein.

VI. CONCLUSION

Multilingual-Tiny as an authoring tool to support indigenous students that will be future teachers of Spanish language when writing texts in Spanish, takes a relevant role in order to help students to improve their writing skills at grammatical level so that they will be proficient teachers of Spanish. Multilingual-Tiny also provides a group of services that allow creating learning objects and design activities in the context of learning Spanish as a second language. This tool can be considered an advance in information and communication technologies to support the training process of indigenous students in the context of bilingual intercultural programs.

The case-based reasoning technique applied to the process of sentence analysis in order to identify grammatical errors mainly in terms of grammatical number and gender, is an efficient technique due to the use of the past user experience. Besides, the similarity algorithm, including the local similarity functions, the global similarity function and the retrieval process based on the K-NN algorithm in JColibri applied in the retrieval step works as it was expected in order to retrieve the most similar cases compared with a new case provided.

Evaluation developed shows that with respect to the algorithms used to retrieval cases that OverlapOrdered and TokensContained are functions with better performance to retrieve cases from the case library, so we can confirm that they are useful when we are dealing with grammatical structures of sentences in Spanish in form of part-of-speech

tags. As a result of this experiment we decided to combine both functions when performing the retrieval phase of the case-based reasoning cycle.

This strategy allows improving the system's performance in order to identify possible grammatical errors in gender and number agreement in Spanish language.

ACKNOWLEDGMENT

This work is supported in part by the European Commission through the funding of the ALTERNATIVA Project (DCI-ALA/19.09.01/10/21526/245- 575/ALFA III (2010)88) and for the Spanish Science and Education Ministry in the ARrELS project (TIN2011-23930). The BCDS group (ref. GRCT40) is part of the DURSI consolidated research group COMUNICACIONS I SISTEMES INTELLIGENTS (CSI) (ref. SGR-1202).

REFERENCES

[1] UNICEF & PROEIB Andes, *Atlas Sociolingüístico de pueblos indígenas en América Latina - Tomo I.* FUNPROEIB Andes, Bolivia, 2009.

[2] UNICEF – *Oficina Regional para America Latina y el Caribe*, [Online], Available: http://www.unicef.org/lac/lenguas_indigenas.pdf

[3] Trillos, M, "Bilingüismo Desigual en las escuelas de la sierra nevada de Santa Marta". THESAURUS. Tomo LI. Núm. 3, pp. 401-486. 1996.

[4] Hamel, R. E., Francis, N, "The Teaching of Spanish as a Second Language in an Indige-nous Bilingual Intercultural Curriculum". Language, Culture and Curriculum. Vol 19:2, pp. 171-188. 2006

[5] Moxiecode Systems AB - TinyMCE. [Online]. Available at: http://www.tinymce.com/

[6] Lier, L. "The Ecology and Semiotics of Language Learning – A Sociocultural Perspective". Springer Netherlands – Kluwer Academic Publishers, Boston. 2004.

[7] Valdiviezo, L. A., "Indigenous worldviews in intercultural education: teachers' construction of interculturalism in a bilingual Quechua–Spanish program. Intercultural Education.". Vol. 21,1, pp. 27-39. 2010.

[8] Arévalo, I., Pardo, K., Vigil, N, "Enseñanza de castellano como segunda lengua en las escuelas EBI del Peru". Dirección Nacional de Educación Bilingüe Intercultural.

[9] University of Toronto, ATutor Learning Management System, [Online], Available: http://atutor.ca/

[10] Padró, L., "Analizadores Multilingües en FreeLing", *Linguamatica*. 3 (2), pp. 13-20. 2011.

[11] Expert Advisory Group on Language Engineering Standards – EAGLES. "EAGLES Guidelines" [Online]. Available: http://www.ilc.cnr.it/EAGLES96/home.html

[12] J.A. Recio-García, B. Díaz-Agudo, M.A Gómez-Martín, and N. Wiratunga, "Extending jCOLIBRI for Textual CBR". In: Muñoz-Ávila, H., Ricci, F. (eds.) Case-Based Reasoning Research and Development. LNCS, vol. 3620, pp. 421-435. Springer, Heidelberg 2011.

[13] K. Yu, L. Ji, and X. Zhang, "Kernel Nearest-Neighbor algorithm", Neural Processing Letters 15, pp. 147-156, 2002.

[14] H. Núñez, M. Sànchez-Marrè, U. Cortés, J. Comas, M. Martínez, I. Rodríguez-Roda and M. Poch," A comparative study on the use of similarity measures in case-based reasoning to improve the classification of environmental system situations, Environmental Modelling & Software", vol. 19, 9, pp. 809-819, Sep. 2004.

[15] N. Elita, M. Gavrila, and C. Vertan, "Experiments with string similarity measures in the ebmt framework". In Proc. RANLP 2007 Conf., Bulgaria, Sep. 2007.

[16] J.J. Astrain, J.R. González de Mendívil and J.R. Garitagoitia, "Fuzzy automata with -moves compute fuzzy measures between strings, Fuzzy Sets and Systems", vol. 157, 11, pp. 1550-1559, Jun. 2006.

[17] M. Hadjieleftheriou and D. Srivastava, "Weighted set-based string similarity", IEEE Data Eng. Bull, 33(1), pp. 25-36, 2010.

[18] P. Achananuparp, X. Hu, and X. Shen, "The Evaluation of Sentence Similarity Measures", In: Song, I., Eder, J., Nguyen, T. (eds.) Data Warehousing and Knowledge Discovery 2008. LNCS, 5182, pp. 305-316. Springer, Heidelberg, 2008.

[19] D. Patterson, N. Rooney, M. Galushka, V. Dobrynin and E. Smirnova, "SOPHIA-TCBR: A knowledge discovery framework for textual case-based reasoning", Knowledge-Based Systems, vol. 21, 5, pp. 404-414, Jul. 2008

Jorge Luis Bacca Acosta is Systems Technologist and Telematics Engineer from the Universidad Distrital Francisco José de Caldas in Colombia, MSc. on Industrial Informatics from University of Girona in Spain, as well as Ph.D Student at University of Girona. He belongs to the BCDS (Broadband Communications and Distributed systems) group from the University of Girona. He is also a developer at the ALTER-NATIVA European Project. His research interest are technology enhanced learning, case-based reasoning, intelligent agents, semantic web and augmented reality.

Silvia Margarita Baldiris Navarro Systems and Industrial Engineer from Industrial University of Santander (UIS), Colombia Master in Industrial Informatics and Automatic from University of Girona and PhD in Technologies from University of Girona. The topics of User Modelling and Adaptations applied to learning management systems, Knowledge management and standardizations in elearning are her research interest. She has participated in several projects: Adaptation based on machine learning, user modelling and planning for complex user-oriented tasks (ADAPTAPlan), EIE-Surveyor: Reference Point for Electrical and Information Engineering in Europe, A2UN@: Accessibility and Adaptation for ALL in Higher Education, and in the European project ALTER-NATIVA.

Ramon Fabregat Gesa Ramón Fabregat is professor at the Computer Architecture and Technology Department at the University of Girona, Spain. He received his degree in computer engineering from the Universitat Autónoma de Barcelona and his doctorate degree in technology from the University of Girona. His research interests are adaptive hypermedia systems, user modeling and e-learning. He has participated to the technical program committees of several conferences. He is director of the master of computer science and automation and leads a research group on BCDS.

Cecilia Avila Garzón is a researcher at BCDS Group from Universitat de Girona (Spain). She graduated in Telematics Engineering from the Universidad Distrital Francisco José de Caldas (Colombia) and has a MSc in Industrial Informatics from the Universitat de Girona. She is a PhD student in the area of technology. His work addresses several topics as web accessibility, collaborative learning, learning objects, and work in communities. She has been researching integrated in the ALTER-NATIVA European Project as a software developer for the last 2 years.

21

O-ODM Framework for Object-Relational Databases

Carlos Alberto Rombaldo Jr, Solange N. Alves Souza, *Department of Computing Engineering and Digital System of University of São Paulo, São Paulo, Brazil.*
Luiz Sergio de Souza, *Faculdade de Tecnologia - Carapicuíba, São Paulo, Brazil*

Abstract —Object-Relational Databases introduce new features which allow manipulating objects in databases. At present, many DBMS offer resources to manipulate objects in database, but most application developers just map class to relations tables, failing to exploit the O-R model strength. The lack of tools that aid the database project contributes to this situation. This work presents O-ODM (Object-Object Database Mapping), a persistent framework that maps objects from OO applications to database objects. Persistent Frameworks have been used to aid developers, managing all access to DBMS. This kind of tool allows developers to persist objects without solid knowledge about DBMSs and specific languages, improving the developers' productivity, mainly when a different DBMS is used. The results of some experiments using O-ODM are shown.

Keywords —Object-Relational Databases. Persistence Framework. Java Annotations. SQL:2008

I. INTRODUCTION

Persistent frameworks, frequently called ORM (Object Relational Mapping) tool [6]-[7]-[8], have been used to aid database projects. This kind of tool maps objects from application to relation (relational databases - RD) [11]. Using ORM tools, developers have advantages; (1) they can persist data in RD without solid knowledge of Relational Database Management System (RDBMS). It allows developers to focus on application development (OO paradigm and language aspects); (2) all data access is made through the tool, since ORM tools are integrated in programming environment; developers can use a single environment to do this work; (3) generally, when more than one DBMS is used, only one instruction is modified. This instruction indicates the new DBMS; then, all the code produced by the tool to one DBMS is automatically changed to another. In case the instruction was not available, the developer would have to produce the new SQL code according to the characteristic of the DBMS chosen. All these aspects aid both: the point of view system maintenance and developers' productivity. Thus, the benefit of using persistent frameworks cannot be ignored. On the other hand, it is necessary to consider the new characteristics of Object-Relation Databases (ORDB).
ORDB allows manipulating objects in databases. Many DBMS offer new resources such as UDTs (User Data Types), composite types, REF types, inheritance and others that can be used to model objects in databases. Besides, using REF types

to represent relationship between objects can result in improvement of performance given that no field needs to be created in an existing or new relation. This characteristic could be more suitable for new applications that have emerged and which present complex objects such as CAD/CAM (Computer Aided Design/ Computer Aided Manufacturing), GIS (Geographic Information System), Genetic, etc [9]. Adding to this, using ORDB, objects from an application must be mapped to objects from databases; thus the impedance mismatch, which has been reported in the literature and in real applications as a problem, can be avoided. Another ORDB advantage is the possibility to use only one conceptual model for both the application and the data tiers [1]. Generally, the entity-relationship model (ERM) and UML class model are built when the relational model is employed. This causes an overhead not only related to mapping class to relation, but also to elaborating the ERM and the need of specific knowledge to generate this model.

Since the strength of the Object-Relational Model might be more explored [4] together with the lack of tools to aid projects and maintenance of ORDB, this paper proposes an O-ODBM (Object- Object Database Mapping) tool, an object-relational persistence framework. O-ODBM maps an object from the application to the ORDB object [16].

However, not all DBMS implement all the resources of objects specified in the SQL standard. Therefore, some elements can be unavailable in some of them. Undoubtedly, this is another important aspect which contributes to ignoring object resources from DBMS and adds complexity to build CASE and Persistent framework tools to ORDB.

An example was used to evaluate the O-ODBM. We here present not only the O-ODBM characteristics, but also an example and the results.

To develop the O-ODBM, characteristics and operations were studied which are defined or implemented in JPA (Java Persistence API) and/or JDO (Java Data Object) standards and in Hibernate and Torque frameworks. Some of those characteristics, which provide benefits and /or facilities to developers, were implemented in our first version of O-ODBM Framework.

This article is organized as follows. In chapter 2, some characteristics of JPA (Java Persistence API) and JDO (Java Data Object), which were incorporated to O-ODBM, are introduced. Chapter 3 introduces the O-ODBM. Chapter 4 presents the example employed to evaluate the O-ODBM tool, and the results. Finally, chapter 5 concludes and presents future works.

II. JPA AND JDO STANDARDS – SOME CHARACTERISTICS

The O-ODBM *Framework* was developed in Java programming language. Some reasons pointed for this choice are (1) many ORM *Frameworks* available are based on Java language. (2) Java language facilitates the interoperability and (3) the number of the OO applications that developed in Java are increasing.

The JDO (*Java Data Object*) [7] e JPA (*Java Persistence API*) [8] standards define mapping from application object to relations of RDB. These standards also include a set of properties that simplify persistence and data access. Some of these properties were highlighted considering the scope of the O-ODBM project:

- all access to data is made only by the framework. As a result, it is no longer necessary to have a solid knowledge about the DB, SQL and DBMS used.

- offers a language for manipulating data that is closer to OO programming language than SQL.

- transaction manage, which allows the developer to define the beginning and end of transactions. The Framework is responsible for the interface with the DBMS used.

- mechanism for performance control to access, insert, delete and update objects. In OO applications, references between objects are very common. These references are mapped to tables and integrity rules, so that when a query is made, more than a table could be accessed. The use of annotations [12] is employed by the developer to indicate which objects must be persisted. Annotations allow adding information to java classes directly. The Framework uses this information to create the SQL code to generate tables, attributes, integrity rules in attributes and between tables, etc.

III. PROJECT OF O-ODBM FRAMEWORK

The rules of mapping defined for RDB are not suitable, since the new data types connected to the OO paradigm available in ORDBMS are not considered. The rules defined for the Framework proposed are summarized in Tables I and II. More details of these rules can be found in [1]-[2], which are a complementation of [4]-[9]-[14] from the point of view of real applications.

Requirements of O-ODBM Framework

A set of requirements, which are detailed as follows, was defined to guide the development of the Framework. In doing so, the characteristics of ORM Frameworks were considered, which are advantages for both application and developers. Then, JPA and JDO standards were studied, as well as Hibernate and torque implementations [6]-[7]-[8]. In view of the ORDBMS, SQL:2008 was also studied, along with Oracle 11g release 2 and BD2 9.7.5 version DBMS. To simplify the reference, the requirements were identified by the R letter and a sequential number, presented as follows.

R1 – to control the referential integrity rule connected to

TABLE I
MAPPING OF OBJECT FROM APPLICATIONS TO ORDBMS OBJECTS - ADAPTED FROM [1]

OO	ORDBMS	Justify
Class	Table UDT Typed table	Classes may be mapped to conventional tables. However, if the intention is to define methods and/or hierarchies, an UDT must be defined and, to store data, a typed table connected to UDT needs be created.
Abstract class	UDT	an UDT should be created whithout a typed table conected to it to represent an abstract class. In this case, the UDT would be used for defining other UDTs and as it does not have a typed table connected to it, instances will not be persisted.
Simple attribute	Build-in type	SQL:2008 presents many built-in types such as integer, real, etc. It is hence possible to find a corresponding type in SQL for each primitive type of Java.
multivalued attribute	Array or Multiset	multidimensional structures are suitable to store attributes of the same type (collections).
Methods	UDT methods	It is possible to define methods connected to UDTs. Thus, developers can choose to define methods in the database or in the application.

REF type. ORDB allows defining the relationship between objects using REF. However, if an object A, which is

TABLE II
MAPPING OF ASSOCIATIONS AND HIERARCHY IN ORDBMS – ADAPTED FROM [1]

Association		Corresponding in ORDBMS	
Bidiretional Association	Composition/ Aggregation/ Association	1..1	a cross reference is defined, i.e., each class maintains a reference (REF) to the other.
		1..*	a cross reference is also used, although the aggregated class will be an Array ou a Multiset of references.
Unidirectional Association		Similarly to the bidirectional associations above presented, though the reference will be only in table.	
Nth Association (three or more classes)		A table or a UDT is defined with the name of the association. The table or the UDT (and the typed table) must maintain references to the classes involved.	
Associative Class		a table or a UDT can be defined for the association class similarly to nth association.	
Generalization/ Specialization		a UDT is defined for each class of the hierarchy. Typed table would be defined later if data need to be persisted.	

referenced by object B, is removed, B gets a null reference.

Then, a rule, similar to the rule that controls foreign key in RDB, needs to be implemented to avoid a null reference.

R2 - Flexibility for multiple platforms of databases. This requirement means that the Framework gives a simple mechanism for a developer to change the DBMS and all SQL code for persistence and data access, which was generated by the framework for the first DBMS, will automatically be replaced by the code for the new DBMS. It is important to highlight, as explained before, there are differences among ORDBMS and some resources for database object can be not available; therefore, this may be the most difficult requirement to be achieved.

R3 – The developer does not need to know the SQL and DBMS employed. Thus, the framework has to present a language or a mechanism for object manipulation very similar to the OO programming language (if compared with the SQL). As a result, the learning process is facilitated, since the developer does not need to know SQL to use a DBMS.

R4 – Managing DBMS connections – including to open, to close and to verify the timeout of connections. If there are unfinished transactions, the Framework will keep the connection open until the commit or rollback of these transactions. The Framework would force itself to interrupt the transactions, despite keeping (ex. doing rollback) the data integrity in the database.

R5 – Managing the execution of transactions. For this, the Framework has to offer an interface for the developer to define his transactions.

R6 – Automatic code generation for object schema in DBMS, including codes for manipulating these objects.

R7 – the framework will be an access point to database; making the direct connection between application and database unnecessary.

R8 – use of annotations for defining which will be persisted in the database, facilitating the configuration of objects schema. The ORM Frameworks studied employs a similar mechanism; however, in the case of O-ODBM Framework, appropriated annotations have to be created.

R9 – Implementation of inheritance in database, according to OO.

R10 – Implementation of unidirectional, bidirectional and multivalued relationship, using reference (REF) to object when possible.

R11 – Application performance is not degraded.

R12 – Data could be retrieved on demand. In other words, according to what is defined by the developer, the Framework will postpone or will not retrieve related data to improve the performance of the data access [6]-[8]. This is an important aspect for performance because one object referenced by another can keep references for others and so on, which would certainly degrade the data access performance. Therefore, when there is no interest in referenced objects, the retrieval of object and its references would cause unnecessary performance degradation.

R13 – Data could be persisted on demand, which is defined as cascade property in JPA [8]. In this case, the Framework would do the persistence of the associated objects, preventing null references from being found, i.e., references for objects that do not exist

A. Architecture of O-ODBM Framework O-ODBM

The tool accepts input in two different formats: Java code, in which annotations are used to declare persistent classes, or XML files, which correspond to SQL code for the DBMS chosen. In the first case, the Framework presents a set of annotations, similarly to the ORM Frameworks. In the second case, the XML file, which represents the logical schema to ORDB, is generated by a case tool [1] for ORDB. The Framework should be part of the development integrated environment, in which from a conceptual model (ex. UML class model), or from a logical schema, the OR database can be automatically implemented in the DBMS chosen and accessed by the Framework.

A XSD (XML Schema Definition) was formalized to register the mapping from Java classes to ORDB objects. In this XSD, according to SQL:2008 [10] the ORDB data types are defined that define database objects, methods, inheritance, collections and other OO concepts. Therefore, in case the input of the Framework is a XML file produced by the modeling tool, the XSD would be used to verify it. In addition, XML documents are also used internally by the tool for describing the necessary information to mapping among different formats produced by the tool

Figure 1 introduces the architecture of the O-ODBM Framework and its components are described as follows.

Configuration Processor: reads the Java class annotated with the annotations introduced by the Framework. Once the Java classes have been interpreted, this module processes the annotations and generates the XML code with OR structure based on SQL:2008. It was decided to first generate the SQL code for SQL:2008 and then translate it to a dialect of specific DBMS. This decision was made due to the differences among DBMS regarding the object resources offered. Some DBMS implement part of these resources only; moreover, the implementation of the specific element can be different among these DBMS. On the other hand, the SQL:2008 not only has all the elements related to objects, but can also be easily

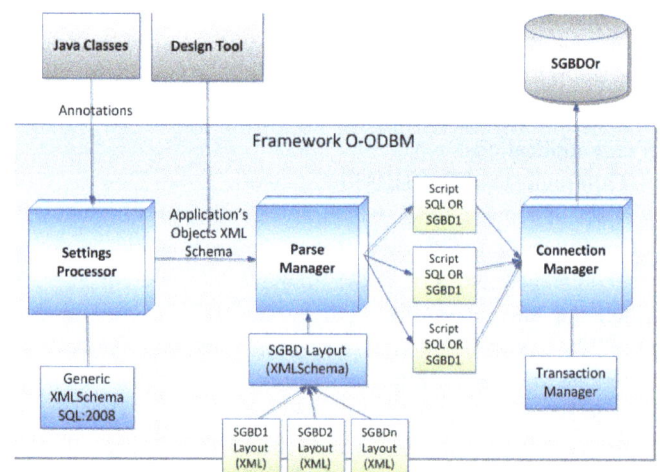

Fig. 1. Architecture of O-ODBM Framework.

translated into another SQL dialect. The XML-SQL schema that represents the database object schema is equivalent to the application object schema. The XML-SQL schema generated by the Configuration Processor is the input of the Conversion Manager Component.

Conversion Manager: generates the SQL scripts to be executed by the DBMS chosen. The Conversion Manager uses the DBMS layout file appropriated for translating the XML-SQL code into an adequate SQL dialect. For this, the Framework uses the XML file (DBMS Layout) that has the specific syntax for each DBMS. The output of this module is the SQL script, which is submitted to the DBMS by the Connection Manager Component.

Connection Manager: all the operations between the Framework and DBMS, for example, execution of SQL script to create structures, persistence and retrieval of objects are made by a connection. This component manages the connections with DBMS and this is transparent for the developer. Connections are automatically opened by the Framework whenever the operation is submitted.

Transaction Controller: manages all the transactions with the DBMS. When a transaction is opened, this component is activated and when the connection needs be closed, this component is consulted to verify/guarantee that there are no transactions open for that connection. In this process, a transaction can be finished (rollback or commit), or the connection is not closed. This component also manages the transaction inactive time and automatically finishes it if the transaction achieves the timeout.

DBMS Layout: Since a XML file, produced by the CASE tool, could be the input for the Framework; a XSD is also used by the Framework, similarly to the SQL schema, for validating this file.

B. Annotations

The API (*Application Programming Interface*) of the Framework is integrated with the programming environment. This way, the developer has the set of annotations, which were produced in this work, available for use and integrated with the development environment. The type of annotation will determine the map from Java class to ORDB element made by the tool. TABLE III introduces the set of annotations. TABLE IV and TABLE V show more annotations that are used for defining parameters and default values, respectively. Experienced developers in Framework and/or in ORDB could redefine default values.

IV. EXAMPLE USED FOR TESTING THE FRAMEWORK

An example, the persistent object schema of which is shown in Figure 2, was used for testing the applicability of the Framework. The main concern was to evaluate the behavior for queries involving objects in hierarchy and the use of reference (REF) for representing association between objects. However, this evaluation is not enough to draw conclusions about the performance of ORDBMS. Therefore, a more careful evaluation must be made in the future.

TABLE III
ANNOTATIONS.

Annotation	Description
@DbObject	indicates the class must be persisted.
@DbField	indicates the attribute must be persisted.
@DbMethod	indicates the object method must be created in DBMS.
@DbInhetitance	indicates the object is part of the hierarchy. Then, the hierarchy must be represented in DBMS. If the parent object has not been annotated with DbObject, only the derived objects would be part of a hierarchy in DBMS, although the characteristics inherited will be part of the derivated objects.
@DbRelation	indicates the attribute represents the association. The associations are represented by the inclusion of the attributes in associated classes. These attributes make references between themselves and, depending on the cardinality of association, this reference may be to an object or to a collection of objects..

In the example, only the annotations shown in Table III

TABLE IV
CONFIGURATIONS FOR @DBFIELD ANNOTATION.S.

PARAMETER	Default value	Description
size	255 for text and numbers.	defines the attribute max size.
isPK	none	indicates the attribute will be a primary key.
autoIncremet	none	indicates the attribute values will be generated by the DBMS.
type	keeps the equivalent data type in the DBMS.	defines the data types that will be used in DBMS

TABLE V
CONFIGURATIONS FOR @DBFIELD ANNOTATION.S.

PARAMETER	Default value	Description
size	255 for text and numbers.	defines the attribute max size.
isPK	none	indicates the attribute will be a primary key.
autoIncremet	none	indicates the attribute values will be generated by the DBMS.
type	keeps the equivalent data type in the DBMS.	defines the data types that will be used in DBMS

were used.

Since the class was annotated, a DAO class for each persistent class was generated. Then, using the O-ODBM Framework, the SQL script of the database schema was generated and executed in DBMS. After that, insert, update, delete and select operations were carried out. First, DB2 DBMS was used, and later Oracle DBMS. It is important to highlight that all these procedures were made by changing the directives of configurations only, i. e., neither class nor annotations were changed. TABLE VI introduces the results of each operation

V. EVALUATION

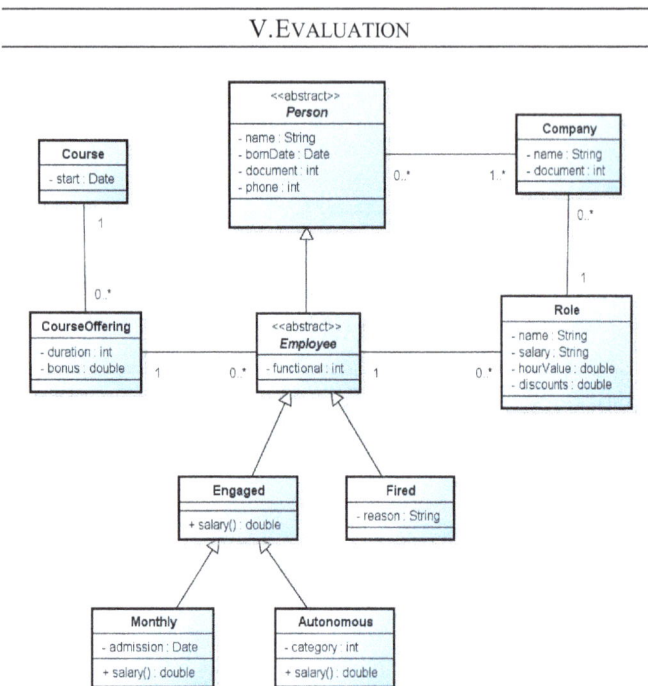

Fig. 2. Class Diagram used in the example to evaluate the Framework.

The set of requirements, defined in section III, and the result of the tests were used for evaluating the o-ODBM Framework.

A. Compliance with the requirements

R1 – the referential integrity rule will be implemented by the Framework if the Java code presenting the appropriate annotations (i.e., the attribute of the object has @DbRelation annotation). Therefore, the Framework will implement an operation to guarantee that null references do not exist.

R2 – the Framework only supports Oracle and DB2, although the modification of one into another is very simple for the developer, since he only declares what DBMS will used and the Framework generates the appropriate code. There are few   DBMS that support OR characteristics and this limits the application of this requirement. However, it can be considered met with the use of these two DBMS.

R3 – the Framework has no data access language. However, the annotations can be used for persistence, queries and updating objects.

R4 – the Framework manages all the connections with the

TABLE VI
TIME OF OPERATION.

	JDBC	O-ODBM
creation of schema	-----	2689 ms.
initialization	512 ms	734 ms
insert	129 ms	141 ms
update	198 ms	216 ms
Select	155 ms	173 ms

DBMS.

R5 – the Framework presents an interface that allows the developer to define the transaction beginning and end. In fact, the control of the transaction is made by the JDBC, which passes this control on to DBMS.

R6 – using the annotated class, the Framework generates the code to interact with the DBMS.

R7 – The Framework is a centralized data access point.

R8 – as stated before, a set of annotations is available and the developer can use it to indicate which must be persisted.

R9 – the Framework generates the code with the structures to represent inheritance as long as the correct annotation has been used.   Then, UDTs hierarchy and typed tables are created in the database.

R10 – since there is the indication of the cardinality of association between the objects, the Framework, by default, creates a list of references in both objects for N:N cardinality. For 1:N, the reference can be to (1) only one object, (2) a list of objects, (3) the reference can be on both sides, in this case, on one side the reference is for an object and on other one, for a list of objects. This is similar to the OO application.

R11 – to evaluate if there is or not performance degradation, the decision was to compare the time spent for database access with and without using the Framework. For this, the OR schema was generated manually, using a JDBC. It was verified that the use of the Framework does not cause performance degradation.

R12 – the capacity of retrieving data on demand (lazy and eager strategy in JPA [6]-[8]) is implemented by the Framework. It allows having fewer unnecessary accesses to DBMS.

R13- cascade strategy (JPA) [6]-[8] is implemented by the Framework.

B. Analysis of Results

Three measures were used for assessing the results that are:

Productivity: here, the productivity is the amount of code the user needs to create to interact with the Framework, as compared with the amount that he has to generate without the Framework. It is worth highlighting that the code generated by the Framework will present a lower number of errors than the code generated by the developer. Another important issue is related to the necessary time for learning to use the Framework. This time will be less than that spent to learn about SQL and ORDB.

Support to OR characteristics: it is the capacity of generating code with structures that allow implementing OO characteristics in DBMS such as object, inheritance, aggregation, composition, references, multivalued structures using the elements available in ORDB [14].

Performance: here, performance is the response time to execute the specific operation in ORDB with and without the use of Framework.

The use of annotations aims to increase productivity, since the use of the Framework is simpler and more intuitive from the developer's point of view. Learning was also considered facilitated by the use of annotations, since the set of annotations are integrated to the programming environment, which the developer interacts with more naturally, similarly to other Frameworks, such as Hibernate.

Another important issue, the use of annotations eliminates the need of more detailed knowledge about the local of persistence and objects there defined. In other words, it is transparent for the developer if UDTs, typed tables, REF types, etc were created in DBMS. This directly affects the developer's productivity, since there are less concepts he/she needs know.

Similarly to other ORM tools, such as Hibernate, an interface was available to allow developers to define transactions.

Without using Framework, it was necessary to generate all the database schema manually in each DBMS and JDBC was employed to make the connection and to access each database. It is not possible, therefore, to compare the performance for database schema generation between these two approaches (with and without the use of Framework). Conversely, the performance considering these two approaches for the insert, update and select operations were really closed, without significant differences. Concerning performance, i.e., response time in data access, few tests were performed with a simple example and with a small number of data. Then, specific work must be done for a real performance evaluation. In direct access (JDBC), the developer needs detailed knowledge about the ORDB, DBMS used and available data types, besides the access language.

As to OR characteristics, the O-ODBM Framework did the mapping using resources of DBMS objects and inheritance, aggregation, composition, references and multivalued structure were employed in this process, i.e., UDTs, REFs, ROWs, MULTSETs and ARRAYs were used. Although there are differences among Oracle, DB2 and SQL:2008, the Framework generated appropriate code to map and to access all of them.

VI. COMMENTS AND CONCLUSIONS

As the ORM Frameworks do not use the new available data type for ORDBM, this article introduced a proposal for a new Framework for ORDBMS, called O-ODBM. As the others, O-ODBM provides a transparent persistence mechanism. The

advantage of the O-ODBM is the use of ORDB, so that the strength of object-relational model is not ignored [4] and its suitability for new applications can be more explored. For example, for scientific applications, it is necessary to deal with a large number of data, which can be related or gotten in groups to obtain information of interest. In this case, the use of RDB could achieve the high level of redundancy of data due to the kind of associations that will be necessary. Besides, to obtain statistic information, not only the existent functions (ex. average, some, etc.) could be necessary. The use of elements, such as UDTs from ORDBMS, allows new solutions to be more easily employed [15].

According to the evaluation made in this work, the O-ODBM was efficient. The advantages are: new concepts are not necessary to use it; the performance remains near the direct access (without Framework); automated generation of code for the persistence of objects; SQL and DBMS do not need be known by the developer; persistence mechanism is transparent for the developer.

Finally, the O-ODBM Framework is still a prototype and for the tool to be effectively used, functionalities need to be implemented or improved. However, the prototype was effective to demonstrate the viability of the proposal.

REFERENCES

[1] T. R. Castro, Projeto Lógico para BDOR de acordo com SQL:2003, proposta de uma ferramenta CASE. 2011. Dissertação Mestrado – Engineering School, University of São Paulo, São Paulo, 2011. Avaiable in <www.teses.usp.br/teses/disponiveis/3/3141/tde-01062011-131450/>. Acessado em janeiro de 2012.

[2] T. R. Castro, S. N.A. Souza, L. S. DeSouza, "Case tool for Object-Relational Database Design", Proceedings of CISTI 2012 – 7ª Conferencia Ibérica de Sistemas y Tecnologías de Información, Madrid, España, p 181-186, 2012. Avaiable in: http://aisti.eu/cisti2012.

[3] G. Feuerlicht ,J. Pokorný, K. Richta, "Object-Relational Database Design: Can Your Application Benefit from SQL:2003?" Galway, Ireland: Springer , p 1-13, 2009.

[4] M. Fotache, C. Strîmbei, "Object-Relational Databases: An Area with Some Theoretical Promises and Few Practical Achievements", Communications of the IBIMA, v. 9, ISSN 1943-7765, 2009. Avaiable in: <www.ibimapublishing.com/journals/CIBIMA/volume9/v9n7.pdf>. Acessado em Janeiro de 2012.

[5] M. F. Golobisky, A. Vecchietti, "Mapping UML Class Diagrams into Object-Relational Schemas". Rosario, Argentina: Proceedings of ASSE, p 65-79, 2005.

[6] D. King, C. Bauer,M. Rydahl, E. Bernard, S. Ebersole, "HIBERNATE - Relational Persistence for Idiomatic Java" Capitulo 14 e 19, 2009, Avaiable in: <docs.jboss.org/Hibernate/core/3.3/reference/en/html>. Acessoado em feveiro de 2012.

[7] Java Data Objects (JDO). Avaiable in: <www.oracle.com/technetwork/java/index-jsp-135919.html>.

[8] The Java Persistence API - A Simpler Programming Model for Entity Persistence. Avaiable in: <www.oracle.com/technetwork/articles/javaee/jpa-137156.html>.

[9] E. Marcos, B. Vela, J. M. Cavero, "A Methodological Approach for Object-Relational Database Design using UML", Heidelberg : Springer Berlin, p 59-72, 2003.

[10] Database languages SQL:, ISO-ANSI WD 9075, ISO, Working Group WG3, 2008.

[11] A. Silberschatz, H. Korth, S. Sudarshan, Sistemas de Banco de Dados; Campus, 1a edição. 2006.

[12] Annotations. 2003. Avaiable in: <download.oracle.com/javase/tutorial/java/javaOO/annotations.html>.

[13] J. M. Vara, B. Vela, J. M. Cavero, E. Marcos, "Model Transformation for Object-Relational Database Development". Proceedings of the ACM symposium on Applied computing, pp. 1012 – 1019. 2007.

[14] C. Calero, F. Ruiz, A. Baroni, F. Brito, M. Piattini, "An Ontological Approach to Describe the SQL:2003 Object-Relational Features." *Journal of Computer Standards & Interfaces - Elsevier*. 2005

[15] E. PARDEDE, J. W. RAHAYU, D. TANIAR, New SQL Stardand for Object-Relational Database Applications. Conference Proceedings on Standardization and Innovation in Information Technology - SIIT2003. PP. 191-198. 2003.

[16] C. A. Rombaldo Jr,,S. N. A. Souza, L. S. De Souza, "Framework de Persistência em Banco de Dados Objeto-Relacional", Proceedings of CISTI 2012 – 7ª Conferencia Ibérica de Sistemas y Tecnologias de Información, Madrid, España, p 341-347, 2012. Avaiable in http://aisti.eu/cisti2012.

22

Performance comparison of hierarchical checkpoint protocols grid computing

Ndeye Massata NDIAYE[1,2], Pierre SENS[1], Ousmane THIARE[2]

(1) Regal team, LIP6, UPMC Paris Jussieu France

(2) Gaston Berger University of Saint-Louis Senegal

Abstract — **Grid infrastructure is a large set of nodes geographically distributed and connected by a communication. In this context, fault tolerance is a necessity imposed by the distribution that poses a number of problems related to the heterogeneity of hardware, operating systems, networks, middleware, applications, the dynamic resource, the scalability, the lack of common memory, the lack of a common clock, the asynchronous communication between processes. To improve the robustness of supercomputing applications in the presence of failures, many techniques have been developed to provide resistance to these faults of the system. Fault tolerance is intended to allow the system to provide service as specified in spite of occurrences of faults. It appears as an indispensable element in distributed systems. To meet this need, several techniques have been proposed in the literature. We will study the protocols based on rollback recovery. These protocols are classified into two categories: coordinated checkpointing and rollback protocols and log-based independent checkpointing protocols or message logging protocols. However, the performance of a protocol depends on the characteristics of the system, network and applications running. Faced with the constraints of large-scale environments, many of algorithms of the literature showed inadequate. Given an application environment and a system, it is not easy to identify the recovery protocol that is most appropriate for a cluster or hierarchical environment, like grid computing. While some protocols have been used successfully in small scale, they are not suitable for use in large scale. Hence there is a need to implement these protocols in a hierarchical fashion to compare their performance in grid computing. In this paper, we propose hierarchical version of four well-known protocols. We have implemented and compare the performance of these protocols in clusters and grid computing using the Omnet++ simulator.**

Keywords — **Grid computing, fault tolerance, checkpointing, message-logging**

I. INTRODUCTION

Molecular biology, astrophysics, high energy physics, those are only a few examples among the numerous research fields that have needs for tremendous computing power, in order to execute simulations, or analyze data. Increasing the computing power of the machines to deal with this endlessly increasing needs has its limits. The natural evolution was to divide the work among several processing units. Parallelism was first introduced with monolithic parallel machines, but the arrival of high-speed networks, and especially Wide Area Network (WAN) made possible the concept of clusters of machines, which were further extended to large scale distributed platforms, leading to a new field in computer science, grid computing.

The first definition of a grid has been given by Foster and Kesselman in [40]. A grid is a distributed platform which is the aggregation of heterogeneous resources. They do an analogy with the electrical power grid. The computing power provided by a grid should be transparently made available from everywhere, and for everyone. The ultimate purpose is to provide to scientific communities, governments and industries an unlimited computing power, in a transparent manner. This raised lots of research challenges, due to the complexity of the infrastructure. Heterogeneity is present at all levels, from the hardware (computing power, available memory, interconnection network), to the software (operating system, available libraries and software), via the administration policies.

From this definition, several kinds of architectures were born. One of the most commonly used architecture, referred to as remote cluster computing, is composed of the aggregation of many networked loosely coupled computers, usually those computers are grouped into clusters of homogeneous and well connected machines. These infrastructures are often dedicated to scientific or industrial needs, and thus provide large amount of computing resources, and a quite good stability.

Today, grid computing technologies make it possible to securely share data and programs for multiple computers, whether desktop or personal supercomputers. These resources are networked and shared through software solutions. In recent years, grid technology has emerged as an important tool for solving compute-intensive problems within the scientific community and in industry. To further the development and adoption of this technology, researchers and practitioners from different disciplines have collaborated to produce standard specifications for creating large-scale, interoperable grid system. The focus of this activity has been the Open Grid Forum (OGF) [8], but other standard development organizations have also produced specifications, such as

[9][10], that are used in grid systems. To fully transition grid technology to operational use and to expand the range and scale of grid applications, grid systems must exhibit high reliability; i.e. they must be able to continuously provide correct service [11]. Moreover, it is important that the specifications used to build these systems fully support reliable grid services. With the increase in use of grid technology, achieving these goals will be made more difficult as grid systems become larger, more heterogeneous in composition, and more dynamic. Many grids are appearing in the sciences, production grids are now being implemented in companies and among agencies: Grid'5000, TeraGrid, Sun Grid, Xgrid ... Grid computing will allow dynamic sharing of resources among participants, organizations and businesses in order to be able to pool, and thus run compute-intensive applications or treatment of very large volumes of data.

Since the failure probability increases with a rising number of components, fault tolerance is an essential characteristic of massively parallel systems. Such systems must provide redundancy and mechanisms to detect and localize errors as well as to reconfigure the system and to recover from error states. A fault tolerant approach may therefore be useful in order to potentially prevent a faulty node affecting the overall performance of the application. Fault tolerance appears as an indispensable element in grid computing. Many protocols for distributed computing have been designed [1]. These protocols are classified into four different classes, namely, coordinated checkpointing, communication induced checkpointing, independent checkpointing and log-based protocols.

We have implemented and compare the performance of these protocols in clusters and grid computing using the Omnet++ simulator [7].

Section II describes the protocols implemented in Omnet++. In section III, we talk about hierarchical checkpointing for grids. The experimental setup and results obtained by executing these protocols are presented in Section IV. In section V, we present the related work and finally section VI concludes.

II. CHECKPOINT AND ROLLBACK-RECOVERY PROTOCOLS

Checkpointing is a standard method for the repair of faults in systems. The idea is to save the state of the system on a stable periodic to prevent breakdowns (Fig. 1). That way when you restart after a power failure, the state saved newest restored and execution resumes its course before the crash. The overall status of a distributed system is defined by the union of local states of all processes belonging to the system.

Taking checkpoints is the process of periodically saving the state of a running process to durable storage. Checkpointing allows a process that fails to be restarted from the point its state was last saved, or its *checkpoint*. If the host processor has not failed, temporal redundancy can be used to *roll back* and restart the process on the same platform. As in other systems, this method is widely used in grids [36][37][38]. Otherwise, if the host has failed, the process may be *migrated*, or

transferred, to a different execution environment where it can be restarted from a checkpoint (a technique also referred to as *failover*). This section begins by discussing checkpoint and process migration methods used in commercial and science grid systems that are based on methods used in high-performance cluster computing. This is followed by discussion of new methods being developed or adapted for scaled grid environments, together with related issues that need to be resolved. Most notable is the issue of finding efficient methods for checkpointing many concurrent, intercommunicating processes, so that in the event of failure, they can resume from a common saved state [39]. Checkpointing can be initiated either from within grid systems or within applications.

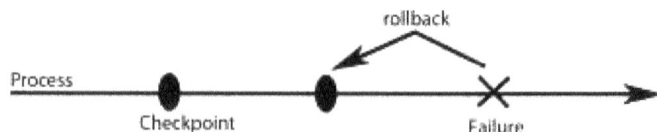

Fig 1: Rollback-Recovery

There are two main classes of protocols: coordinated checkpointing and message logging.

A. Coordinated checkpointing

Coordinated checkpointing is an attractive approach for transparently adding fault tolerance to distributed applications without requiring additional programmer ef- forts. In this approach, the state of each process in the sys- tem is periodically saved on stable storage, which is called a checkpoint of the process. To recover from a failure, the system restarts its execution from a previous error-free, consistent global state recorded by the checkpoints of all processes. More specifically, the failed processes are re-started on any available machine and their address spaces are restored from their latest checkpoints on stable storage. Other processes may have to rollback to their checkpoints on stable storage in order to restore the entire system to a consistent state. Coordinated checkpointing simplifies failure recovery and eliminates domino effects in case of failures by preserving a consistent global checkpoint on stable storage. However, the approach suffers from high overhead associated with the checkpointing process. Two approaches are used to reduce the overhead: First is to minimize the number of synchronization messages and the number of checkpoints, the other is to make the checkpointing process nonblocking.

The protocol requires processes coordinate their checkpoints to form a consistent global state. A global state is consistent if it does not include any orphan messages (i.e, a message received but not already sent). This approach simplifies the recovery and avoids the domino effect, since every process always restarts at the resume point later. Also, the protocol requires each process to maintain only one permanent checkpoint in stable storage, reducing the overhead due to storage and release of checkpoints (garbage collection) [1].

Its main drawback however is the large latency that require interaction with the outside world, in this case the solution is to perform a checkpoint after every input / output. To improve the performance of the backup coordinated, several techniques have been proposed. We have implemented as non-blocking coordinated checkpointing.

1) Non-blocking coordinated checkpointing

A nonblocking checkpointing algorithm does not require any process to suspend its underlying computation. When processes do not suspend their computations, it is possible for a process to receive a computation message from an other process which is already running in a new checkpoint interval. If this situation is not properly dealt with, it may result in an inconsistency. For example, in Fig. 2, P2 initiates a checkpointing process. The example of coordinated checkpoint non-blocking is that of Chandy and Lamport algorithm [2]. This algorithm uses markers to coordinate the backup, and operates under the assumption of FIFO channels. In [3], a comparison of protocols for coordinated checkpoint blocking and non-blocking has been made. Experiments have shown that the synchronization between nodes induced by the protocol blocking further penalize the performance of the calculation with a non-blocking protocol. However, using frequencies of taken checkpoints usual performance of the blocking approach is better on a cluster to high-performance communications.

2) Communication induced checkpointing

This protocol defines two types of checkpoints [1]: local checkpoints taken by processes independently, to avoid the synchronization of coordinated backup and forced checkpoints based on messages sent and received and dependency information carried 'piggyback' on these posts, so to avoid the domino effect of uncoordinated backup, ensuring the advancement of online collection. Unlike coordinated checkpoint protocols, the additional cost due to the medium access protocol disappears because the protocol does not require any message exchange to force a checkpoint: this information is inserted piggyback on the messages exchanged.

B. Message-Logging protocols

Message logging (for example [12] [13] [14] [15] [16] [17] [18] [19] [20]) is a common technique used to build systems that can tolerate process crash failure. These protocols required that each process periodically record its local state and log the messages it received after having recorded that state. When a process crashes, a new process is created in its place: the new process is given the appropriate recorded local state, and then it is sent the logged messages in the order they were originally received. Thus, message logging protocols implement an abstraction of a resilient process in which the crash of a process is translated into intermittent unavailability of that process.

All message logging protocols require that the state of a recovered process be consistent with the states of the other processes. This consistency requirement is usually expressed in terms of *orphan processes*, which are surviving processes whose states are inconsistent with the recovered state of crashed process. Thus, in the terminology of message logging, message logging protocols must guarantee that there are no orphan processes, either through careful logging of through a somewhat complex recovery protocol.

The logging mechanism uses the fact that a process can be modeled as a sequence of deterministic state intervals, each event begins with a non-deterministic. An event may be receiving a message, or issued or other event in the process. It is deterministic if from a given initial state, it always happens at the same final state. [1]

The principle of Logging is to record on a reliable storage any occurrences of non-deterministic events to be able to replay them in recovering from a failure. During execution, each process performs periodic backups of their states, and recorded in a log information about messages exchanged between processes. There are three message-logging categories: optimistic, pessimistic and causal.

1) Pessimistic message-logging

This protocol was designed under the assumption that a failure may occur after any nondeterministic event (i.e. message reception). Then, each message is saved on a stable storage before to be delivering to the application.

These protocols are often made reference to the synchronized because when logging process logs an event of non-deterministic stable memory, it waits for an acknowledgment to continue its execution.

In a pessimistic logging system, the status of each process can be recovered independently. This property has four advantages:

- Process can send messages to the outside without using a special protocol

- The process restarted at the most recent checkpoint.

- Recovery is simple because the effects of a failure are limited only on the fail process

- The garbage collector is simple

The main drawback is the high latency of communications, which results in degradation of the applications response time. Several approaches have been developed to minimize synchronizations:

- The use of semiconductor memories such as non-volatile stable support

- The sender based message logging (SBML) [14] which preserves the determinant or the message in the volatile memory of the transmitter, instead of a remote memory

2) Optimistic message-logging

This protocol uses the assumption that the logging of a message on reliable support will be complete before a failure occurs. Indeed, during the process execution, the determinants of messages are stored in volatile memory, before being saved periodically on stable support. The storage stable memory is asynchronous: the protocol does not require the application to be blocked during the backup memory stable. Induced latency is then very low.

However, a failure may occur before the messages are saved on stable storage. In this case, the information stored in volatile memory of the process down is lost and the messages sent by this process are orphaned. This can produce a domino effect of rollbacks, which increases the recovery time.

3) Causal message-logging

This protocol combines the advantages of both previous methods. As optimistic logging, it avoids the synchronized access to stable, except during the input / output. As pessimistic logging, it allows the process to make interactions with the outside world independently, and does not create process orphan. Causal logging protocols piggyback determinants of messages previously received on outgoing messages so that they are stored by their receivers.

III. HIERARCHICAL CHECKPOINTING FOR GRIDS

The architecture of a grid can be defined as a set of clusters connected by a WAN-type network. The cluster consists of multiple nodes connected by a broadband network. We adopt a hierarchical scheme. In each cluster, there is one leader connected to all other nodes of its cluster. All leaders are connected together (Fig. 2).

The leader assumes the role of intermediary in the inter-cluster communications. The backup takes place in four phases:

1) *Initialization*: an initiator sends a checkpoint-request to its leader

2) *Coordination of leaders*: the leader transfers the checkpoint request to the other leaders

3) *Local checkpointing* : Each leader initiates a checkpoint in its cluster

4) *Termination*: When local checkpoint is over, each leader sends an acknowledgement to the initial leader.

The recovery follows the same rules as the backup: coordination phase of the leaders, and a phase of recovery limited to the cluster.

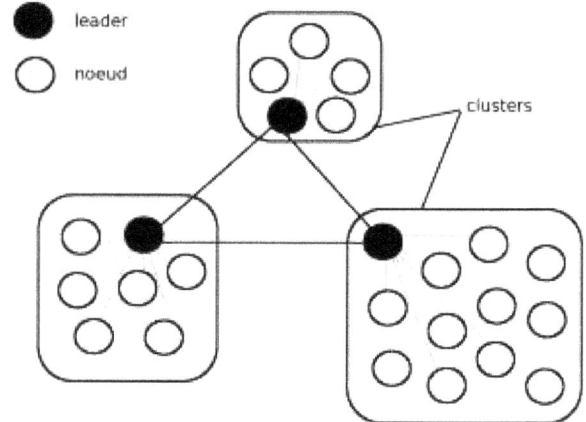

Fig 2: Hierarchical checkpointing for grids

IV. PERFORMANCE EVALUATION

In the most previous studies, fault tolerance algorithms were tested in flat architectures, namely in a cluster. The aim of our study is to determine which algorithm best suits the architectural grid. To this aim, we implement the seven checkpoint algorithms described in Section 2: the 3 main messages logging protocols (represented as "ML" in the figures), Chandy-Lamport, Communication induced protocol (CIC in figures), and blocking coordinated checkpointing.We compare the performance of these algorithms in cluster and grid environments. We use the Omnet++ simulator [7]. The cluster is configured with 25 nodes. For the grid configuration, 25 nodes were uniformly spread in 5 clusters. The intra-cluster delay is fixed to 0.1 ms and the inter-cluster delay is fixed to 100ms. Our tests were carried out with 50 application processes. Messages between processes were randomly generated.

A. Failure free performance

Fig. 3 presents the performance of the algorithms in both configurations. It is obvious that the time taken to run an application with checkpointing is greater than the time taken for it to run without checkpoint. Protocol overhead checkpoint coordinated non-blocking is less compared to other approaches to that phase synchronization is limited to the cluster and the second concerns only the leaders of each cluster. The additional cost of communications-driven approach is due to the forced checkpoints during execution. Logging protocols are sensitive to characteristics of the application, especially in communications-intensive applications. Indeed, they produce a large overhead due to the backup of messages on stable storage and the increasing size of messages to piggyback determinants.

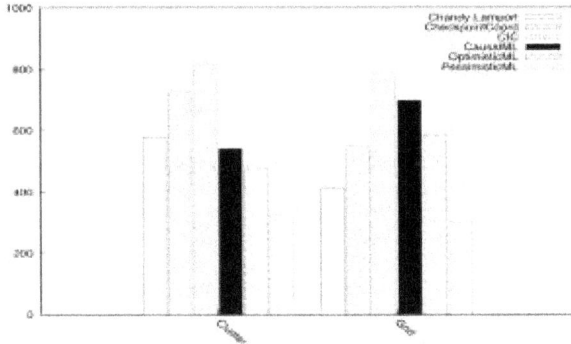

Fig. 3: Failure free performance, Checkpoint interval=180s, Execution time=900s

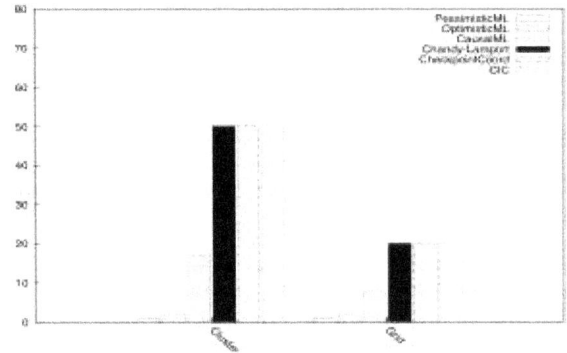

Fig. 5: Number of process, Checkpoint interval=180s, Execution time=900s, Numbers of fault= 1

B. Recovery time

The recovery time depends on the number of checkpoints maintained by the protocol and the number of rollbacks. In coordinated checkpointing and pessimistic logging, recovery is simplified because the system is rolled back only to the last recent checkpoint. In the grid approach, the additional cost of recovery decreases slightly. Indeed, if the faulty node has no dependencies with nodes of other cluster nodes, the fault is confined to the cluster node's fault. So all the nodes of the grid do not perform the recovery procedure. By cons, if the inter-cluster communications are intensive, the overhead increases as in the case of causal and optimistic logging.

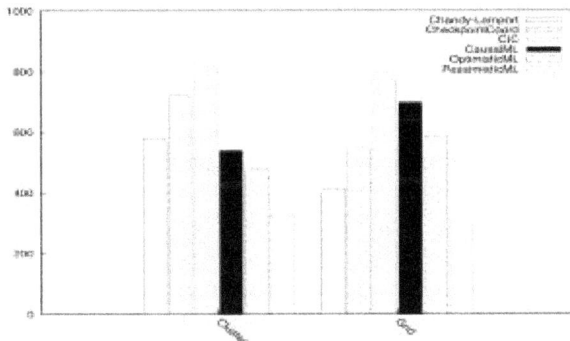

Fig.4: Overhead of recovery, checkpoint interval=180s, execution time=900s, numbers of fault=10

C. Number of rollbacks

For coordinated checkpoint protocols, all processes must resume during recovery. The logging protocol reduces the number of rollback. This number is minimal in pessimistic approach since only faulty processes need to be rolled back. For the other logging protocol, this number depends on the information stored in backups and in the main memory of correct processes.

V. RELATED WORK

Paul et a.l [4] proposes a hierarchical protocol based on coordinated checkpoint. This protocol is designed for hierarchical networks like the Internet. The experiments were made on a network of four clusters of eight nodes. Authors consider three roles of the different processes. *Initiator* is the process that initiates checkpoint sessions. One *Leader* process coordinates the activities within each cluster, in line with the instructions of the *Initiator*. *Follower* are the rest of the system processes, they follow the instructions of their *Leader*. The checkpoint protocol is hierarchical in two phases. The first phase is the execution of the algorithm coordinated checkpoint limited to the cluster. During this phase the processes are blocked and establish a consistent checkpoint. The second phase is a coordinated checkpoint but the leaders are the only participants, with the initiator, which acts as a coordinator. The experiments showed that the overhead of checkpointing in the hierarchical approach is lower than in the standard "flat" coordinated protocol. However the protocol hierarchy is sensitive to the frequency of messages between clusters. Indeed the extra cost of checkpoint increases progressively as the frequency of messages increases, and tends towards that of the checkpoint protocol standard.

Bhatia et al. [5] propose a hierarchical causal logging protocol that addresses the scalability problems of causal logging. Indeed, the traditional causal logging algorithms are used successfully in small-scale systems. They are known to provide a low overhead during failure-free executions sending no extra messages. But they are not scalable since each application process needs to maintain a data structure, which grows quadratically with the number of processes in the system.

Authors reduce the data structure by an exponential amount. They propose a hierarchical approach using a set of proxies spread on the network that act as a distributed cache. This approach highly reduces the amount of information piggybacked on each messages. However, the use of proxies decreases the performance of recovery since the recovery information is spread on the proxies.

Monnet et al. [6] propose a hierarchical checkpointing protocol, which combines coordinated checkpointing inside clusters and a checkpoint induced by communications between clusters. Simulation of the protocol shows that it generates a high number of forced checkpoints when the communication rate between clusters increases. Then, this approach is more suitable for code coupling applications where communications are mainly local inside clusters.

Several techniques are used to implement fault tol- erance in message-passing systems. Simple replication is not relevant for such systems, since if the system is designed to tolerate n faults, every component must be replicated n times and the computation resources are thus divided by n. The two main techniques used are message- logging and coordinated checkpoints. A review of the different techniques can be fount in [2].

Message-logging consists in saving the messages sent between the computation nodes, and replay them if a failure occurs. It is based on the *piecewise deterministic assumption*: the execution of a process is a sequence of deterministic events separated by non deterministic ones [14]. With this assumption, replaying the same sequence of non-deterministic events at the same moment makes possible the recovering of the state preceding a failure. Thus these protocols consist for every process to save

all its non-deterministic events in a reliable manner and to checkpoint regularly. When a failure occurs, only the crashed process is restarted from its last checkpoint, and it recovers its last state after having replayed all saved events. There is no need to coordinate the checkpoints of the different processes. No orphan processes (*i.e.* processes that are waiting for a message that will never come, since the expected sender is more advanced into its execution) are created. The recover mechanism is more complex than with coordinated checkpoints as a process shall obtain its past events and be able to replay them. Moreover the overhead induced during failure-free execution decreases the performances in not very faulty environments, such as clusters [23]. Furthermore, it can lead to the domino effect [24]: a process that rollbacks and that need a message to be replayed, asks another process to rollback. This process does, and asks another one to do so, etc. The execution can be restarted from the beginning because of cascading rollbacks and so the benefits of fault tolerance are lost.

Message-logging protocols are classified into three categories : optimistic, pessimistic and causal proto- cols. Optimistic protocols assume that no failure will occur between the moment a process executes a non- deterministic event and the moment this event is saved on a reliable storage support. So when a process executes a non deterministic event, it sends it to the reliable storage support then continues its computation without waiting any acknowledgment [22]. The induced overhead during failure-free execution is then quite small, but the optimistic hypothesis introduces the risk to get an incoherent state if it is not realized. Pessimistic protocols do not make this hypothesis, and the processes wait for an acknowledgment from the reliable storage support to continue their execution [23]. The induced overhead during fault-free

execution is then important. The third category of message-logging protocols tries to gather the advantages of both optimistic and pessimistic protocols: low overhead during fault-free execution, and no risk to recover into an incoherent state. It consists in saving the causality information on a reliable storage, but does not need to wait for the acknowledgment from this medium by piggybacking these information in the messages until the acknowledgments are received. A description can be found in [24], and another causal protocol based on dependencies graphs is described in [25]. A metric to evaluate the performances of message-logging protocols can be found in [26].

Coordinated checkpointing has been introduced by Chandy and Lamport in [27]. This technique requiresthat at least one process sends a marker to notify the other ones to take a snapshot of their local state and then form a global checkpoint. The global state obtained from a coordinated checkpoint is coherent, allowing the system to recover from the last full completed checkpoint wave. It does not generate any orphan processes nor domino effect, but all the computation nodes must rollback to a previous state. The recover process is simple, and a simple garbage collection reduces the size needed to store the checkpoints.

In blocking checkpointing protocols, the processes stop their execution to perform the checkpoint, save it on a reliable storage support (that can be distant), send an acknowledgment to the checkpoint initiator and wait for its commit. They continue the execution only when they have received this commit. The initiator sends the commit only when it has received all the acknowledg- ments from all the computing nodes to make sure that the global state that has been saved is fully completed. As claimed in [28], blocking checkpoints induce an important latency and non-blocking checkpoints are then more efficient.

Non-blocking coordinated checkpoints with dis- tributed snapshots consists in taking checkpoints when a marker is received. This marker can be received from a centralized entity, that initiates the checkpoint wave, or from another computation node which has itself received the maker and transmits the checkpoint signal to the other nodes. This algorithm assumes that all the communication channels comply with the FIFO property. Therefore the computation processes do not have to wait for the other ones to finish their checkpoint, and then the delay induced by the checkpoint corresponds only to the local checkpointing.

Communication-induced checkpoint protocols (CIC) perform uncoordinated checkpoints but avoid the domino effect [29]. Unlike coordinated checkpoints, it does not require additional messages for a process to know when it has to perform a local checkpoint. The information about when a local checkpoint must be performed are piggybacked in the messages exchanged between the processes. Two kinds of checkpoints are defined: local and forced. Local checkpoints are decided by the local process, forced ones are decided by the process accord- ing to the information piggybacked in the messages. The forced ones avoid the domino effect and ensure then the progress of the recovery line, i.e. the set of checkpoints of all the processes describing a coherent global state. When a

failures occurs, all the processes rollback to their last stored local checkpoint and then to the last recovery
line. CIC is an interesting theoretical solution but it has been shown in [30], using NPB 2.3 benchmark suite [31], that it is not relevant for typical cluster applications.
Several MPI libraries are fault tolerant. A review can be found in [32]. Coordinated checkpointing has been implemented in several MPI implementations on different levels of the application.

LAM/MPI [33], [34] implements the Chandy-Lamport algorithm for a system-initiated global checkpointing. When a checkpoint must be performed, the *mpirun* process receives a checkpoint request from a user or from the batch scheduler. It propagates the checkpoint request to each MPI process to initiate a checkpoint wave. As in our blocking Chandy-Lamport implementation, each MPI process then coordinates itself with all the others, flushing every communication channel, in order to reach a consistent global state. If a failure occurs, mpirun restarts all the processes from their last stored state. Finally, processes rebuild their communication channels with the other ones and resume their execution.

VI. CONCLUSION

In this paper, we compare checkpoint protocols and message logging in grid computing. We propose a hierarchical approach to combine different algorithms. We find that the protocols that require the recovery of all processes in case of single failure are poorly suited to systems with many processes. The message logging protocols are more suitable for large configuration with the exception of some causal logging approach, which induces communications to all processes during the recovery. Non-blocking coordinated checkpoint are not sensitive to the rate of communications. They therefore represent an attractive solution for applications and highly interconnected grid architectures by reducing the number of markers sent during the synchronization phase.

REFERENCES

[1] E. N. (MOOTAZ) ELNOZAHY, LORENZO ALVISI, YI-MIN WANG, DAVID B. JOHNSON, A Survey of Rollback-Recovery Protocols in Message-Passing Systems, ACM Computing Surveys, Vol. 34, No. 3, September 2002, pp. 375–408.
[2] Distributed snapshots: Determining global states of distributed systems. ACM Trans. Comput. Syst. 31, 1, 63–75.
[3] C. Coti, T. Herault, P. Lemarinier, L. Pilard, A. Rezmerita, E. Rodriguez, and F. Cappello. Blocking vs. non-blocking coordinated checkpointing for large-scale fault tolerant MPI. In SC '06: Proceedings of the 2006 ACM/IEEE conference on Supercomputing, page 127, New York, NY, USA, 2006. ACM.
[4] Himadri S. Paul, Arobinda Gupta R. Badrinath, Hierarchical Coordinated Checkpointing Protocol. In International Conference on Parallel and Distributed Computing Systems, pages 240-245, November 2002
[5] K. Bhatia, K. Marzullo, and L. Alvisi. "Scalable causal Message Logging for Wide-Area Environments," Concurrency and Computation: Practice and Experience, 15(3), pp. 873-889, Aug. 2003.
[6] S. Monnet, C. Morin, R. Badrinath, "Hybrid Checkpointing for Parallel Applications in cluster Federations", Proc. 4th IEEE/ACM International Symposium on Cluster Computing and the Grid, Chicago, IL, USA, pp. 773-782, April 2004.
[7] http://www.omnetpp.org

[8] Open Grid Forum. http://ogf.org
[9] Organization for the Advancement of Structured Information Standards (OASIS). http://www.oasis-open.org
[10] Internet Engineering Task Force. http://www.ietf.org
[11] A. Avizienis, J. Laprie, B. Randell and C. Landwehr. "Basic concepts and taxonomy of dependable and secure computing", IEEE Transactions on Dependable and Secure Computing, 1(1), pp. 11-33, 2004
[12] A. Borg, J. Baumbach, and S. Glazer, "A message system supporting fault-tolerance", In Proceedings of the Symposium on Operating Systems Principles, ACM SIGOPS, pp. 90-99, Oct. 1983
[13] M. L. Powell, and D. L. Presotto, "Publishing: A reliable broadcast communication mechanism", In Proceedings of the Ninth Symposium on Operating System Principle, ACM SIGOPS, pp. 100-109, Oct. 1983
[14] R. B. Strom and S. Yemeni, "Optimistic recovery in distributed system" ACM Transactions on Computer Systems, 3(3), pp. 204-226, April 1985
[15] D. B. Johnson and W. Zwaenepoel, "Sender-based message logging", In Digest of Papers: 17 Annual International Symposium on Fault-Tolerant Computing, IEEE Computer Society, pp. 14-19, June 1987
[16] R. E. Strom, D. F. Bacon and S. A. Yemeni, "Volatile logging in n-fault-tolerant distributed systems", In Proceedings of the Eighteenth Annual International Symposium on Fault-Tolerant Computing, pp. 44-49, 1988
[17] A. P. Sistla and J. L. Welch, "Efficient distributed recovery using message logging", In Proceedings of the Eighth Symposium on Principles of Distributed Computing, ACM SIGACT/SIGOPS, pp. 223-238, Aug. 1989
[18] D. B. Johnson, and W. Zwaenepoel, "Recovery in distributed systems using optimistic message logging and checkpointing", Journal of Algorithm, 11: pp. 462-491, 1990
[19] S. Venkatesan, and T. Y. Juang, "Efficient algorithms for optimistic crash recovery", Distributed Computing, 8(2): pp. 105-114, June 1994
[20] E. N. Elnozahy and W. Zwaenepoel, "In the use and implementation of message logging", In Digest of Papers: 24 Annual International Symposium on Fault-Tolerant Computing, IEEE Computer Society, pp. 298-307, June 1994
[21] P. Lemarinier, A. Bouteiller, T. Herault, G. Krawezik, and F. Cappello, "Improved message logging versus improved coordinated checkpointing for fault tolerant MPI," in *IEEE International Conference on Cluster Computing (Cluster 2004)*. IEEE CS Press, 2004
[22] B. Randell, "System structure for software fault tolerance," *IEEE Transactions on Software Engineering*, vol. SE-1, no. 2, pp. 220–232, 1975
[23] A . Bouteiller, F. Cappello, T. Herault, G. Krawezik, P. Lemarinier, and F. Magniette, "MPICH-V2: a fault tolerant MPI for volatile nodes based on pessimistic sender based mes- sage logging," in *High Performance Networking and Computing (SC2003)*, Phoenix USA. IEEE/ACM, November 2003
[24] A. Bouteiller, P. Lemarinier, G. Krawezik, and F. Cappello, "Coordinated checkpoint versus message log for fault tolerant MPI," in *IEEE International Conference on Cluster Computing (Cluster 2003)*. IEEE CS Press, December 2003
[25] E. N. Elnozahy and W. Zwaenepoel, "Replicated distributed processes in manetho," in *22nd International Symposium on Fault Tolerant Computing (FTCS-22)*. Boston, Massachusetts: IEEE Computer Society Press, 1992, pp. 18–27
[26] L. Alvisi and K. Marzullo, "Message logging: Pessimistic, optimistic, causal, and optimal," *IEEE Trans. Software Eng*, vol. 24, no. 2, pp. 149–159, 1998
[27] K. M. Chandy and L.Lamport, "Distributed snapshots : Determining global states of distributed systems," in *Transactions on Computer Systems*, vol. 3(1). ACM, February 1985, pp. 63–75
[28] E. N. Elnozahy, D. B. Johnson, and W. Zwaenepoel, "The performance of consistent checkpointing," in *Symposium on Reliable Distributed Systems*, 1992, pp. 39–47
[29] J.-M. Helary, A. Mostefaoui, and M. Raynal, "Communication- induced determination of consistent snapshots," *IEEE Transac- tions on Parallel and Distributed Systems*, vol. 10, no. 9, pp. 865–877, 1999
[30] L. Alvisi, E. Elnozahy, S. Rao, S. A. Husain, and A. D. Mel, "An analysis of communication induced checkpointing," in *29th Symposium on Fault-Tolerant Computing (FTCS'99)*. IEEE CS Press, june 1999

[31] N. A. R. center, "Nas parallel benchmarks," 1997http://science.nas.nasa.gov/Software/NPB/

[32] W. Gropp, and E. Lusk, "Fault tolerance in MPI Program", Special issue of the Journal of High Performance Computing Applications (IJHPCA), 2002

[33] G. Burns, R. Daoud, and J. Vaigl, "LAM: An Open Cluster Environment for MPI," in *Proceedings of Supercomputing Symposium*, 1994, pp. 379–386

[34] S. Sankaran, J. M. Squyres, B. Barrett, A. Lumsdaine, J. Duell, P. Hargrove, and E. Roman, "The LAM/MPI checkpoint/restart framework: System-initiated checkpointing," in *Proceedings, LACSI Symposium*, Sante Fe, New Mexico, USA, October 2003

[35] S. Zanikolas, and R. Sakellariou, "A taxonomy of Grid monitoring systems", Future Generation Computer System, 21(1), pp. 163-188, 2005

[36] H. Jitsumoto, T. Endo, and S. Matsuoka, "ABARIS : An adaptable fault detection/recovery component framework for MPI", Proceedings of the IEEE International Parallel and Distributed Processing Symposium, IEEE Computer Society Press : Los Alamitos, CA, pp. 1_8, 2007

[37] H. Jin, W. Qiang, and D. Xou, "DRIC : Dependable Grid Computing framework", IEICE Transactions on Information and System : E89-D(2), pp. 612-623, 2006

[38] E. Elnozahy, D. Johnson, and Y. Wang, "A survey of rollback recovery protocols in message passing systems", ACM Computing Surveys, 34(3), pp. 375-408, 2002

[39] I. Foster and C. Kesselman. The Grid 2: Blueprint for a New Computing Infrastructure. Morgan Kaufmann Publishers Inc., San Francisco, CA, USA, ISBN 1558609334, 2003

Ndeye Massata Ndiaye received his B.Sc and M. Sc in Computer science from Gaston Berger University of Saint and M.Phil in Computer Science from Cheikh Anta Diop university of Dakar in Senegal. She is now an assistant professor in university of Bambey Senegal.

Pierre Sens is a full professor at the University of Paris 6 since 2003. He received his PhD in 1994 and the "Habilitation à Diriger des Recherches" in 2000. Since 2002, he leads Regal project which is a joint research team between LIP6 and INRIA, France. He has been author and co-author of published papers in several books, journals and recognized international conferences and symposiums.

Ousmane Thiare. Received a PhD in computer science (Distributed systems) at 2007 from the university of Cergy Pontoise, France. He is an associate professor in Gaston Berger University of Saint-Louis Senegal. He has been co-author of published papers in several journals and recognized international conferences and symposiums.

23

Design of a Mutual Exclusion and Deadlock Algorithm in PCBSD – FreeBSD

Libertad Caicedo Acosta, Camilo Andrés Ospina Acosta, Nancy Yaneth Gelvez García, Oswaldo Alberto Romero Villalobos.

Universidad Distrital Francisco José de Caldas, Bogotá, Colombia

Abstract — **This paper shows the implementation of mutual exclusion in PCBSD-FreeBSD operating systems on SMPng environments, providing solutions to problems like investment priority, priority propagation, interlock, CPU downtime, deadlocks, between other. Mutex Control concept is introduced as a solution to these problems through the integration of the scheduling algorithm of multiple queues fed back and mutexes.**

Keywords — **Mutex, PcBSD, SMPng, FreeBSD, Operating Systems.**

I. INTRODUCTION

OVER time operating systems have evolved to reach the progress that can be seen today: starting batch processing, which involved planning the next job to run on a treadmill until multiprogramming systems in which many users waited to be served. With the advent of personal computers has been generally allowing one active process and more resources to which access, then with the integration of more than one processor on a machine, appeared multiprocessing and therefore the concept of parallelism, which involves making one or many processes running on different processors at the same time, being assigned a process per processor. Such evolution is generated from finding that a perceived performance and user satisfaction is optimal.

One of the main functions of the operating system is making decisions about allocating resources to the various processes are in ready state and require access to the same resource; process scheduler uses the scheduling algorithm to make such decisions. Scheduling algorithms implemented in the kernel of the system depending on the environment in which they are seeking to improve the response time, proportionality, predictability, fairness and prevent data loss. [1]

In environments such as real-time or interactive problems may be found when concurrency occurs one or more processors; where processes wish to share the same resource difficulties are encountered when defining the time and the conditions under which each process makes use of the resource, looking in critical section only able to stay a process, ie, that the final result depends on who is running and when it does. This situation leads to problems usually involving shared memory, files, and resources in general (a resource is a hardware device or a piece of information) are generated, which leads to data loss or downtime CPU.

II. MUTUAL EXCLUSION AND DEADLOCK

Mutual exclusion is born from the generation of the problems listed above with concurrent programming, seeking to ensure that if a process makes use of a shared resource processes exclude others do the same. However, sometimes the processes are performing internal calculations and other things that do not involve access to the critical section, ie, the part of the program that accesses the shared memory. What is desired is that the processes can operate in parallel to data sharing is optimized over time, as long as only one is in critical section. There are some considerations when performing a mutual exclusion algorithm using critical regions:

- There can only be a process critical section at a time.
- Must know the speeds or the number of CPU's.
- Only the process is in critical section may block other.
- There should be no downtime CPU, because no process can wait infinitely to be executed.

Another mechanism that avoids mutual exclusion is partially disabling interrupts on a CPU; however do user processes and may not be re-enabled that would kill the system, or if the CPU multiprocessing disabling performed cease to function. In the same way the method operates lock variables, in which when the lock is 0, the process can access critical section and when no one is 1; however has problems as to the mechanism mentioned above.

One of the latest implementations of mutual exclusion algorithms are the *mutex*, which allow to manage a resource or piece of code; is very helpful for thread sets that are implemented in user space. The mutex variable is a padlock that can be open or closed, which is represented by 0 if it is open and any other value if it is closed. When a process requires entering critical section checks whether the padlock is open and if so hard to run if it is not blocked until the process in that section is released, ie, the padlock opens. If several blocked threads, then one is selected at random to be the next to access critical section.

The mutex can be viewed from its behavior. As shown in Figure 1, a mutex is a padlock variable that can be open or closed and which may have one of the following behaviors:

- Sppining: When you constantly look at the state of the lock to see if the resource is already released. In this case there are no interruptions, then time is wasted waiting for lengthy processes that took control of the resource.
- Blocking: When the resource is not changed to block and awaits the call of the appeal process had, by the time it releases.
- Sleeping: If the resource is not available, it puts the process to sleep until the resource is enabled.

For exceptional cases, the programmer can create a few extra conditions under which a process sends to sleep while the presented problem is solved.

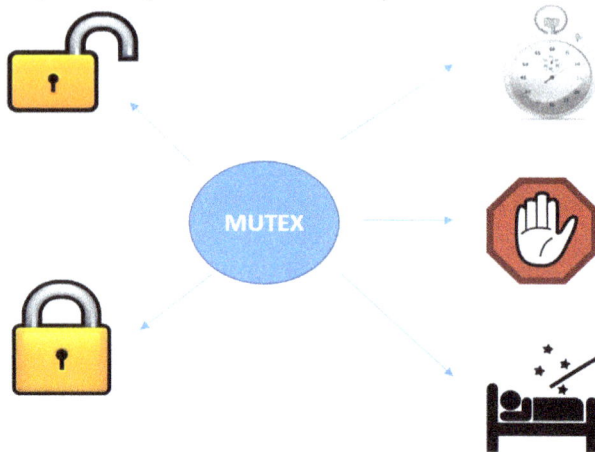

Fig. 1 - Representation of mutual exclusion.

In many cases computers have resources that can be used by only one process in the same space of time; if they run more, inconsistencies or errors occur in the information manipulated. Operating systems temporarily attach to a process exclusive access to certain resources; in cases in which a process one needs more than one resource, first makes the request, but when you need to access another and find that it is occupied by a process 2 which requires use of the resource using process 1; as both are waiting for the other release to free the resource you are running a deadlock occurs. This phenomenon could also occur between machines in the same network with shared devices such as scanners, printers, external drives, etc. [2]. Figure 2 shows more clearly the deadlock problem, with both processes P1 and P2, and two R1 and R2 resources that are left in a standby cycle, and retention of the release of resources.

Deadlocks are usually not preemptive resource linked, ie they may leave without being run over. In some cases there is a quantum that allows equality between processes and subtract the lifetime of the running process and then leave critical section; if it has been completed is deleted, if not, back to the tail of "ready" to run below the remaining time. Such resources may also have an associated priority, which means that in case a higher priority process needs the resource that is being used, use it and send the running process to a suspended list.

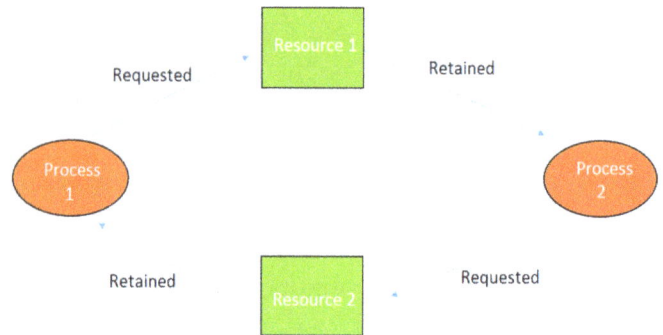

Fig. 2 - Representation of deadlock.

Currently the FreeBSD kernel supports symmetric multiprocessing (SMP), in which all Central Processing Units (CPU's) have a single connection to a non-uniform memory, implementing the Mutex Synchronization strategy as the primary method to manage short-term threads synchronization [3]. One of the desired characteristics for the mutexes design is that acquiring and releasing uncontested mutexes should be as fast as possible, which is one of the reasons for designing an algorithm that works in junction with the FreeBSD-PCBSD mutexes.

III. DESIGN

On operating systems such as FreeBSD PcBSD have been implementing mutual exclusion algorithms increasingly trying to minimize the problems that arise concerning the allocation of resources to processes.

PcBSD is a desktop operating system based on FreeBSD, which provides stability and security in server environments; makes use of window managers and open source application installers of the same type. It is currently in version 10 call PcBSD Joule, which is used in the following analysis and implementation to be discussed later. [4]

A. Mutual Exclusion Algorithm in PcBSD

The implementation of the scheduling algorithms in pcbsd involves evolution in terms of versioning, as currently implemented in some types of which a mutex some extra functionality integrated avoid problems such as investment priority, priority propagation and interlock.

The current implementation emerges from the problems presented with KSE (Kernel Scheduling Entity) in multiprocessing environments that generated downtime CPU and deadlocks [5]; therefore, from version 5.0 of FreeBSD kernel restructuring is done in the way of working threads in such environments, implementing mutexes that lead to a kernel SMPng according to [6].

As mentioned in section 2, there are three types of behaviors associated with mutexes which are kept in the model proposed by the creators of FreeBSD development, ie that there are shared mutex as Spinning, Blocking or Sleeping. Additionally mutexes define four types of [7], which are defined below:

- Mutex: When access to data is located on 1 CPU and accessed by a single thread.
- RW Lock: When access to data is made with several threads on several CPU's, in which reading and writing is permitted, however, only one process can be in write mode, while many in read mode.
- Lock RM: is equal to the RW Lock, only varies in the fact that the reading time is optimized.
- Waitchannel: When a thread requires the use of another thread is assigned a stop expected to sleeping.

It is preferable to use a mutex Blocking that a mutex Spinning in most cases, there are only a few exceptions where the other is better.

Below in Figures 3, 4 and 5 shows the implementation of mutexes, which are encoded in the kernel of PCBSD.

```
struct mtx {
        struct lock_object     lock_object;    /* Common lock properties. */
        volatile uintptr_t     mtx_lock;       /* Owner and flags. */
};
```

Figure 3 - Defining the mutex structure.

The above structure is the /usr/include/sys/_mutex.h system directory.

```
/* Lock a normal mutex. */
#define __mtx_lock(mp, tid, opts, file, line) do {              \
        uintptr_t _tid = (uintptr_t)(tid);                      \
                                                                \
        if (!_mtx_obtain_lock((mp), _tid))                      \
                _mtx_lock_sleep((mp), _tid, (opts), (file), (line));    \
        else                                                    \
                LOCKSTAT_PROFILE_OBTAIN_LOCK_SUCCESS(LS_MTX_LOCK_ACQUIRE, \
                    mp, 0, 0, (file), (line));                  \
} while (0)
```

Fig. 4 - Definition of a mutex lock.

The lock and unlock functions are in the C++ library mutex.h, which are located in the /usr/include/sys filesystem.

```
/* Unlock a normal mutex. */
#define __mtx_unlock(mp, tid, opts, file, line) do {           \
        uintptr_t _tid = (uintptr_t)(tid);                     \
                                                               \
        if (!_mtx_release_lock((mp), _tid))                    \
                _mtx_unlock_sleep((mp), (opts), (file), (line)); \
} while (0)
```

Fig. 5 - Defining unlock a mutex.

B. Design of mutexes in multiple queues fedback to an environment PCBSD-FreeBSD

In conjunction with the scheduling algorithm of multiple queues fed back [8], we propose an extra control for handling mutual exclusion and deadlock through mutexes which we call Mutex Control. For the specific case of pcbsd-FreeBSD, this control within each scheduling algorithm was implemented as shown in Figure 6, allowing at the time of assessment step in a process for critical section, a mutex is assigned to it. All this for the purpose to have a better management and resource allocation to avoid problems CPU timeouts, deadlocks, interlocking, among others.

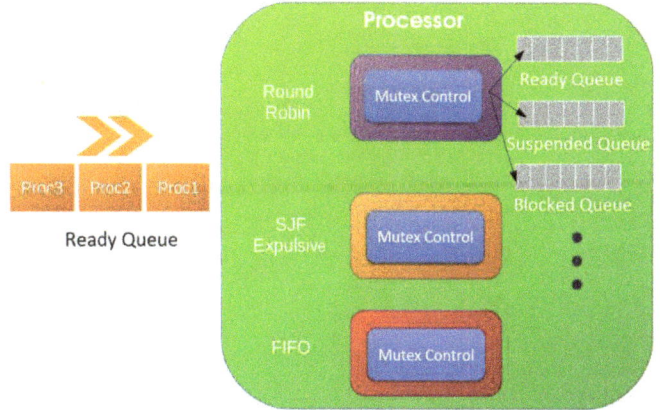

Fig. 6 - Environment and location Mutex Control

Design of Control made Mutex involves a simulated environment algorithm fed back tails, specifically built to allow the integration of mutexes proposed PCBSD-FreeBSD. In Figure 7, was able to visualize the proposed multiple queues fed back flowchart Mutex Control behavior.

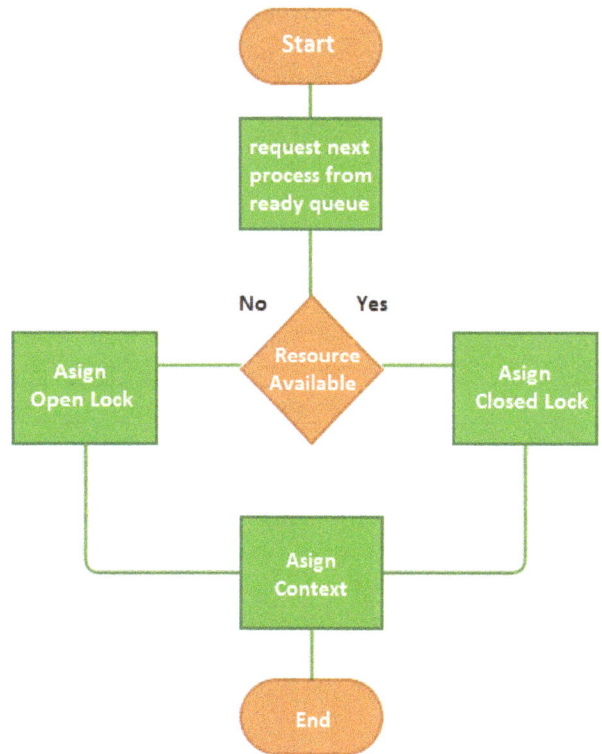

Fig.7 - Mutex Control Operation

The structure mutex proposal to integrate itself to queues fed back algorithm, is given as an adaptation of the libraries implemented in C++ Kernel PCBSD-FreeBSD, reflecting the main behavior that they exhibit a SMPng environment. Representation is in terms of Objects Oriented Programming and structural, in which a Mtx class, mtx_lock function and mtx_unlock function similar to structures in C++, mtx, __mtx_lock and __mtx_unlock respectively is created and seen previously in Figures 3, 4 and 5.

```
class mtx():
    '''
    Simple mutex class for locking resources
    '''
    def __init__(self, resource=None, owner=None, lock=0):
        self.lock_object={"resource":resource}
        self.mtx_lock={"owner":owner, "lock":lock} #0 lock else unlock}


def mtx_lock(mtx):
    if mtx.mtx_lock["lock"]:
        mtx.mtx_lock["lock"]=0


def mtx_unlock(mtx):
    if not mtx.mtx_lock["lock"]:
        mtx.mtx_lock["lock"]=1
```

Fig. 8 - Mtx code implementation.

Integrating mutex controls the scheduling algorithms is performed just after receiving the next process in the ready queue and is given by an assessment of the resources needed by each process, allocation and release locks as can be seen in Figure 8.

```
proc=self.core1.roundRobinAlgo.readyProcesses.next()

#Create New Mutex
new_mtx=mtx(proc.resource, proc.name)
#Eval locking
if bIP.resource["type"] not in (core1resource,core2resource,core3resource):
    mtx_lock(new_mtx)
else:
    mtx_unlock(new_mtx)
#Assign Mutex to Process
proc.mtx= new_mtx

self.core1.roundRobinAlgo.blockedProcesses.append(proc)
```

Fig. 9 - Code implementing the Mutex Control

IV.   MUTEX SIMULATION

In order to simulate the coupled behavior of the mutex controller proposed, it was necessary to create an application by multiple queues fed back to incorporate into their calling from the processes queue on each core to mutex control in order to make the decision to change the context of each process (blocked, suspended and critical section).

Then several screens showing action working together mutex control through multiple queues fed back (multiprocessor) can be observed. It should be noted that although the simulation was designed thinking of ways to perform mutual exclusion in PcBSD-FreeBSD, an implementation of a mutex control style could be proposed in other operating systems (improving the effectiveness of the algorithms for SMPng environments).

Fig. 10 - Assigning each core processes

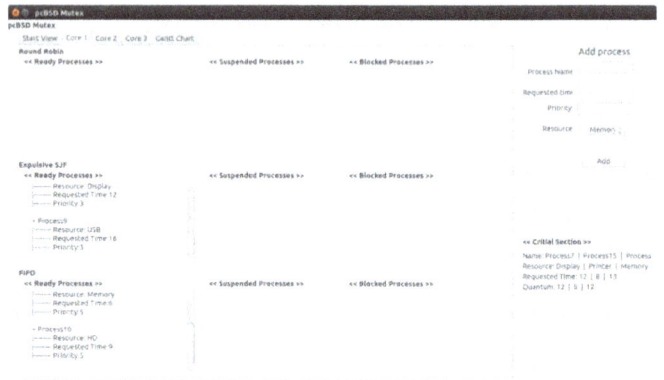

Fig. 11 - Core 1 running

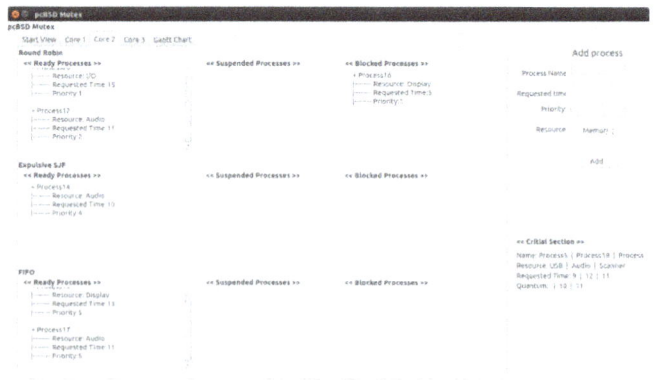

Fig. 12 - Core 2 running

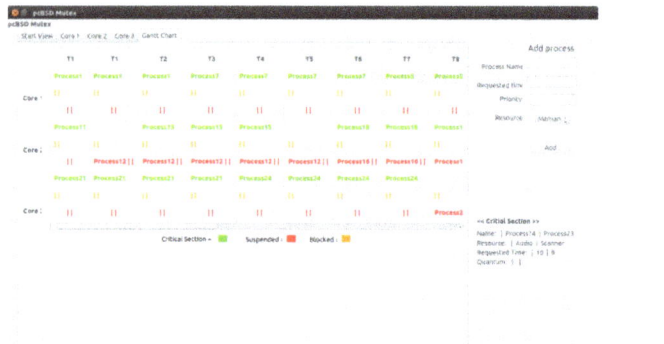

Fig.13 - Gantt execution of algorithms

V. CONCLUSIONS

As previously mentioned, FreeBSD-PCBSD implements the Mutex Synchronization strategy as the primary method to manage short-term threads synchronization which in junction with the proposed algorithm improves the OS desired characteristics for the mutexes.

For each operating system there is a way to implement mutual exclusion that is best suited to your operation. Not always the most complex has better benefits.

Scheduling algorithms and mutual exclusion, require adaptations to environments smpng because it worked very well on a single processor environment, tends to have problems or inefficiencies in the management of resources and response times Multiprocessor.

The mutex control minimizes the problems that are presented to the planning algorithms in environments SMPng such as investment priority, priority propagation, interlock, CPU timeouts, and deadlocks, among others.

Integrating mutexes to implement multiple exclusion in PcBSD-FreeBSD operating systems for SMPng environments, represents major advantages implementation over other mutual exclusion algorithms.

REFERENCES

[1] A. S. Tanenbaum, "Sistemas Operativos Modernos", Pearson Educación, 3ra edición. pp. 146-160. México 2009.

[2] Zobel, "Operating Systems Review", Automatica. Vol. 6. Number 1/2, June, 1972.

[3] Marshall Kirk McKusick, George V. Neville-Neil, Robert N.M, Watson, "The Design and Implementation of the FreeBSD Operating System".

[4] PcBSD. (2014, Julio 13). iXsystems, Inc. [En línea]. Disponible en: http://www.pcbsd.org/es/.

[5] FreeBSD. (2014, Julio 13)." KSE". [En línea]. Disponible en: http://www.freebsd.org/cgi/man.cgi?query=kse&sektion=2&manpath=FreeBSD+5.0-RELEASE.

[6] Baldwin John. (2014, Julio 13)." How SMPng Works and Why It Doesn't Work The Way You Think ". [En línea]. Disponible en: http://people.freebsd.org/~jhb/papers/smp/slides.pdf.

[7] Rao, Attilio. "FreeBSD src/ committer since 2007." AsiaBSDCon 2009 Paper Session. The first part of the two. [En línea] Disponible en: https://www.youtube.com/watch?v=a3XLROUjXic

[8] JRA, "Planificación de Procesos", Departamento de Informática, Facultad de Ingeniería.Universidad Nacional de la Patagonia "San Juan Bosco" 2010.

24

A Probability-based Evolutionary Algorithm with Mutations to Learn Bayesian Networks

Sho Fukuda, Yuuma Yamanaka, Takuya Yoshihiro

Wakayama University, Sakaedani, Wakayama, Japan

Abstract — **Bayesian networks are regarded as one of the essential tools to analyze causal relationship between events from data. To learn the structure of highly-reliable Bayesian networks from data as quickly as possible is one of the important problems that several studies have been tried to achieve. In recent years, probability-based evolutionary algorithms have been proposed as a new efficient approach to learn Bayesian networks. In this paper, we target on one of the probability-based evolutionary algorithms called PBIL (Probability-Based Incremental Learning), and propose a new mutation operator. Through performance evaluation, we found that the proposed mutation operator has a good performance in learning Bayesian networks.**

Keywords — **Bayesian Networks, PBIL, Evolutionary Algorithms**

I. INTRODUCTION

BAYESIAN network is a well-known probabilistic model that represents causal relationships among events, which has been applied to so many areas such as Bioinformatics, medical analyses, document classifications, information searches, decision support, etc. Recently, due to several useful tools to construct Bayesian networks, and also due to rapid growth of computer powers, Bayesian networks became regarded as one of the promising analytic tools that help detailed analyses of large data in variety of important study areas.

To learn a near-optimal Bayesian network structure from a set of target data, efficient optimization algorithm is required that searches an exponentially large solution space for near-optimal Bayesian network structure, as this problem was proved to be NP-hard [1]. To find better Bayesian network structures with less time, several efficient search algorithms have been proposed so far. Cooper et al., proposed a well-known deterministic algorithm called K2 [2] that searches for near-optimal solutions by applying a constraint of the order of events. As for the general cases without the order constraint, although several approaches have been proposed so far, many of which uses genetic algorithms (GAs), which find good Bayesian network structures within a reasonable time

[3][4][5]. However, because recently we are facing on large data, more efficient algorithms to find better Bayesian network models are expected.

To meet this requirement, recently, a new category of algorithms so called EDA (Estimation of Distribution Algorithm) has been reported to provide better performance in learning Bayesian Networks. EDA is a kind of genetic algorithms that evolves statistic distributions to produce individuals over generations. There are several types of EDA such as UMDA (Uni-variate Marginal Distribution Algorithm) [12], PBIL (Population-Based Incremental Learning) [7], MIMIC (Mutual Information Maximization for Input Clustering) [13], etc. According to the result of Kim et al. [11], PBIL-based algorithm would be the most suitable for learning Bayesian networks.

The first PBIL-based algorithm for Bayesian networks was presented by Blanco et al. [9], which learns good Bayesian net- works within short time. However, because this algorithm does not include mutation, it easily falls into local minimum solution. To avoid converging at local minimum solutions, Handa et al. introduced a *bitwise mutation* into PBIL and showed that the mutation operator improved the quality of solutions in four-peaks problem, Fc4 function, and max-sat problem[10]. Although this operator was not applied to Bayesian networks, Kim et al. later proposed a new mutation operator transpose mutation specifically for Bayesian networks, and compares the performance of EDA-based Bayesian network learning with several mutation variations including bitwise mutation [11].

In this paper, we propose a new mutation operator called *probability mutation* for PBIL-based Bayesian Network learning. Through evaluation, we show that our new mutation operator is also efficient to find good Bayesian network structures.

The rest of this paper is organized as follows: In Section 2, we give the basic definitions on Bayesian networks and also describe related work in this area of study. In Section 3, we propose a new mutation operator called probability mutation to achieve better learning performance of Bayesian networks. In Section 4, we describe the evaluation results, and finally we conclude this paper in Section 5.

X_1	$P(X_1)$
0	0.999
1	0.001

X_2	$P(X_2)$
0	0.998
1	0.002

X_1	X_2	$P(X_3 \mid X_1, X_2)$	
		0	1
0	0	0.999	0.001
0	1	0.710	0.290
1	0	0.060	0.840
1	1	0.050	0.950

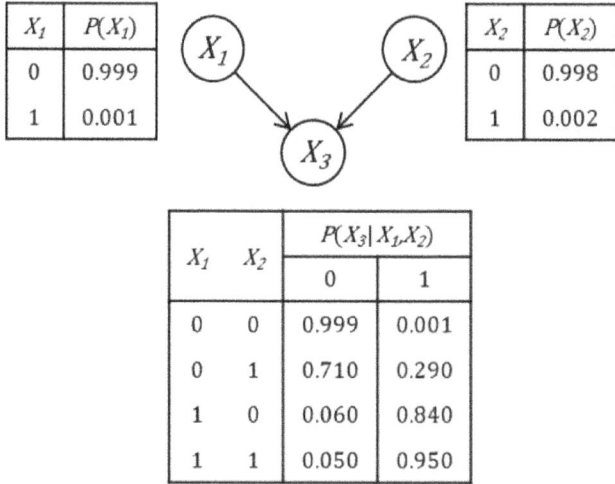

Fig. 1. A Bayesian Network Model

II. LEARNING BAYESIAN NETWORKS

A. Bayesian Network Models

A Bayesian network model visualizes the causal relationship among events through graph representation. In a Bayesian network model, events are represented by nodes while causal relationships are represented by edges. See Figure 1 for example. Nodes X_1, X_2, and X_3 represent distinct events where they take 1 if the corresponding events occur, and take 0 if the events do not occur. Edges $X_1 \rightarrow X_3$ and $X_2 \rightarrow X_3$ represent causal relationships, which mean that the probability of $X_3 = 1$ depends on events X_1 and X_2. If edge $X_1 \rightarrow X_3$ exists, we call that X_1 is a parent of X_3 and X_3 is a child of X_1. Because nodes X_1 and X_2 do not have their parents, they have own prior probabilities $P(X_1)$ and $P(X_2)$. On the other hand, because node X_3 has two parents X_1 and X_2, it has a conditional probability $P(X_3 \mid X_1, X_2)$. In this example, the probability that X_3 occurs is 0.950 under the assumption that both X_1 and X_2 occur. Note that, from this model, Bayesian inference is possible: if X_3 is known, then the posterior probability of X_1 and X_2 can be determined, which enables us to infer events that causes the child event.

The Bayesian networks can be learned from the data obtained through the observation of events. Let $O = \{o_j\}, 1 \le j \le S$ be a set of observations, where S is the number of observations. Let $o_j = (x_{j1}, x_{j2}, ..., x_{jN})$ be a j-th observation, which is a set of observed values x_{ji} on event X_i for all $i (1 \le i \le N)$, where N is the number of events. We try to learn a good Bayesian network model θ from the given set of observations. Note that the model θ should be able to explain the observation O, i.e., O should be likely to be observed under θ. As an evaluation criterion to measure the

level of fitting between θ and O, we use AIC (Akaike's Information Criterion) [6], which is one of the best known criterion used in Bayesian networks. Formally, the problem of learning Bayesian networks that we consider in this paper is defined as follows:

Problem 1: From the given set of observations O, compute a Bayesian network model θ that has the lowest AIC criterion value.

B. K2 Algorithm

K2 [2] is one of the best-used traditional algorithms to learn Bayesian network models. Note that searching good Bayesian network models is generally time consuming because the problem to learn Bayesian networks is NP-hard [1]. K2 avoids the problem of running time by limiting the search space through the constraint of totally order of events. Namely, for a given order of events $X_1 < X_2 < ... < X_N$, causal relationship $X_k \rightarrow X_l$, where $k > l$ is not allowed. Note that this constraint is suitable for some cases: if events have their time of occurrence, an event X_k that occurred later than X_l cannot be a cause of X_l. Several practical scenes would be the case.

The process of K2 algorithm applied to a set of events $X_1, X_2, ..., X_N$ with the constraint $X_1, X_2, ..., X_N$ is described as follows:

(1) Select the best structure using two events X_N and X_{N-1}. Here, the two structures, i.e., $X_{N-1} \rightarrow X_N$ and the independent case, can be the candidates, and the one with better criterion value is selected.

(2) Add X_{N-2} to the structure. Namely, select the best structure from every possible cases where X_{N-2} has edges connected to X_{N-1} and X_N. Namely, from the cases (i) $X_{N-2} \rightarrow X_{N-1}$ and $X_{N-2} \rightarrow X_N$, (ii) $X_{N-2} \rightarrow X_{N-1}$ only, (iii) $X_{N-2} \rightarrow X_N$ only, and (iv) where X_{N-2} has no edge.

(3) Repeat step (2) to add events to the structure in the order $X_{N-3}, ..., X_2, X_1$.

P		Parent Node					
		X_1	X_2	...	X_i	...	X_N
Child node	X_1	0.0	0.5	...	p_{i1}	...	0.5
	X_2	0.5	0.0	...	p_{i2}	...	0.5
	⋮	⋮	⋮	⋱	⋮	...	⋮
	X_j	p_{1j}	p_{2j}	...	p_{ij}	...	p_{Nj}
	⋮	⋮	⋮	⋮	⋮	⋱	⋮
	X_N	0.5	0.5	...	p_{iN}	...	0.0

Fig. 2. A Probability Vector

(4) Output the final structure composed of all events. Although K2 requires low computational time due to the constraint

of event order, many problems do not allow the constraint. In such cases, we require to tackle the NP-hard problem using a heuristic algorithm for approximate solutions.

C. Related Work for Un-ordered Bayesian Network Models

Even for the cases where the constraint of order is not allowed, several approaches to learn Bayesian network models has been proposed. One of the most basic method is to use K2 with random order, where randomly generated orders are applied repeatedly to K2 to search for good Bayesian network models.

As more sophisticated approaches, several ideas have been proposed so far. Hsu, et al. proposed a method to use K2 algorithm to which the orders evolved by genetic algorithms are applied [3]. Barrière, et al. proposed an algorithm to evolve Bayesian network models based on a variation of genetic algorithms called co-evolving processes [4]. Tonda, et al. proposed another variation of genetic algorithms that applies a graph-based evolution process [5]. However, with these approaches, the performance seems to be limited, and a new paradigm of the algorithm that learn Bayesian networks more efficiently is strongly required.

D. Population-Based Incremental Learning

Recently, a category of the evolutionary algorithms called EDA (Estimation Distribution Algorithm) appears and reported to be efficient to learn Bayesian network models. As one of EDAs, PBIL [7] is proposed by Baluja et al. in 1994, which is based on genetic algorithm, but is designed to evolve a probability vector. Later, Blanco et al. applied PBIL to the Bayesian network learning, and showed that PBIL efficiently works in this problem [9].

In PBIL, an individual creature s is defined as a vector $s = (v_1, v_2, ..., v_L)$, where $v_i (1 \le s \le L)$ is the i-th element that takes a value 0 or 1, and L is the number of elements that consist of an individual. Let $P = (p_1, p_2, ..., p_L)$ be a probability vector where $p_i (1 \le i \le L)$ represents the probability to be $v_i = 1$. Then, the algorithm of PBIL is described as follows:

(1) As initialization, we let $p_i = 0.5$ for all $i = 1, 2, ..., L$.

(2) Generate a set S that consists of C individuals according to P. Namely, element v_i of each individual is determined according to the corresponding probability p_i .

(3) Compute the evaluation value for each individual $s \in S$.

(4) Select a set of individuals S' whose members have evaluation values within top C' in S , and update the probability vector according to the following formula:

$$p_i^{new} = ratio(i) \cdot \alpha + p_i \cdot (1.0 - \alpha) \quad (1)$$

where p_i^{new} is the updated value of the new probability

vector P^{new} (P is soon replaced with P^{new}), $ratio(i)$ is

$$P = (0.0, \ 0.5, \ 0.8, \ 0.1, \ 0.0, \ 0.5, \ 0.3, \ 0.4, \ 0.0)$$

P		Parent Node		
		X_1	X_2	X_3
Child Node	X_1	0.0	0.1	0.3
	X_2	0.5	0.0	0.4
	X_3	0.8	0.5	0.0

Determine v_{ij}
According to P_{ij}

$$s_1 = (\ 0, \quad 0, \quad 1, \quad 0, \quad 0, \quad 1, \quad 0, \quad 0, \quad 0)$$

s_1		Parent Node		
		X_1	X_2	X_3
Child Node	X_1	0	0	0
	X_2	0	0	0
	X_3	1	1	0

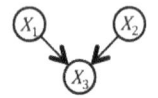

Structure of model s_1

Fig. 3. Step (2): Generating Individuals

the function that represents the ratio of individuals in S' that include link i (i.e., $v_i = 1$), and α is the parameter called learning ratio.

(5) Repeat steps (2)-(4).

By merging top-C' individuals, PBIL evolves the probability vector such that the good individuals are more likely to be generated. Different from other genetic algorithms, PBIL does not include "crossover" between individuals. Instead, it evolves the probability vector as a "parent" of the generated individuals.

III. PBIL-BASED BAYESIAN NETWORK LEARNING

In this section, we present a PBIL-based algorithm to learn Bayesian network models to which we apply a new mutation operator. Since our problem (i.e., Problem 1) to learn Bayesian networks is a little different from the general description of PBIL shown in the previous section, a little adjustment is required.

In our algorithm, individual creatures correspond to each Bayesian network model. Namely, with the number of events N, an individual model is represented as $s = (v_{11}, v_{12}, ..., v_{1N}, v_{21}, v_{22}, ..., v_{N1}, v_{N2}, ..., v_{NN})$, where v_{ij} corresponds to the edge from events X_i to X_j, i.e., if $v_{ij} = 1$ the edge from X_i to X_j exists in s, and if $v_{ij} = 0$ it does not exist. Similarly, we have the probability vector P to generate individual models as $P = (p_{11}, p_{12}, ..., p_{1N}, p_{21}, p_{22}, ..., p_{N1}, \quad p_{N2}, ..., p_{NN})$ where

A set of models S $(|S|=C)$

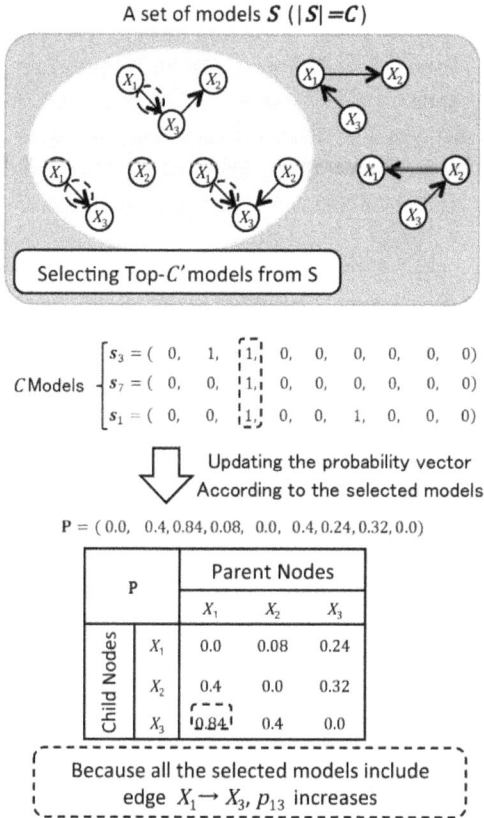

Selecting Top-C' models from S

C Models $\begin{cases} s_3 = (\ 0,\quad 1,\quad 1,\quad 0,\quad 0,\quad 0,\quad 0,\quad 0,\quad 0) \\ s_7 = (\ 0,\quad 0,\quad 1,\quad 0,\quad 0,\quad 0,\quad 0,\quad 0,\quad 0) \\ s_1 = (\ 0,\quad 0,\quad 1,\quad 0,\quad 0,\quad 1,\quad 0,\quad 0,\quad 0) \end{cases}$

Updating the probability vector
According to the selected models

$P = (\,0.0,\quad 0.4, 0.84, 0.08,\quad 0.0,\quad 0.4, 0.24, 0.32, 0.0)$

P		Parent Nodes		
		X_1	X_2	X_3
Child Nodes	X_1	0.0	0.08	0.24
	X_2	0.4	0.0	0.32
	X_3	0.84	0.4	0.0

Because all the selected models include
edge $X_1 \rightarrow X_3$, p_{13} increases

Fig. 4. Step (3)(4): Updating the Probability Vector

p_{ij} is the probability that the edge from X_i to X_j exists. A probability vector can be regarded as a table as illustrated in Fig. 2. Note that, because Bayesian networks do not allow self-edges, p_{ij} is always 0 if $i = j$.

The process of the proposed algorithm is basically obtained from the steps of PBIL. Namely, the basic steps are described as follows:

(1) Initialize the probability vector P as $p_{ij} = 0$ if $i = j$ and $p_{ij} = 0.5$ otherwise.

(2) Generate S as a set of C individual models according to P. (This step is illustrated in Fig. 3.)

(3) Compute values of the evaluation criterion for all individual models $s \in S$.

(4) Select a set of individuals S' whose members have top-C' evaluation values in S, and update the probability vector according to the formula (1). (These steps (3) and (4) are illustrated in Fig. 4.)

(5) Repeat steps (2)-(4).

Same as PBIL, the proposed algorithm evolves the

probability vector to be likely to generate better individual models. However, there is a point specific to Bayesian networks, that is, a Bayesian network model is not allowed to have cycles in it. To consider this point in our algorithm, step 2 is detailed as follows:

(2a) Create a random order of pairs (i, j), where $1 \le i, j \le N$ and $i \ne j$.

(2b) Determine the values of v_{ij} according to P, with the

$P = (\,0.0,\quad 0.4, 0.84, 0.08,\quad 0.0,\quad 0.4, 0.24, 0.32, 0.0)$

P		Parent Nodes		
		X_1	X_2	X_3
Child Nodes	X_1	0.0	0.08	0.24
	X_2	0.4	0.0	0.32
	X_3	0.84	0.4	0.0

Permutation on edge $X_2 \rightarrow X_1$

$P = (\,0.0,\quad 0.4, 0.84, 0.54,\quad 0.0,\quad 0.4, 0.24, 0.32, 0.0)$

P		Parent Nodes		
		X_1	X_2	X_3
Child Nodes	X_1	0.0	0.54	0.24
	X_2	0.4	0.0	0.32
	X_3	0.84	0.4	0.0

Fig. 5. Probability Mutation (PM)

ordercreated in step (2a); every time v_{ij} is determined, if v_{ij} is determined as 1, we check whether this edge from X_i to X_j creates a cycle with all the edges determined to exist so far. If it creates a cycle, let v_{ij} be 0.

(2c) Repeat steps (2a) and (2b) until all pairs (i, j) in the order are processed.These steps enable us to treat the problem of learning good Bayesian network models within the framework of PBIL. Note that checking the cycle creation

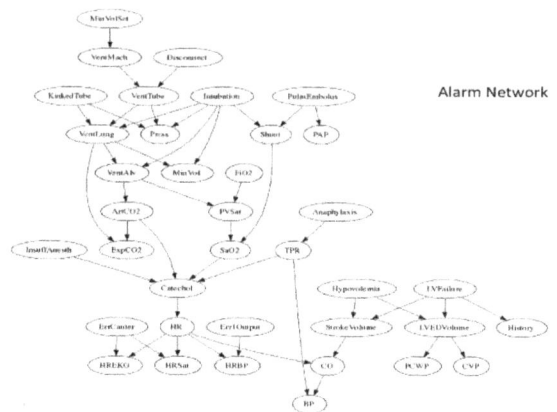

Fig. 6. The Alarm Network

in step (2b) can be done efficiently using a simple table that manages the taboo edges that create cycles when they are added to the model.

A. Mutation Operators

Note that the algorithm introduced in the previous section does not include mutation operator. Thus, naturally, it is easy to converge to a local minimum solution. Actually, PBIL-based algorithm to learn Bayesian networks proposed by

Blanco et al. [9] stops when the solution converges to a minimal solution, i.e., when score does not improve for recent k generations. However, local minimum solutions prevent us to search for better solutions, thus it should be avoided.

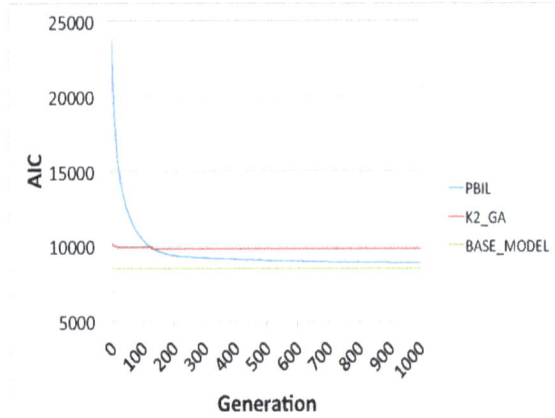

Fig. 7. Performance of the PBIL-based Algorithm

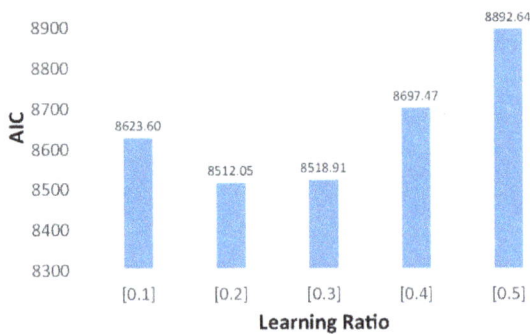

Fig. 8. AIC Scores under Variation of Learning Ratio

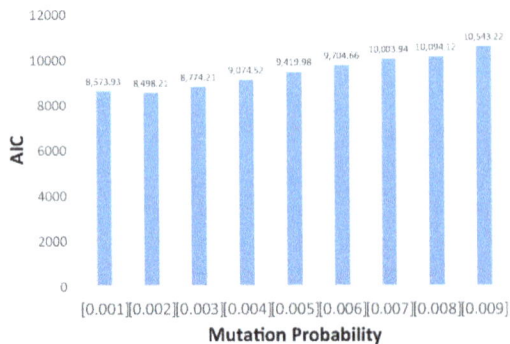

Fig. 9. AIC Scores under Variation of Mutation Probability

To avoid converging to the local minimum solution and to improve the performance of the algorithm, typically several mutation operations are inserted between steps (2) and (3). The most popular mutation operator is called *bitwise mutation* (BM) introduced by Handa [10], which apply mutations to each link in each individual, as described in the following step:

BM: For each individual in S generated in step (2), we flip each edge with probability p_{mut}. Namely, for each pair of nodes i and $j (1 \le i, j \le N)$, $v_{ij} \leftarrow 1$ if $v_{ij} = 0$, and $v_{ij} \leftarrow 0$ otherwise, with probability p_{mut}.

The other mutation operator we try in this paper is called *transpose mutation* (TM) introduced by [11]. This operation is proposed based on the observation that that reverse-edges frequently appear in the solutions. To avoid this, transpose mutation changes the direction of edges in the individuals produced in each generation. The specific operation inserted between steps (2) and (3) is in the following.

TM: For each individual in S generated in step (2), with probability p_{mut}, we do the following operation: we reverse all edges in the individual with probability p_{mut}, namely, $v_{ij} \leftarrow v_{ji}$ for all i and j.

In contrast to these conventional mutations shown above, our new mutation operator called *probability mutation* (PM) does not manipulate individuals produced in each generations. Instead, we manipulate the probability vector P to generate better individuals in the next generation, which is inserted between steps (4) and (5). The specific operation of this mutation is shown and in the following (See also Fig. 5):

PM: Apply mutations on the new probability vector P : For all pairs of events $(X_i, X_j), i \ne j$, we apply the following formula with probability p_{mut}, where the function $rand()$ generates a random value from range $[0,1]$.

$$p_i^{new} = rand() \leftarrow \beta + p_i \leftarrow (1 - \beta) \qquad (2)$$

IV. EVALUATION

A. Methods

In order to reveal the effectiveness of PBIL-based algorithms, we first evaluate the PBIL-based algorithm with probability mutation in comparison with K2 with its constraint (i.e., the order of events) evolved with genetic algorithms, which is a representative method among traditional approaches to learn Bayesian networks. In this conventional algorithm, we repeat creating Bayesian network models, in which its constraints (i.e., order of nodes) are continuously evolved with a typical genetic algorithm over generations, and output the best score among those computed ever. The results are described in Sec. IV-B. We next compare the performance of three mutation operators BM, TM, and PM applied to the PBIL-based algorithm. With this evaluation, we show that the new mutation operator PM proposed in this paper has good performance. The results are described in Sec. IV-C. In our experiment, we use Alarm Network [8] shown in Fig. 6, which is a Bayesian network model frequently used as a benchmark problem in this area of study. We create a set of 1000 observations according to the structure and the conditional probability of Alarm Network, and then learn Bayesian network models from the observations using those two algorithms. As the evaluation criterion, we use AIC, one of the representative criterion in this area. Namely, we compare the AIC values in order to evaluate how good is the Bayesian network models obtained by these two algorithms. As for parameters, we use $C = 1000, C' = 1, \alpha = 0.2, \beta = 0.5,$ and $p_{mut} = 0.002$.

B. Result 1: Performance of PBIL-based Algorithms

The first result is shown in Fig. 7, which indicates the AIC score of the best Bayesian network model found with the growth of generations. In this figure, the AIC score of the original Alarm Network, which is the optimal score, is denoted by "BASE MODEL." The proposed algorithm with *probability mutation* (represented as PBIL in the figure) converges to the optimal score as time passes, whereas K2-GA stops improving in the early stage. We can conclude that the performance of the PBIL-based algorithm is better than the conventional algorithm in that the PBIL-based algorithm computes better Bayesian network models according to time taken in execution. Note that the running time per generation in the proposed method is far shorter than K2-GA; the difference is more than 250 times in our implementation.

Fig. 8 and 9 show the performance of the proposed algorithm with variation of learning ratio α and mutation probability p_{mut} in 10,000 generations. These results show that the performance of the proposed method depends on α and p_{mut}, which indicates that we should care for these values

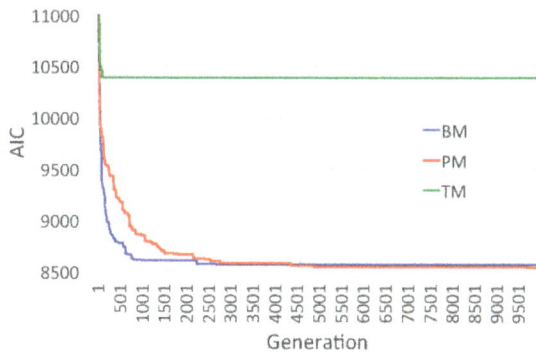

Fig. 10. Performance Comparison in Mutation Variations

to improve the performance of the proposed algorithm. Note that, from these results, we have the best-performance values $\alpha = 0.2$ and $p_{mut} = 0.002$, which are used as the default values in our experiment.

C. Result 2: Comparison of Mutation Variations

We further compared the performance of the PBIL-based algorithm with three mutations, bitwise mutation (BM), transpose mutation (TM), and probability mutation (PM). Facing on this experiment, we carefully choose the mutation probability of each method through preliminary experiments. For BM, we examined the performance of the mutation probability in range [0.001:0.2], and chose the value of the best performance, 0.005. For TM, we similarly tried the performance of the mutation probability in range [0.05:0.5],

and chose 0.1 as the best value. For PM, from the result shown in Fig. 9, we chose the mutation probability 0.002, which is the same value as our first result shown in Fig. 7.

The result is shown in Fig. 10. We see that BM and PM continue improving as generation passes, whereas TM stops improving at the early stage of generation. Also, we see that the curve of BM and PM are slightly different where BM reach better scores in the early stage while PM outperforms BM in the late stage. This result shows that the newly proposed mutation operator PM is also useful especially in long-term learning of Bayesian network models under PBIL-based algorithms.

V. CONCLUSION

In this paper, we introduced the literature of PBIL-based learning of Bayesian network models, and proposed a new mutation operator called probability mutation that manipulates probability vector of PBIL. Through evaluation of these algorithms, we found that (i) the PBIL-based algorithm outperforms K2-based traditional algorithms with the long-term continuous improvement, and (ii) probability mutation works well under PBIL-based algorithms especially in long-term computation to obtain high-quality Bayesian network models. Designing more efficient search algorithms based on EDA is one of the most attractive future tasks.

REFERENCES

[1] D.M. Chickering, D. Heckerman, C. Meek, "Large-Sample Learning of Bayesian Networks is NP-Hard," Journal of Machine Learning Research, Vol.5, pp.1287—1330, 2004.

[2] Cooper, G. F., and Herskovits, E. A Bayesian method for the induction of probabilistic networks from data. Machine Learning, 9, 309-347, 1992

[3] W.H. Hsu, H. Guo, B.B. Perry, and J.A. Stilson, A permutation genetic algorithm for variable ordering in learning Bayesian networks from data In Proceedings of the Genetic and Evolutionary Computation Conference (GECCO), 2002.

[4] O.Barrière, E. Lutton, P.H. Wuillemin, "Bayesian Network Structure Learning using Cooperative Coevolution," In Proceedings of the Genetic and Evolutionary Computation Conference (GECCO), pp.755-762, 2009.

[5] A.P Tonda, E. Lutton, R. Reuillon, G. Squillero, and P.H. Wuillemin, Bayesian network structure learning from limited datasets through graph evolution, In Proceedings of the 15th European conference on Genetic Programming (EuroGP'12), pp.254-265, 2012.

[6] Akaike, H., "Information theory and an extension of the maximum likelihood principle", Proceedings of the 2nd International Symposium on Information Theory, pp.267-281 (1973).

[7] Shumeet Baluja. Population-based incremental learning: A method for integrating genetic search based function optimization and competitive learning. Technical Report No. CMU-CS-94-163, Carf Michigan, Ann Arbor, 1994.

[8] Beinlich, I,A., Suermondt, H.J., Chavez R.M., Cooper G.F. The ALARM monitoring System: A Case Study with Two Probabilistic Inference Techniques for Belief Networks" In: Second European Conference on Artificial Intelligence in Medicine. Volume 38., London, Great Britain, Springer-Verlag, Berlin 247—256, 1989.

[9] R. Blanco, I. Inza, P. Larrañaga, "Learning Bayesian Networks in the Space of Structures by Estimation of Distribution Algorithms," International Journal of Intelligent Systems, Vol.18, pp.205–220, 2003.

[10] H. Handa, "Estimation of Distribution Algorithms with Mutation," Lecture Notes in Computer Science, Vol.3448, pp.112-121, 2005.

[11] D.W. Kim, S. Ko, and B.Y. Kang, "Structure Learning of Bayesian Networks by Estimation of Distribution Algorithms with Transpose Mutation," Journal of Applied Research and Technology, Vol.11, pp.586– 596, 2013.
[12] H. Muhlenbein, "The Equation for Response to Selection and Its Use for Prediction," Evolutionary Computation, Vol.5, No.3, pp. 303–346, 1997.
[13] J.S. De Bonet et al., "MIMIC: Finding Optima by Estimating Probability Densities," Advances in Neural Information Processing Systems, Vol.9, pp.424–430, 1997.

Sho Fukuda received his B.E. and M.E. degrees from Wakayama University in 2012 and 2014, respectively. He is currently working with Intec Hankyu Hanshin Co.Ltd. He is interested in Data Analytics with large data sets.

Yuuma Yamanaka is currently pursuing his Bachelor's degree in Faculty of Systems Engineering, Wakayama University. He is interested in Data Analytics and Machine Learning.

Takuya Yoshihiro received his B.E., M.I. and Ph.D. degrees from Kyoto University in 1998, 2000 and 2003, respectively. He was an assistant professor in Wakayama University from 2003 to 2009. He has been an associate professor in Wakayama University from 2009. He is currently interested in the graph theory, distributed algorithms, computer networks, wireless networks, medical applications, bioinformatics, etc. He is a member of IEEE, IEICE, and IPSJ.

25

Evaluation of Shelf Life of Processed Cheese by Implementing Neural Computing Models

Sumit Goyal, Gyanendra Kumar Goyal

National Dairy Research Institute, Karnal, India.

Email - thesumitgoyal@gmail.com, gkg5878@yahoo.com

Abstract — **For predicting the shelf life of processed cheese stored at 7-8° C, Elman single and multilayer models were developed and compared. The input variables used for developing the models were soluble nitrogen, pH; standard plate count, Yeast & mould count, and spore count, while output variable was sensory score. Mean Square Error, Root Mean Square Error, Coefficient of Determination and Nash - Sutcliffo Coefficient were applied in order to compare the prediction ability of the developed models. The Elman models got simulated very well and showed excellent agreement between the experimental data and the predicted values, suggesting that the Elman models can be used for predicting the shelf life of processed cheese.**

Keywords— **Artificial Neural Network, Artificial Intelligence, Elman, Processed Cheese, Shelf Life**

I. INTRODUCTION

ARTIFICIAL neural network (ANN), usually called neural network is a mathematical model or computational model that is inspired by the structure and functional aspects of ANN. ANN based computing method is an adaptive system that changes its structure based on external or internal information that flows through the network during the learning phase. In ANN based intelligent computing, simple artificial nodes called "neurons", "neurodes", "processing elements" or "units" are connected together to form a network of nodes mimicking the biological neural networks. Generally, ANN involves a network of simple processing elements that exhibit complex global behavior determined by connections between processing elements and element parameters. While an ANN does not have to be adaptive, its practical use comes with algorithms designed to alter the weights of the connections in the network to produce a desired signal flow [1].Elman models are two layered backpropagation networks, with the addition of a feedback connection from the output of the hidden layer to its input. This feedback path allows Elman model to learn to recognize and generate temporal patterns, as well as spatial patterns. The Elman ANN model has *tansig* neurons in its hidden layer, and *purelin* neurons in its output layer. This combination is special in that two layered networks with these

transfer functions can approximate any function (with a finite number of discontinuities) with arbitrary accuracy. The only requirement is that the hidden layer must have enough neurons. More hidden neurons are needed as the function being fitted increases in complexity. Elman model differs from conventional two layer networks in that the first layer has a recurrent connection. The delay in this connection stores values from the previous time step, which can be used in the current time step. Therefore, even if two Elman models, with the same weights and biases, are given identical inputs at a given time step, their outputs can be different because of different feedback states. Because the network can store information for future reference, it is able to learn temporal patterns as well as spatial patterns. The Elman models can be trained to respond to, and to generate, both kinds of patterns [2].Shelf life studies can provide important information to product developers enabling them to ensure that the consumer will see a high quality product for a significant period of time after production. Of course, long shelf life studies do not fit with the speed requirement and therefore, accelerated studies have been developed as part of innovation [3]. Goyal and Goyal [4] implemented brain based artificially intelligent scientific computing models for shelf life detection of cakes stored at 30°C. The potential of simulated neural networks for predicting shelf life of soft cakes stored at 10°C was highlighted by Goyal and Goyal [5]. Cascade single and double hidden layer models were developed for predicting the shelf life of Kalakand, a desiccated sweetened dairy product [6]. For forecasting the shelf life of instant coffee drink, radial basis artificial neural engineering and multiple linear regression models were suggested [7]. Cascade forward and feedforward backpropagation artificial intelligence models for prediction of sensory quality of instant coffee flavoured sterilized drink have been evolved [8]. Artificial neural networks for predicting the shelf life of milky white dessert jeweled with pistachio were applied by Goyal and Goyal [9]. The shelf life of brown milk cakes decorated with almonds was predicted by developing artificial neural network based radial basis (exact fit) and radial basis (fewer neurons) models [10]. Also, the time-delay and linear layer (design) intelligent computing expert system models have been developed for predicting the shelf life of soft mouth melting milk cakes stored at 6°C [11]. The computerized models have been suggested for predicting the shelf life of post-harvest coffee sterilized milk drink [12]. Neuron based artificial intelligent scientific computer engineering models estimated the shelf life of instant coffee sterilized drink [13]. The aim of the present study is to develop Elman ANN models with single layer and multilayer, and to compare them with each other, for predicting the shelf life of processed cheese stored at 7-8ºC.

II. METHOD MATERIAL

The input variables used in the network were the processed cheese experimental data relating to soluble nitrogen, pH; standard plate count, Yeast & mould count, and spore count. The sensory score assigned by the trained panelists was taken as output variable for developing computing models (Fig.1). Experimentally obtained 36 observations for each input and output variables were used for developing the models. The dataset was randomly divided into two disjoint subsets, namely, training set having 30 (80% for training) observations, and validation set (20% for testing) consisting of 6 observations.

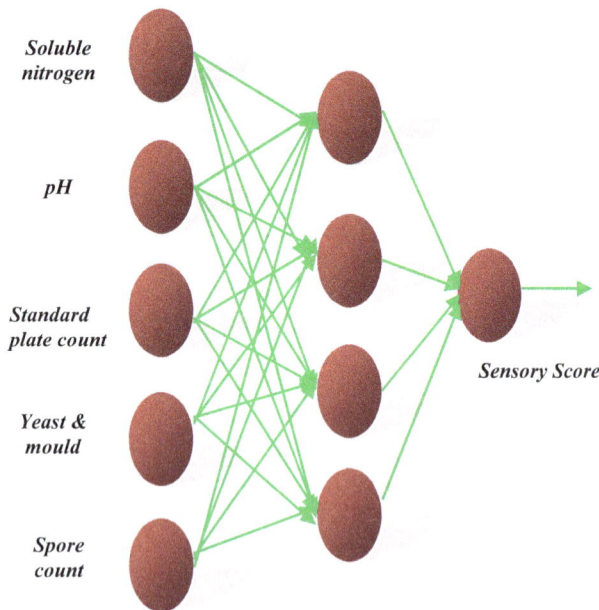

Fig. 1: Input and output parameters for elman models

$$MSE = \left[\sum_{1}^{N} \left(\frac{Q_{exp} - Q_{cal}}{n} \right)^2 \right]$$
(1)

$$RMSE = \sqrt{\frac{1}{n} \left[\sum_{1}^{N} \left(\frac{Q_{exp} - Q_{cal}}{Q_{exp}} \right)^2 \right]}$$
(2)

$$R^2 = 1 - \left[\sum_{1}^{N} \left(\frac{Q_{exp} - Q_{cal}}{Q_{exp}^2} \right)^2 \right]$$
(3)

$$E^2 = 1 - \left[\sum_{1}^{N} \left(\frac{Q_{exp} - Q_{cal}}{Q_{exp} - \overline{Q_{exp}}} \right)^2 \right]$$
(4)

Where,

Q_{exp} = Observed value;

Q_{cal} = Predicted value;

$\overline{Q_{exp}}$ =Mean predicted value;

n = Number of observations in dataset.

Mean Square Error MSE (1), Root Mean Square Error RMSE (2), Coefficient of Determination R^2 (3) and Nash - Sutcliffe Coefficient E^2 (4) were applied in order to compare the prediction ability of the developed models. *Gradient Descent algorithm with adaptive learning rate, Powell Beale restarts conjugate gradient algorithm, Levenberg Marquardt algorithm, Fletcher Reeves update conjugate gradient algorithm,* and *Bayesian regularization* algorithms were tried. *Bayesian regularization* mechanism was finally selected for training ANN models, as it exhibited the best results. The network was trained up to 100 epochs, and neurons in each hidden layers varied from 1 to 20. The network was trained with single as well as multiple hidden layers, and transfer function for hidden layer was *tangent sigmoid,* while for the output layer it was *pure linear* function. MALTAB software was used for performing experiments.

III. RESULTS AND DISCUSSION

Elman single layer (Table 1) and multilayer (Table 2) ANN models were developed and compared with each other for predicting the shelf life of processed cheese stored at 7-8º C.

TABLE I
RESULTS FOR SINGLE LAYER ELMAN MODEL

Neurons	MSE	RMSE	R^2	E^2
3	9.13178E-05	0.009556034	0.990443966	0.999908682
4	0.000314749	0.01774116	0.98225884	0.999685251
5	0.000449704	0.021206231	0.978793769	0.999550296
6	9.14141E-05	0.009561074	0.990438926	0.999908586
7	0.00039364	0.019840363	0.980159637	0.99960636
8	0.00039364	0.019840363	0.980159637	0.99960636
9	9.15588E-05	0.009568634	0.990431366	0.999908441
10	9.1607E-05	0.009571154	0.990428846	0.999908393

11	0.000330042	0.018167049	0.981832951	0.999669958
12	3.34199E-05	0.005780997	0.994219003	0.99996658
13	0.000188492	0.013729239	0.986270761	0.999811508
14	4.20792E-06	0.002051322	0.997948678	0.999995792
15	9.18484E-05	0.009583755	0.990416245	0.999908152
16	9.18967E-05	0.009586275	0.990413725	0.999908103
17	9.1945E-05	0.009588795	0.990411205	0.999908055
18	0.000115845	0.010763139	0.989236861	0.999884155
19	9.20417E-05	0.009593835	0.990406165	0.999907958
20	**1.87878E-07**	**0.000433449**	**0.999566551**	**0.999999812**

TABLE 2
RESULTS FOR MULTILAYER ELMAN MODEL

Neurons	*MSE*	*RMSE*	*R²*	*E²*
3:3	9.1366E-05	0.009558554	0.990441446	0.999908634
4:4	0.000561383	0.023693521	0.976306479	0.999438617
5:5	9.14141E-05	0.009561074	0.990438926	0.999908586
6:6	9.14141E-05	0.009561074	0.990438926	0.999908586
7:7	9.14623E-05	0.009563594	0.990436406	0.999908538
8:8	9.14623E-05	0.009563594	0.990436406	0.999908538
9:9	9.15105E-05	0.009566114	0.990433886	0.999908489
10:10	4.78872E-05	0.006920061	0.993079939	0.999952113
11:11	0.000535418	0.02313911	0.97686089	0.999464582
12:12	0.000554478	0.023547358	0.976452642	0.999445522
13:13	9.16552E-05	0.009573674	0.990426326	0.999908345
14:14	9.17035E-05	0.009576194	0.990423806	0.999908296
15:15	9.17518E-05	0.009578715	0.990421285	0.999908248
16:16	8.80247E-05	0.009382151	0.990617849	0.999911975
17:17	3.93431E-05	0.006272407	0.993727593	0.999960657
18:18	9.18484E-05	0.009583755	0.990416245	0.999908152
19:19	0.000711004	0.026664661	0.973335339	0.999288996
20:20	**1.17981E-05**	**0.00343483**	**0.99656517**	**0.999988202**

Elman single layer and multilayer computerized models were developed for predicting the shelf life of processed cheese stored at 7-8° C. Single layer model with 5-20-1 combination (**MSE: 1.87878E-07; RMSE: 0.000433449; R² : 0.999566551; E²: 0.999999812**) gave the best result among single layer experiments (Table 1); and for multilayer Elman models, the best result was with 5-20-20-1 combination (**MSE: 1.17981E-05; RMSE: 0.00343483; R² : 0.99656517; E² : 0.999988202**) (Table 2). The comparison of these two results showed that the multilayer model with a combination of 5-20-20-1 performed better for predicting the shelf life of processed cheese. The comparison of Actual Sensory Score (ASS) and Predicted Sensory Score (PSS) for Elman single layer and multilayer models are illustrated in Fig.2 and Fig.3, respectively.

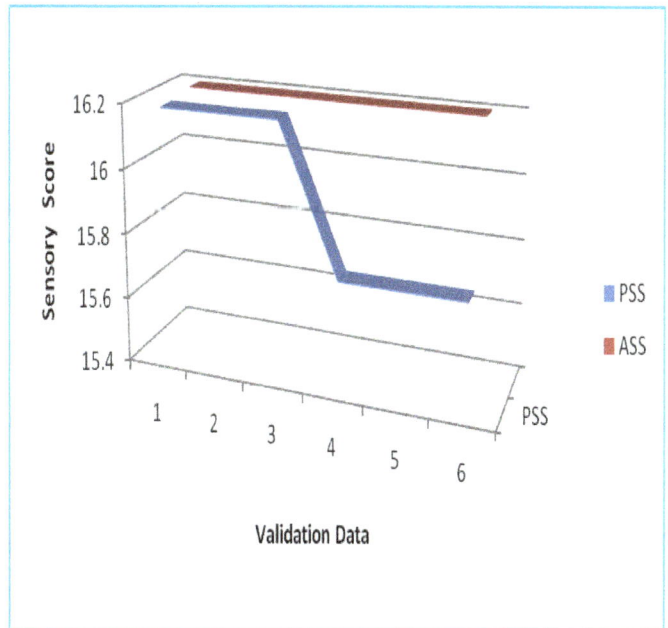

Fig. 2: Comparison of ASS and PSS for Elman single layer model

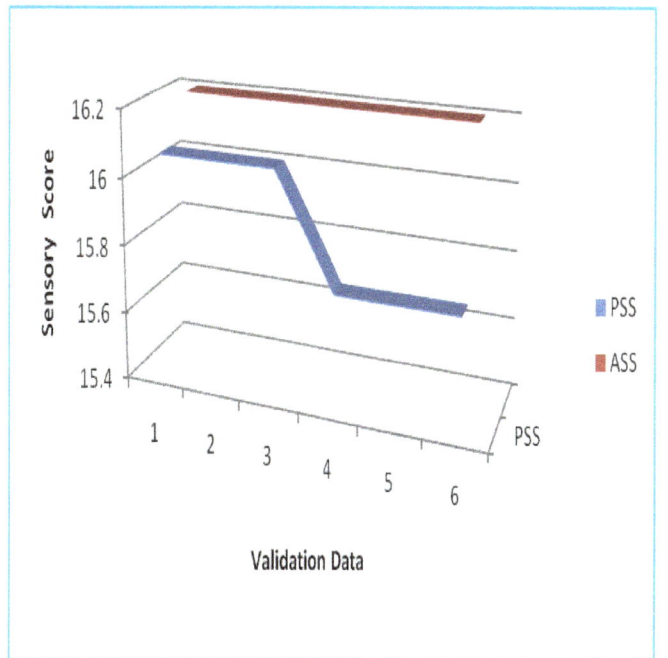

Fig. 3: Comparison of ASS and PSS for Elman multilayer model

From the results, it is observed that Elman models got simulated exceedingly well, and are very effective in predicting the shelf life of processed cheese stored at 7-8 ° C.

IV. CONCLUSION

Elman single and multilayer ANN models were developed and compared with each other. The inputs variables of the network consisted of soluble nitrogen, pH; standard plate count, yeast & mould count, and spore count. The output variable was sensory score of the processed cheese stored at 7-8°C. The modelling results revealed very good agreement between the experimental data and the predicted values, with a high determination coefficient, establishing that the developed Elman ANN models were able to analyze non-linear multivariate data with excellent performance, fewer parameters, and shorter calculation time. This Elman model might be an alternative low cost and less time consuming method for determining the expiration date of stored processed cheese, shown on labels and provide consumers with a safer food supply [14-20].

V. REFERENCES

[1] Artificial Nerural Network http://en.wikipedia.org/wiki/Artificial_neural_network (accessed on 4.7.2011)

[2] H. Demuth , M. Beale and M. Hagan. Neural network toolbox user's guide". The MathWorks Inc., Natrick, USA , (2009).

[3] www.medlabs.com/Downloads/food_product_shelf_life_web.pdf (accessed on 7.3.2011).

[4] Sumit Goyal and G.K. Goyal, "Brain based artificial neural network scientific computing models for shelf life prediction of cakes", Canadian Journal on Artificial Intelligence, Machine Learning and Pattern Recognition, 2(6), 73-77,2011.

[5] Sumit Goyal and G.K. Goyal, "Simulated neural network intelligent computing models for predicting shelf life of soft cakes", Global Journal of Computer Science and Technology, 11(14), Version 1.0, 29-33,2011.

[6] Sumit Goyal and G.K. Goyal, "Advanced computing research on cascade single and double hidden layers for detecting shelf life of kalakand: An artificial neural network approach", International Journal of Computer Science & Emerging Technologies, 2(5), 292-295, 2011.

[7] Sumit Goyal and G.K. Goyal, "Application of artificial neural engineering and regression models for forecasting shelf life of instant coffee drink", International Journal of Computer Science Issues, 8(4), No 1, 320-324, 2011.

[8] Sumit Goyal and G.K. Goyal, "Cascade and feedforward backpropagation artificial neural networks models for prediction of sensory quality of instant coffee flavoured sterilized drink", Canadian Journal on Artificial Intelligence, Machine Learning and Pattern Recognition, 2(6), 78-82, 2011.

[9] Sumit Goyal and G.K. Goyal, "A new scientific approach of intelligent artificial neural network engineering for predicting shelf life of milky white dessert jeweled with pistachio", International Journal of Scientific and Engineering Research, 2(9), 1-4, 2011.

[10] Sumit Goyal and G.K. Goyal, "Radial basis artificial neural network computer engineering approach for predicting shelf life of brown milk cakes decorated with almonds", International Journal of Latest Trends in Computing, 2(3), 434-438, 2011.

[11] Sumit Goyal and G.K. Goyal, "Development of intelligent computing expert system models for shelf life prediction of soft mouth melting milk cakes", International Journal of Computer Applications, 25(9), 41-44, 2011.

[12] Sumit Goyal and G.K. Goyal, "Computerized models for shelf life prediction of post-harvest coffee sterilized milk drink", Libyan Agriculture Research Center Journal International, 2 (6), 274-278, 2011.

[13] Sumit Goyal and G.K. Goyal, "Development of neuron based artificial intelligent scientific computer engineering models for estimating shelf life of instant coffee sterilized drink", International Journal of Computational Intelligence and Information Security, 2(7), 4-12,2011.

[14] Sumit Goyal and G.K. Goyal, "Soft computing single hidden layer models for shelf life prediction of burfi", Russian Journal of Agricultural and Socio-Economic Sciences, 5(5), 28-32, 2012.

[15] Sumit Goyal and G.K. Goyal, "Predicting shelf life of dairy product by using artificial neural networks (ANN) and statistical computerized methods", International Journal of Computer Engineering Research, 3(2), 20-24, 2012.

[16] Sumit Goyal and G.K. Goyal, "Time – delay single layer artificial neural network models for estimating shelf life of burfi", International Journal of Research Studies in Computing, 1(2), 11-18, 2012.

[17] Sumit Goyal and G.K. Goyal, "Study on single and double hidden layers of cascade artificial neural intelligence neurocomputing models for predicting sensory quality of roasted coffee flavoured sterilized drink", International Journal of Applied Information Systems, 1(3), 1-4, 2012.

[18] Sumit Goyal and G.K. Goyal, "Shelf life determination of kalakand using soft computing technique", Advances in Computational Mathematics and its Applications, 1(3), 131-135, 2012.

[19] Sumit Goyal and G.K. Goyal, "Radial basis (exact fit) and linear layer (design) ANN models for shelf life prediction of processed cheese", International Journal of u- and e- Service, Science and Technology, 5(1), 63-69, 2012.

[20] Sumit Goyal and G.K. Goyal, "Central nervous system based computing models for shelf life prediction of soft mouth melting milk cakes", International Journal of Information Technology and Computer Science, 4(4), 33-39, 2012.

Author's Biodata

Sumit Goyal: is M.Phil. in Computer Science, Master of Computer Applications and Bachelor of Information Technology. His research interests have been in the area of soft computing, artificial neural networks and prediction of shelf life of food products. His research has appeared in Canadian Journal on Artificial Intelligence, Machine Learning and Pattern Recognition, Int. J. of Computer Applications, Int. J. of Computational Intelligence and Information Security, Int. J. of Latest Trends in Computing, Int. J. of Scientific and Engineering Research, Int. J. of Computer Science Issues, Int. J. of Computer Science & Emerging Technologies, Global Journal of Computer Science and Technology, Int. J. of Artificial Intelligence and Knowledge Discovery, amongst others. He is member of IDA.

Gyanendra Kumar Goyal: obtained his Ph.D. degree in 1979 from Panjab University, Chandigarh, India. He was recipient of United Nations fellowship award and World Bank's fellowship award. He was also awarded Belgian Government's fellowship award. In 1985-86, he did specialized research work on Dairy and Food Packaging at Michigan State University, East Lansing, U.S.A.; and in the year 1999 he received advanced training in Education Technology at Cornell University, New York, U.S.A. His research interests include dairy & food packaging and shelf life determination of food products. He has published more than 150 research papers in national and international journals, and presented his work in national and international conferences. His research work has been published in Int. J. of Food Sci. Technol. and Nutrition, Nutrition and Food Science, Milchwissenschaft, American Journal of Food Technology, British Food Journal, Canadian Journal on Artificial Intelligence, Machine Learning and Pattern Recognition, Int. J. of Computer Applications, Int. J. of Computer Science Issues, Int. J. of Computer Science & Emerging Technologies, Int. J. of Artificial Intelligence and Knowledge Discovery, amongst others. He is life member of AFST (I) and IDA.

26

Wireless Sensor Networks and Real-Time Locating Systems to Fight against Maritime Piracy

Óscar García, Ricardo S. Alonso, Dante I. Tapia, Fabio Guevara, *R&D Department, Nebusens, S.L.*
Fernando de la Prieta, Raúl A. Bravo, *Department of Computer Science and Automation, University of Salamanca*

Abstract — There is a wide range of military and civil applications where Wireless Sensor Networks (WSNs) and Multi-Agent Systems (MASs) can be used for providing context-awareness for troops and special corps. On the one hand, WSNs comprise an ideal technology to develop Real-Time Locating Systems (RTLSs) aimed at indoor environments, where existing global navigation satellite systems do not work properly. On the other hand, agent-based architectures allow building autonomous and robust systems that are capable of working on highly dynamic scenarios. This paper presents two piracy scenarios where the n-Core platform can be applied. n-Core is a hardware and software platform intended for developing and deploying easily and quickly a wide variety of WSNs applications based on the ZigBee standard. In the first scenario a RTLS is deployed to support boarding and rescue operations. In the second scenario a multi-agent system is proposed to detect the unloading of illegal traffic of merchandise at ports.

Keywords— Wireless Sensor Networks, Real-Time Locating Systems, Multi-Agent Systems, Maritime Piracy.

This work has been supported by the
Spanish Ministry of Science and Innovation (Subprograma Torres Quevedo).

Óscar García is with the R&D Department of Nebusens, S.L.. Scientific Park of the University of Salamanca, Building M2, calle Adaja s/n, 37185, Villamayor de la Armuña, Salamanca, Spain (corresponding author to provide e-mail: oscar.garcia@nebusens.com).

Ricardo S. Alonso is with the R&D Department of Nebusens, S.L.. Scientific Park of the University of Salamanca, Building M2, calle Adaja s/n, 37185, Villamayor de la Armuña, Salamanca, Spain (e-mail: ricardo.alonso@nebusens.com).

Dante I. Tapia is with the R&D Department of Nebusens, S.L.. Scientific Park of the University of Salamanca, Building M2, calle Adaja s/n, 37185, Villamayor de la Armuña, Salamanca, Spain (e-mail: dante.tapia@nebusens.com).

Fabio Guevara is with the R&D Department of Nebusens, S.L.. Scientific Park of the University of Salamanca, Building M2, calle Adaja s/n, 37185, Villamayor de la Armuña, Salamanca, Spain (e-mail: fabio.guevara@nebusens.com).

Fernando de la Prieta is with the Department of Computer Science and Automation, University of Salamanca. Plaza de la Merced, s/n, 37008, Salamanca, Spain (e-mail: fer@usal.es).

Raúl A. Bravo is with the Department of Computer Science and Automation, University of Salamanca. Plaza de la Merced, s/n, 37008, Salamanca, Spain (e-mail: raulabel@usal.es).

I. Introduction

WIRELESS Sensor Networks (WSNs) are used for gathering the information useful to build context-aware environments, whether in home automation, industrial applications or smart hospitals [1]. Nevertheless, the information obtained by Wireless Sensor Networks must be managed by intelligent and self-adaptable technologies to provide an adequate interaction between the users and their environment. In this sense, agents and Multi-Agent Systems (MASs) [2] comprise one of the areas that contribute expanding the possibilities of Wireless Sensor Networks.

One of the most interesting applications for WSNs is Real-Time Locating Systems (RTLSs). Although outdoor locating is well covered by systems such as the current GPS or the future Galileo, indoor locating needs still more development, especially with respect to accuracy and low-cost and efficient infrastructures [3]. Therefore, it is necessary to develop RTLSs that allow performing efficient indoor locating in terms of precision and optimization of resources. In this sense, the use of optimized locating techniques allows obtaining more accurate locations using even fewer sensors and with less computational requirements [3].

In this sense, Nebusens and the BISITE Research Group of the University of Salamanca have developed n-Core [4], a hardware and software platform intended for developing and deploying easily and quickly a wide variety of WSN applications based on the ZigBee standard [5]. n-Core consists of several modules, fully integrated among them, which provide all the functionalities of the platform through an Application Programming Interface (API), including two engines to develop specific applications, one to build automation applications and another intended for creating Real-Time Locating Systems [4].

This paper, which is an extension of the work published in the proceedings of DCAI 2012 [6], proposes two maritime piracy case studies where n-Core can be applied. The first one consists of a RTLS that can be deployed to support maritime boarding and rescue operations. This system, called n-Core Polaris [4] [7] and also developed by Nebusens and BISITE, is

Fig. 1. n-Core Sirius devices: Sirius B (left), Sirius A (center), Sirius D (right).

based on WSNs and MASs and includes the n-Core's innovative set of locating and automation engines. n-Core Polaris is an especially useful tool in environments where it is needed to locate people or assets in real-time with a fast deployment, such as natural or nuclear disasters. This way, the system will support special corps when performing rescue operations that involve a hostile boarding. In this regard, the system will provide them with real-time information, facilitating the coordination of the operation and avoiding casualties.

The second case study consists of a MAS intended for detecting the unloading of illegal traffic of merchandise at ports. On the one hand, the MAS uses the Global Positioning System (GPS) to keep track of the location of ships at a global scale. On the other hand, the system makes use of the WSNs and the automation and locating engines provided by n-Core. This way, the system can detect the proximity of a certain ship to a port, identifying automatically the merchandise that is unloaded. This information is compared with the assets that were loaded at the port of origin, thus detecting and preventing illegal traffic situations.

The rest of the paper is structured as follows. The next section explains the problem description, as well as the research areas involved in the development of the n-Core platform and the n-Core Polaris RTLS. Then, the main characteristics of the innovative n-Core platform and n-Core Polaris system are described. After that, two case studies where the n-Core platform is proposed to be applied to fight against maritime piracy are described. Finally, the conclusions and future lines of work are presented.

II. PROBLEM DESCRIPTION

In recent years, the problem of maritime piracy has become worryingly well-known all over the world due to attacks against fishing ships and oil tankers in Indian Ocean's waters near Somalia coast [8]. These attacks imply substantial human, social and economic costs for the fishing and merchant countries due to military expenses, ransoms, as well as the reduction of the international commerce and fishing. In this

regard, the use of technology can help civilian and military personnel both at sea and at ground to face emergency situations, reducing drastically the costs derived from an eventual rescue intervention, as well as the expenses in preventive measures.

Nevertheless, technology should help users to perform surveillance and rescue tasks without distracting them. In addition, technology should increase the knowledge about the environment by users, have a steep learning curve, as well as be non-invasive, context-aware, efficient and inexpensive. Some of the research areas and technologies proposed in this work to fight against maritime piracy are Wireless Sensor Networks, Multi-Agent Systems and Real-Time Locating Systems.

One of the most important technologies used for providing context-awareness for systems and applications is Wireless Sensor Networks (WSN) [1]. Context-aware technologies allow civil and military developments to automatically obtain information from users and their environment in a distributed and ubiquitous way. The context information may consist of many different parameters such as location, the ambient status (e.g., temperature), vital signs (e.g., heart rhythm), etc. Sensor networks need to be fast and easy to install and maintain. In this regard, Wireless Sensor Networks are more flexible and require less infrastructural support than wired sensor networks, existing plenty of technologies for implementing WSNs, such as RFID, UWB, ZigBee, Wi-Fi or Bluetooth [1].

Moreover, the information obtained by WSNs can be managed by intelligent and self-adaptable technologies to provide an adequate interaction between the users and their environment. In this sense, the development of agents is an essential piece in the analysis of data from distributed sensors and gives them the ability to work together and analyze complex situations, thus achieving high levels of interaction with humans [2]. Furthermore, agents can use reasoning mechanisms and methods in order to learn from past experiences and to adapt their behavior according to the context [9].

Tracking the real-time position of people and assets can make the difference in a maritime piracy scenario. One of the most interesting applications for WSNs is Real-Time Locating Systems (RTLSs). The most important factors in the locating process are the kinds of sensors used and the techniques applied for the calculation of the position based on the information recovered by these sensors. Besides, there is a need to develop Real-Time Locating Systems that perform efficient indoor locating in terms of precision and resource optimization [3] [10]. This optimization of resources includes the reduction of the costs and size of the sensor infrastructure involved on the locating system. Real-Time Locating Systems can be categorized by the kind of its wireless sensor infrastructure and by the locating techniques used for calculating the position of the tags (i.e., the locating engine). This way, there is a combination of several wireless technologies, such as RFID, Wi-Fi, UWB and ZigBee, and also a wide range of locating techniques that can be used for determining the position of the tags. Among the most widely used locating techniques we have *signpost*, *fingerprinting*, *triangulation*, *trilateration* and *multilateration* [3] [11] [12]. However, each of these must deal with important problems when trying to develop a precise locating system that uses WSNs in its infrastructure, especially for indoor environments.

III. THE N-CORE PLATFORM AND THE N-CORE POLARIS RTLS

Nebusens and the BISITE (Bioinformatics, Intelligent Systems and Education Technology) Research Group of the University of Salamanca have developed the n-Core platform [4]. The n-Core platform is based on the IEEE 802.15.4/ZigBee international standard, which operates in the 868/915MHz and 2.4GHz unlicensed bands. Unlike Wi-Fi or Bluetooth, ZigBee is designed to work with low-power nodes and allows up to 65,534 nodes to be connected in a star, tree or mesh topology network [5]. The n-Core platform consists of several modules, fully integrated among them, which provide all the functionalities of the platform.

At the hardware level, the n-Core platform provides a set of radio-frequency devices, called n-Core Sirius A, Sirius B and Sirius D (Figure 1). Each n-Core Sirius device includes an 8-bit RISC (Atmel ATmega 1281v) microcontroller with 8KB RAM, 4KB EEPROM and 128KB Flash memory and an IEEE 802.15.4/ZigBee 2.4GHz (AT86RF230) or 868/915MHz (AT86RF212) transceiver, and several communication ports (GPIO, ADC, I2C, PWM and UART through USB or DB-9 RS-232) to connect to distinct devices, such as computers, sensors and actuators [4].

At the software level, all n-Core Sirius devices are provided with a specific firmware that offers all its functionalities. This way, developers do not have to write embedded code. They can either simply configure the devices functionalities from a specific tool or write high-level code using the n-Core Application Programming Interface (API) from a computer. The n-Core API allows creating easily end-user applications from any compatible language and Integrated Development Environment (IDE), for example, C/C++, .NET, Java, or Python, among many others. n-Core also offers through this API different modules/engines to develop specific applications, including an automation engine (for controlling sensors and actuators), a locating engine (includes innovative algorithms to calculate the position of any n-Core device) and a data engine (for transmitting general-purpose data frames among devices).

Therefore, the functionalities provided by the n-Core Platform allow building systems in a wide range of application areas, including home automation (control of lighting and HVAC, control of electronic devices, security), healthcare

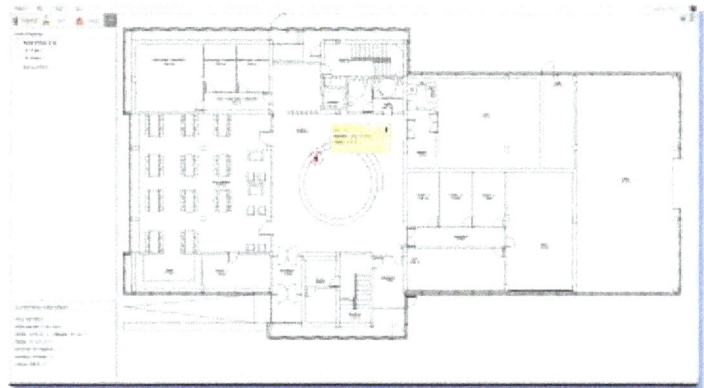

Fig. 2. The Web Services based architecture of the n-Core Polaris RTLS.

(location of patients and medical staff, access control and wander prevention), industry (location of workers, monitoring of assets and dangerous materials, process automation), environment (monitoring of environmental data, irrigation systems, animal tracking) or energy (control of energy costs, monitoring of consumption patterns), among others.

Besides, n-Core Polaris is an innovative Real-Time Locating System also developed by Nebusens and BISITE over the n-Core platform, and features a tested accuracy, flexibility and automation integration [4] [7] [13]. Therefore, the wireless infrastructure of n-Core Polaris is made up of a set of n-Core Sirius devices. In the n-Core Polaris RTLS, n-Core Sirius B devices are used as tags, while n-Core Sirius D devices are used as readers (i.e., position references). This way, n-Core Sirius B devices are carried by users and objects to be located, whereas n-Core Sirius D devices are placed throughout the environment to detect the tags. Finally, n-Core Sirius A devices are used for connecting sensors and actuators through their communication ports.

Figure 2 shows the basic architecture of the n-Core Polaris RTLS. The kernel of the system is a computer that is connected to a ZigBee network formed by n-Core Sirius devices. That is, the computer is connected to an n-Core Sirius D device through its USB port. This device acts as coordinator of the ZigBee network. The computer runs a web server module that offers the innovative locating techniques provided by the n-Core API. On the one hand, the computer gathers the detection information sent by the n-Core Sirius D devices acting as readers to the coordinator node. One the other hand, the computer acts as a web server offering the location info to a wide range of possible client interfaces. In addition, the web server module can access to a remote database to obtain

information about the users and register historical data, such as alerts and location tracking.

The operation of the system is as follows. Each user or object to be located in the system carries an n-Core Sirius B acting as tag. Each of these tags broadcasts periodically a data frame including, amongst other information, its unique identifier in the system. The rest of the time these devices are in a sleep mode, so that the power consumption is reduced. This way, battery lifetime can reach even several months, regarding the parameters of the system (broadcast period and transmission power). A set of n-Core Sirius D devices is used as readers throughout the environment. The broadcast frames sent by each tag are received by the readers that are close to them. This way, readers store in their memory a table with an entry per each detected tag. Each entry contains the identifier of the tag, as well as the RSSI (Received Signal Strength Indication) and the LQI (Link Quality Indicator) gathered from the broadcast frame reception. Periodically, each reader sends this table to the coordinator node connected to the computer. The coordinator forwards each table received from each reader to the computer through the USB port (or using any other data transmission link, such as a military RF/SAT link). Using these detection information tables, the n-Core API applies a set of locating techniques to estimate the position of each tag in the environment. These locating techniques include signpost, trilateration, as well as an innovative locating technique that takes into account different confidence levels when estimating the distances between tags and readers from the detected RSSI values (due to multipath effects, some detected RSSI intervals are less reliable than others).

Then, the web server module offers the location data to remote client interfaces as web services using SOAP (Simple

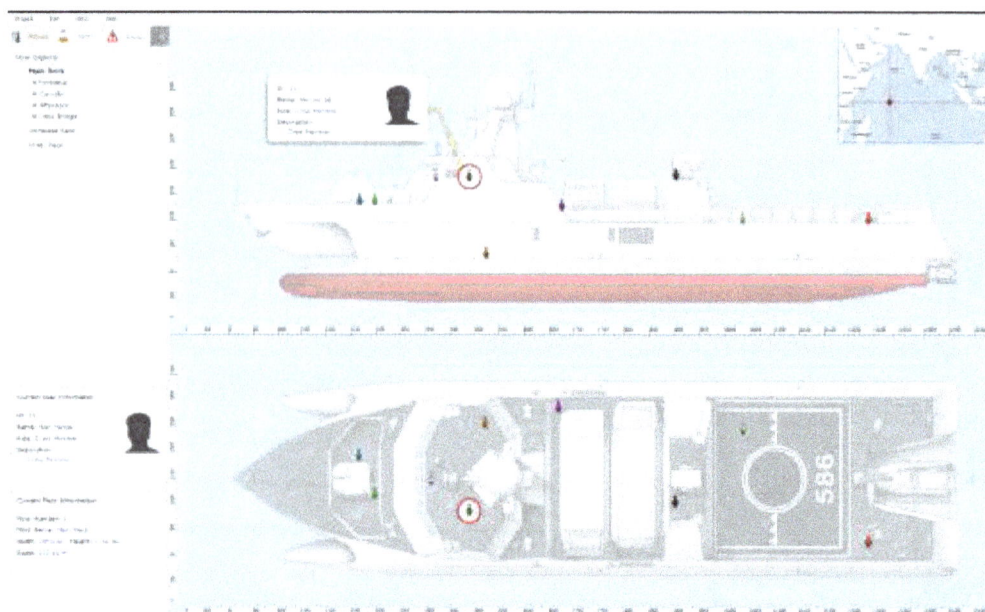

Fig. 3. Web client Graphical user Interface (GUI) of the of the n-Core Polaris RTLS for a maritime scenario.

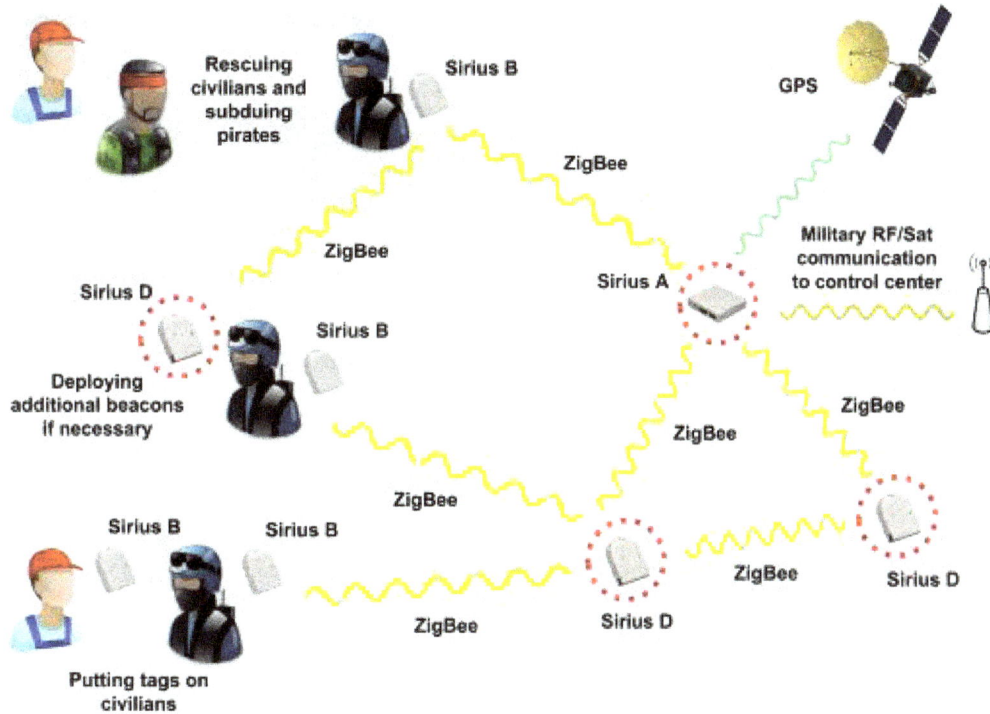

Fig. 4. Schema of the n-Core Polaris RTLS in a boarding and rescue scenario.

Object Access Protocol) over HTTP (Hypertext Transfer Protocol). Figure 3 shows a screenshot of the web client Graphical user Interface (GUI). This client interface has been designed to be simple, intuitive and easy-to-use. Through the different interfaces, administrator users can watch the position of all users and objects in the system in real-time. Furthermore, administrators can define restricted areas according to the users' permissions. This way, if a user enters in an area that is forbidden to him according to his permissions, the system will generate an alert that is shown to the administrator through the client interfaces. In addition, such alerts are registered into the database, so administrators can check anytime if any user violated his permissions. Likewise, administrators can query the database to obtain the location track of a certain user, obtaining statistical measurements about its mobility or the most frequent areas where it moves.

Furthermore, users can use one of the general-purpose buttons provided by the n-Core Sirius B devices to send an alert to the system. Similarly, administrators can send alerts from the system to a user or a set of users, which can confirm the reception using other of the buttons. The system not only provides locating features, but also scheduling and automation functionalities. The system can be easily integrated with a wide range of sensors and actuators using the variety of communication ports included in the n-Core Sirius A devices. By means of the automation engine provided by the n-Core API, the n-Core Polaris system can schedule automation tasks, as well as monitor all sensors in the environment in real-time.

IV. PROPOSED SCENARIOS

This section describes two case studies where the n-Core Platform is proposed to be applied to fight against maritime piracy and illegal traffic. First one presents a scenario where the n-Core Polaris RTLS is proposed to be applied to fight against maritime piracy and illegal traffic. The second one proposes a system that allows monitoring automatically ships' merchandise containers.

A. Real-Time Locating System for Boarding Support and Rescue

The system proposed in this case study consists of a Real-Time Locating System that can be deployed to support maritime boarding and rescue operations. In this scenario, the main objective is to avoid casualties, as well as avoid ransom payments and discourage further hijackings. As n-Core Polaris can be deployed in just few minutes throughout the area of interest and works properly indoors, such as buildings or tunnels, achieving an indoor accuracy with just 1m error [4] [7] [13], these features make it suitable for military applications where is required to monitor the position of people and objects in real-time and with minimal installation and deployment times.

Figure 4 shows the basic schema of n-Core Polaris running in a boarding and rescue scenario. Each member of the rescue military troops carries an n-Core Sirius B device so that the system can locate him in a certain area at all times, both indoors and outdoors. In addition, some soldiers can carry additional n-Core Sirius B devices to be used as tags by

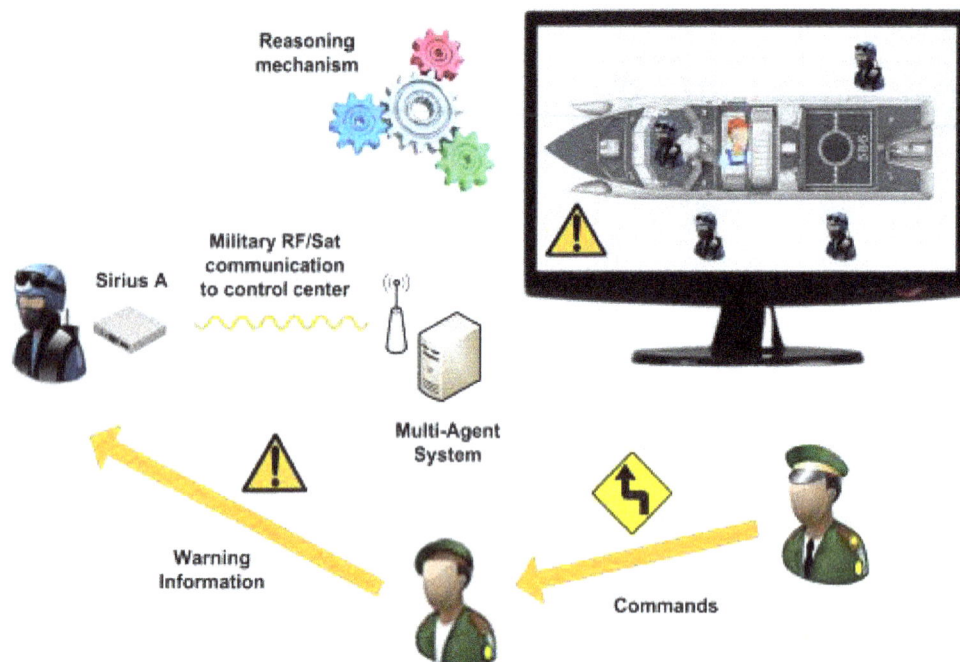

Fig. 5. Multi-Agent System and reasoning mechanisms in the boarding and rescue scenario.

civilians. This way, soldiers put an additional n-Core Sirius B device on each civilian, activating it to be tracked by the system. Likewise, some soldiers carry a set of n-Core Sirius D devices to be placed as additional beacons in the environment and acting as distances references in the system.

Furthermore, there is an n-Core Sirius A device acting as coordinator node in the system and that can be carried by a soldier or by a boat close to the rescue area. This n-Core Sirius A device is connected to a GPS receiver to obtain its global position as the main reference. In addition, this device is connected to a remote control center through a military radio-frequency or satellite link. The remote control center runs a Multi-Agent System that includes reasoning mechanisms and makes use of the locating techniques provided by the n-Core API, as can be seen in Figure 5. The remote control center gathers the detection data sent by the n-Core Sirius D acting as readers to the coordinator node.

B. Multi-Agent System for Controlling the Unloading of Illegal Traffic of Merchandise

The second case study consists of a system that allows monitoring automatically the containers that are transported from port to port by cargo ships. This system makes use of the n-Core platform and agent technology. This way, each of the containers that are transported by sea carries an n-Core Sirius A device that alerts the system if a container is manipulated improperly during a travel. The load of each container in each ship is controlled when both leaving the source port and arriving at the destination port. By means of GPS technology and radio and satellite communications, the system can know where a container is globally at all times. Thus, the loading and unloading tasks are facilitated, preventing in the same way

the traffic of illegal merchandise, such as drug, arms or contraband.

Figure 6 shows the schema and main components of the system. As can be seen, each ship container includes an n-Core Sirius A device to identify it, locate it, as well as control if it is opened, stolen, lost, or if it should not be in that ship. All these features are possible by means of the automation and locating engines provided by the n-Core platform. In addition, each ship includes an intelligent system with reasoning mechanisms running on a local server in order to locate and keep track of all containers in the ship (through ZigBee), locate globally the ship (using GPS) and communicate with ports and control centers (via satellite link communications). Furthermore, every ship is registered when leaving and arriving at a port, by means of a MAS that makes use of the n-Core API features and an n-Core Sirius A device for each dock. This way, all containers and their seals are also checked at port. Finally, all ships' load is tracked from port to port using a global MAS aimed at processing massive data.

V. CONCLUSIONS AND FUTURE WORK

Piracy and illegal traffic imply human, economic, social and political costs. In this sense, it is necessary to apply non-invasive, context-aware, efficient and inexpensive technology to minimize these costs. Systems based on Multi-Agent Systems, Wireless Sensor Networks and Real-Time Locating Systems can give support to military and civil authorities to deal with these problems.

There are different wireless technologies that can be used on RTLSs. The ZigBee standard offers interesting features over the rest of technologies, as it allows the use of large mesh

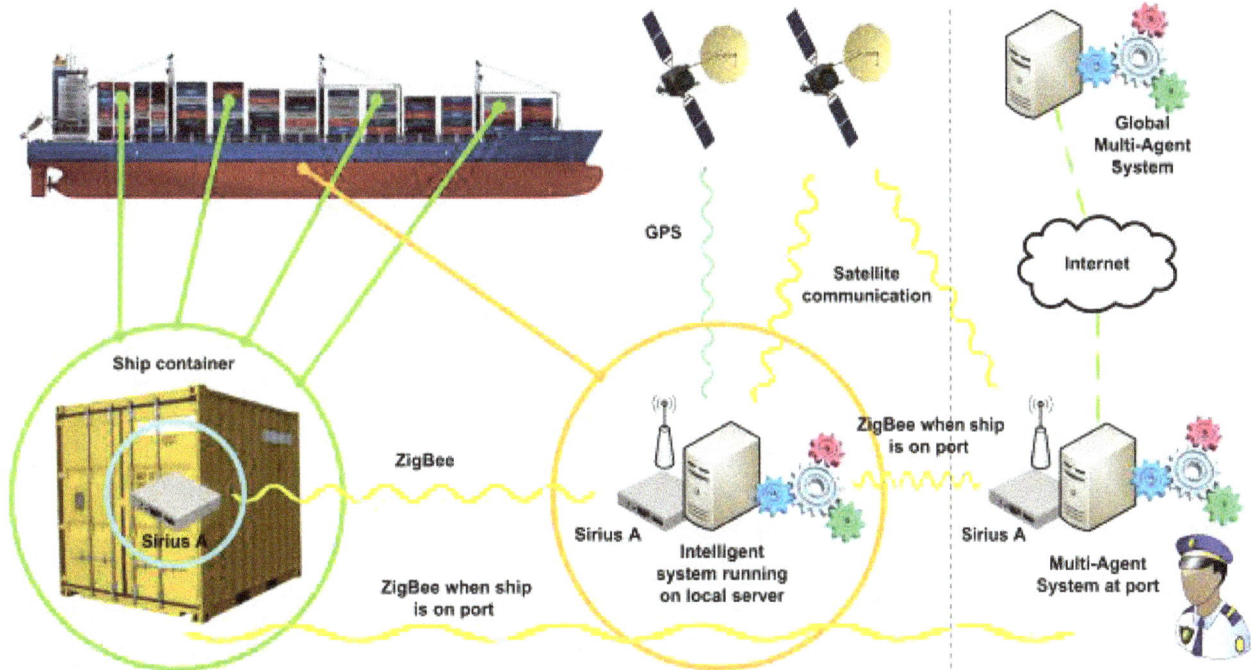

Fig. 6. Multi-Agent System for controlling the unloading of illegal traffic of merchandise.

networks of low-power devices and the integration with many other applications.

In this regard, the n-Core platform and the n-Core Polaris RTLS can provide an important competitive advantage to applications where it is necessary to gather sensing data, automate tasks and know the location of people or objects. Among its multiple application areas are the healthcare, the industrial or the agricultural sectors, as well as those related to security. Its optimal indoor and outdoor functioning makes the n-Core platform and the n-Core Polaris RTLS flexible, powerful and versatile solutions.

Regarding its performance, the n-Core Polaris indoor locating system has been awarded as the winner of the first international competition on indoor localization and tracking, organized by the Ambient-Assisted Living Open Association (AALOA) [13]. These results demonstrate that n-Core Polaris is a robust system suitable to be used in indoor environments and that can locate users and assets with up to 1m accuracy without interfering in the daily-life of people.

Future lines of work include obtaining ideas from specialized military and civilian users to get feedback and improve the proposed case study. Then, it will be performed a detailed analysis and design process to develop and deploy prototypes to test performance and get additional feedback.

REFERENCES

[1] J. Sarangapani, *Wireless Ad hoc Sensor Networks: Protocols, Performance and Control*. CRC Press, 2007.

[2] M.J. Wooldridge, *An Introduction to MultiAgent Systems*. Wiley New York, 2009.

[3] H. Liu, H. Darabi, P. Banerjee, J. Liu, *Survey of Wireless Indoor Positioning Techniques and Systems*. IEEE Transactions on Systems, Man, and Cybernetics, Part C: Applications and Reviews, vol.37, no.6, Nov. 2007, pp.1067-1080.

[4] Nebusens. (2012, June 10). n-Core®: A Faster and Easier Way to Create Wireless Sensor Networks [Online]. Available http://www.n-core.info.

[5] P. Baronti, P. Pillai, V.W.C. Chook, *et al. Wireless sensor networks: A survey on the state of the art and the 802.15.4 and ZigBee standards*. Computer Communications, vol.30, 2007, pp. 1655-1695.

[6] Ó. García, R.S. Alonso, D.I. Tapia, F. Guevara, F. de la Prieta, R.A. Bravo, *A Maritime Piracy Scenario for the n-Core Polaris Real-Time Locating System*. In: Distributed Computing and Artificial Intelligence, vol. 151, S. Omatu, J.F. de Paz, S.R. González, J.M. Molina, A.M. Bernardos, J. M. C. Rodríguez, Eds. Springer Berlin / Heidelberg, 2012, pp. 601-608.

[7] D.I. Tapia, R.S. Alonso, S. Rodríguez, F. de la Prieta, J.M. Corchado, J. Bajo, *Implementing a Real-Time Locating System Based on Wireless Sensor Networks and Artificial Neural Networks to Mitigate the Multipath Effect*. In: Information Fusion (FUSION), 2011 Proceedings of the 14th International Conference on. IEEE/ISIF, Chicago, USA, 2011, pp. 1-8

[8] A. Maouche, *Piracy along the Horn of Africa: An Analysis of the Phenomenon within Somalia*. PiraT Arbeitspapier zur Maritimen Sicherheit, 6. Hamburg, 2011.

[9] D.I. Tapia, J.F. de Paz, S. Rodríguez, J. Bajo, J.M. Corchado, *Multi-Agent System for Security Control on Industrial Environments*. International Transactions on System Science and Applications Journal, vol.4, num.3, 2007, pp. 222-226.

[10] C. Nerguizian, C. Despins, S. Affès, *Indoor Geolocation with Received Signal Strength Fingerprinting Technique and Neural Networks*. In: Telecommunications and Networking - ICT 2004, vol. 3124, Springer Berlin / Heidelberg, 2004, pp. 866-875.

[11] B. Ding, L. Chen, D. Chen, H. Yuan, *Application of RTLS in Warehouse Management Based on RFID and Wi-Fi*. In: 4th International Conference on Wireless Communications, Networking and Mobile Computing (WiCOM'08), 2008, pp 1-5.

[12] M.A. Stelios, A.D. Nick, M.T. Effie, *et al, An indoor localization platform for ambient assisted living using UWB*. In: Proceedings of the 6th International Conference on Advances in Mobile Computing and Multimedia ed. ACM, Linz, Austria, 2007, pp. 178-182.

[13] AAL Open Association (2012, June 10). Evaluating AAL Systems through Competitive Benchmarking [Online]. Available http://evaal.aaloa.org.

27

Graph-based Techniques for Topic Classification of Tweets in Spanish

Héctor Cordobés[1], Antonio Fernández Anta[1], Luis F. Chiroque[1],
Fernando Pérez[2], Teófilo Redondo[3], Agustín Santos[1]

1. IMDEA Networks Institute, Madrid, Spain
2. U-Tad, Madrid, Spain
3. Factory Holding Company 25, Madrid, Spain
4. Universidad Carlos III de Madrid, Spain

Abstract — **Topic classification of texts is one of the most interesting challenges in Natural Language Processing (NLP). Topic classifiers commonly use a bag-of-words approach, in which the classifier uses (and is trained with) selected terms from the input texts. In this work we present techniques based on graph similarity to classify short texts by topic. In our classifier we build graphs from the input texts, and then use properties of these graphs to classify them. We have tested the resulting algorithm by classifying Twitter messages in Spanish among a predefined set of topics, achieving more than 70% accuracy.**

Keywords — **Topic classification, text classification, graphs, natural language processing**

I. INTRODUCTION

Topic classification of texts is one of the most interesting challenges in Natural Language Processing (NLP). In the field of the happiness research it is important to combine sentiment analysis with topic classification techniques, in order to determine the reasons why a subject expresses happiness or sadness. The problem is to assign to every input text to be classified one topic chosen from a collection of predefined topics. Topic classifiers have commonly used a bag-of-words approach, in which the classifier uses (and is trained with) selected terms from the input texts. In these types of approaches the biggest issue is that the set of potential terms used is huge, and has to be reduced to have a practical classifier. Hence, the preprocessing of the texts and the selection of the most important terms to be used becomes fundamental.

In this work, we present classification techniques that are not based on the bag-of-words paradigm. Instead, they generates graphs from the texts, and use graph similarity to classify them by topic. The resulting classifier uses much fewer attributes than bag-of-words classical classifiers.

A prototype classifier was developed using the techniques proposed here, and was used to participate in the topic classification challenge of the Workshop on Sentiment

Analysis at SEPLN - 2013, known as TASS 2013 (*Taller de Análisis de Sentimientos en SEPLN 2013*). As in previous years, the challenge organizers prepared and made available a data set for evaluation. For topic classification, a set of Twitter messages (tweets) in Spanish were provided. Some of these tweets had been previously classified among predefined categories (politics, economy, music, sports, etc.), and the rest was to be classified by the systems developed by the challenge participants. The classifier we developed ended in 3rd position (with respect to the F1 characteristic), very close to the systems that ended first and second, which used classical techniques.

Additionally, we have also tested different configurations of our classifier using the whole data set of tweets provided by the TASS organizers (including the ones used for evaluation), and found that our classifier achieves accuracies above 70%, using very few attributes. In the classifier developed and tested in this work, we have also explored pre-processing alternatives, such as simple Named-Entity Recognition, Thesauri and specific dictionaries (e.g., SMS abbreviations) to account for the special medium Twitter is. We believe that thorough work on this pre-made knowledge data bases could greatly improve the results of the classification.

The rest of the paper is structured as follows. We revise graph-based approaches for NLP in Section II. In Section III we describe the basic techniques used by our classifier, while in Section IV we describe how these techniques have been transformed into an operational system. In Section V we present the evaluation results that have been obtained and discuss their significance and implications.

II. STATE OF THE ART

The great representational power of graphs, in terms of element relationships, and the extensive mathematical work in graph theory, have been useful for text processing. Graph techniques have been successfully exploited for many tasks such as text summarization and information retrieval.

In fact, a number of scientific works use graph techniques for text summarization of big documents, such as [2] or [16]. Similarly, the TextRank method [10], which is the application

Partially funded by the SOCAM research project, Spanish Ministry of Industry, Energy and Tourism.

of the well-known PageRank metric [3] to text graphs, has been used with remarkable success [7] to extract good representatives in text-related graphs by using a random-walk approach. The method is based on the assumption that well-connected nodes (e.g., terms or sentences), would be good representatives of a graph. These works also use an additional set of techniques in order to exploit the relation between sentences in the same document. For this matter, methods such as tf-idf [14], combined with mutual information, information gain, Helmholtz principle [4], and other weighting mechanisms, have been developed to fine-tune the importance of the terms, mainly towards a subsequent bag-of-words scheme. For example, for classification tasks, it is common to describe documents within a Vector Space Model (VSM), and classify them with Rocchio or SMO classifiers, in which each feature is a weighted term. These methods relay in calculating centroid representatives of the text to summarize. Unfortunately, they may sometimes fall in a multi-centroid problem, for which good decision borders determination can be difficult to solve.

In this work, we propose a system where very short text classification is possible by using a vector classification model for which the features are not terms, but graph metrics, thus significantly reducing the training and exploitation computational requirements, while retaining reasonable accuracy. As mentioned, this work makes use of the TASS2013 corpus, managed by SEPLN (Spanish Society for Natural Language Processing) for its TASS sentiment classification challenge. This corpus is in Spanish, which prevents us from using well-known baselines for the English language, such as Reuters-21578 [9]. Instead, we will compare ourselves with other participants in the same task.

Nevertheless, this work is a first step in the application of graph techniques to topic classification of short texts, so it must be taken as a proof of concept. More advanced techniques can be used in conjunction with this classification scheme, such as PoS tagging and dependency trees [17], or sophisticated text normalization [13].

III. BASIC GRAPH-BASED CLASSIFICATION TECHNIQUES

The basic principle for all our techniques is that every piece of text (tweets in this case, and in general a sentence) can be represented as a graph. Essentially, for a given text our proposal uses the words in the text as graph vertices (we usually work only with the word lemmas, and optionally with named entities), and creates weighted edges between the words. We have considered different ways of assessing weights on the edges. A simple option is that the weight represents the frequency with which both words occur together in the text. Another more sophisticated (and complex) choice is that this frequency is weighted by the distance between the words in the syntactic tree of the text. There are other alternatives for building the graph that we deem of great interest in future work (especially those based in directed graphs).

Knowing how to build a graph for each tweet, the first hypothesis for our system is that graphs belonging to the same topic have a common representative structure (topic reference graph). For the text classification, we look for the similarities between the graph generated for a given text and different topic reference graphs. Hence, our work uses a technique of graph similarity in order to detect the topic of a piece of text.

Hence, for our experiments, we have built a reference graph for each topic. This graph is the union of all the graphs generated from all the texts of the same topic. In the resulting reference graph, the weights of the same edge in different graphs are added. This decision is based on the second hypothesis of our work, that is, all words relate to each other with different intensities depending on the topic. For instance, when the topic is Politics, the words *Presidencia* and *Congreso* will show a strong relationship. These same words may not appear or have a weak relationship in other topics (e.g., Football). Therefore, the reference (union) graph created for every topic is expected to be very different. The overall process of building the reference graphs is shown in fig. 1.

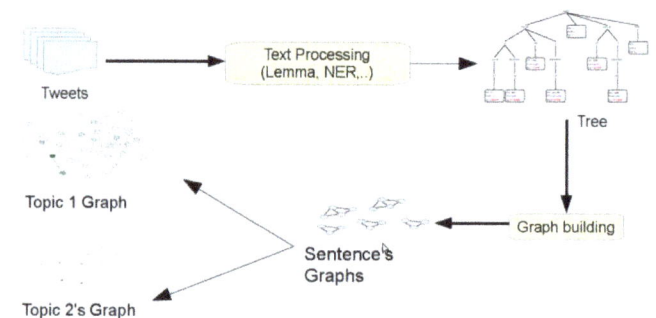

Fig. 1: Graph building process

Hence, using a pre-classified set of tweets for training, our system builds the reference graph for each of the different topics. When a new tweet needs to be classified, its graph is generated. Then, we search for the reference graph with the highest similarity with the tweet graph we want to classify. Fig. 2 shows this process.

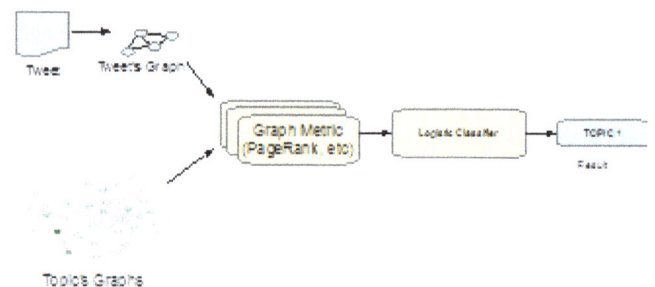

Fig. 2: Tweet classification process

The basic mechanism previously described opens up a wide spectrum of choices and approaches that can be combined in multiple ways. The first step in the mechanism is to build the graph for the tweet. As we have already mentioned, in our work we have explored several options for selecting nodes and assigning weights to the edges. Similarly, we have used several criteria to measure the similarity of a given graph to a reference graph. In the following sections we go into greater detail about the methods we have employed.

IV.   IMPLEMENTING THE CLASSIFIER

In this section we describe how the classifier has been developed, and particularly how the techniques described in the previous section have been implemented. In Section 4.1 we describe the preprocessing that all the tweets go through before using them to build the associated graphs. Section 4.2 describes how the reference graphs get built. Finally, in Section 4.3 we describe how the topic of a new tweet is identified.

A.   Preprocessing of the Text

As a step prior to building and analysing the graphs, we run a preprocessing phase on the texts. This is a typical step in many natural language processing techniques. In this phase, the text is corrected, analysed, and separated into simple elements. In our work we have used the Hunspell dictionaries to obtain an orthographically correct text. We have also used the dictionary of SMS abbreviations and symbols (SMS dictionary) that we already used in the system we developed for TASS 2012 [5]. In addition, we have used Freeling [12] for word lemmatization, taking always into account the automatic disambiguation of lemmas according to the syntactic function. Freeling is also used to parse the syntactic tree of the tweets, which is used for calculating the distances between words. These distances will be used in the following sections.

Another step in the preprocessing phase has been identifying the Named Entities (Named Entity Recognition or NER process). The objective in this step has been to have mechanisms available in order to unify in a single term collections of words that refer to the same concept (e.g., *Real Madrid, Real Madrid C.F.*). To this end, and as a proof of concept, we have used a small manually-created catalog of slightly less than 100 entity names, with several variations for each one. For the creation of the catalog, the texts in the training set have been separated into n-grams, with no limit as to their length, using the technique described in [11]. After the extraction of statistically significant n-grams, the catalog was manually extended both in similar concepts (for instance, the name of a media provider) and in the different ways these concepts may be present.

For the NER we have used a search in the catalog for every single occurrence of the n-gram in the text in order to verify if it refers to one of the entities in the catalog. If so, the n-gram gets substituted by a given canonical name. For instance, the bigram *Mariano Rajoy* has been considered as one such entity, in this case with canonical name *mariano_rajoy_brey*. The whole process has been executed as an experiment, and we believe that broadening its use and having a more complete catalog could improve significantly the quality of the results.

In summary, the preprocessing of each tweet goes through the following phases: first, all URL's are deleted from the tweet; second, using the SMS dictionary, the abbreviations and symbols present are replaced by their textual equivalent; third, orthography is corrected using the Hunspell dictionaries; fourth, the tweet language is detected using Cybozy Labs Language Detection Library [15] and, if it is not Spanish, it is discarded; fifth, NER is applied, substituting the entities found for their canonical name (this phase can be removed at will to check how effective it is in the overall result); sixth, lemmatization is performed using Freeling; seventh and last, all the stop words are removed.

B.   Reference Graphs

The key process to build a reference graph per topic is the process of building a graph for each text, since the reference graph is the union of these graphs. We have tried several options to build text graphs, described below, some of them very involved. The differences are on the set of nodes included in the graph or the way weights are assigned to the edges of the graph.

The simplest option considered for building text graphs has been using as nodes of the graph the words of the text (or the named entities, if used). Then, two nodes are connected with an edge whose weight is the product of their respective number of appearances in the text. (For instance, if in a text the word *concierto* appears twice and the word *guitarra* appears three times, the nodes of these two words are connected with a link of weight 6.) The reference graph obtained with this option has as nodes all the words that appear in the tweets of the topic, and the weight of a link between two words is the number of instances of both words occurring together in the same tweet.

A second option explored assigns to the link between two words a weight that is inversely proportional to the distance between the two words in the text. The intuition is that two words occurring together in a text have larger affinity, and hence should have a stronger link, than words occurring at opposite ends in a sentence. This distance is derived from the syntactic parsed tree as produced by Freeling. To calculate the distance between two words we count the number of jumps in the parsed tree from one word to the other. Our experiments revealed that the results obtained with this option are similar to those with the previous one. Hence, this option was discarded, due to the additional complexity.

Another option that has been explored is using as node set not only the words that appear in the text but also all its synonyms provided by a thesaurus. The intuition is that this will increase the information of the resulting graph. In order to introduce a difference, the weight of the links involving synonyms was slightly below one, while the links connecting words in the text had weight one. In the tests run, the use of synonyms decreased the quality of the results, possibly because they interfered with the use of centrality measures for graph topic. We also tested the use of synonyms when trying to benefit from the graph information (not at the time of creating it). In this case we did not detect any significant improvement either.

As mentioned, none of the options explored was sensibly better that the first option, which is also the simplest one. Hence, this is the type of text graph that is considered in the rest of the paper. However, we think that the use of weights based on distance and synonyms must be addressed in future work since we expect that the augmented graph obtained can improve the reference graphs, and consequently yield a higher rate of successful classifications. In fact, other works such as [1] have already benefited from using thesaurus information.

C. Text Classification

We describe now how the classification of an input text has been done. One of the main questions in our approach is related with the problem of detecting graph similarity. Electing the measure of similarity is a complex decision since there are a great variety of measures and it is not clear which one would be the most appropriate for our problem. In our work we have used several measures, but all of them use the subgraphs of the reference graphs obtained after filtering out the words that do not occur in the text to be classified. That is, for each reference graph we have extracted the words occurring in the text, and we keep the links between them (i.e., we obtain the subgraph induced by the words of the text). Thus, for each topic we obtain a topic subgraph that can even be empty if no word in the text is found within the reference graph.

The following step is to determine one or several topology measures that, when applied to the topic subgraphs, would allow us to choose the topic(s) of the text. We have used two large types of measures: those based in node metrics and those based in relations metrics. The node metrics have mainly been just two: PageRank [3] and HITS [8]. For the computation of these metrics we have used the variants for undirected graphs with weighted links, and applied them to the topic reference graphs. As a result, each node of the reference graph is assigned a measure (its PageRank or HITS values).

Unfortunately, the size of the reference graphs is heavily influenced (biased) by the training set (i.e., number of tweets for each topic), and the centrality measure assigned to the nodes are influenced by the size of the graph. Hence, we attempt to compensate this deviation by means of a normalization of the centrality measures. Following a simple hypothesis, we assume that, given equal representation, the values for the centrality measures would decrease according to the number of graph nodes. Hence, we have normalized the number depending on the size of the reference graph of a given topic. On the other hand, since these values are also dependent on the graph topology in an unpredictable way, we have tried using non-linear operations (particularly, powers like 0.5 or 1/3), in order to give more representation capability to the system.

Then, once the topic subgraph has been extracted for a text, the topic is assigned a value that is the sum of the measures of the nodes of the subgraph (for instance, the normalized sum of PageRank for all the nodes in the subgraph). Computing this value is fast and simple from the precomputed reference graphs. These centrality measures (PageRank and HITS) have been very useful in determining the text topic, as we show later in Section V (see Table I). We observed no big differences between using PageRank and HITS.

As a first approach the value assigned to each topic could be directly used for classification. After adding up the centrality measures for each word in a topic subgraph, the text is classified to the topic with the highest value. With this methodology we achieve nearly a 60% of correct classifications. However, using more sophisticated classifiers (provided in Weka) we achieve a higher rate of accuracy, as we show below.

In addition to the centrality measures, our work has also contemplated links measures. Since every link has a weight, we can compute metrics using those values. We have tried several techniques, but all of them are based on the density of the topic subgraphs (a weighted sum of the links weights). This technique by itself has not rendered better results, but during the evaluation with the training set the technique has proved to be fundamental when combined with the other techniques described before.

In order to combine all the measures described, we have used classifiers included in the Weka system [6]. Each tweet was represented by a vector formed by all the available metrics (PageRank's sum, HITS' sum, graph density, etc.) for every topic reference graph. All in all we have a vector with up to 70 numeric values at our disposal. Of all the classification methods available in Weka, we found that the family of Logistic produced a higher rate of correct classifications. Especially the Logistic MultiClass Classifier method, appeared to give better results in a consistent way over the training set. Hence, all the results shown below use this classifier.

V. RESULTS AND DISCUSSION

We have evaluated our system with different configurations. In all the runs we have trained Weka with the full training set of TASS 2013 (approximately 7,000 tweets) and we have assessed the resulting model with slightly less than the 60,000 tweets of the test set (leaving out some tweets we could not obtain). Weka's algorithm in use has always been SimpleLogistic, as mentioned above.

In Table I we show the results in all the runs. The column "Configuration" shows the text attributes used: PageRank (PR), HITS, graph density (GD), and the modifications applied. These attributes have been generated for every single tweet both during training and evaluation. The column NER shows whether entity recognition has been used or not. As mentioned before, we have disabled this feature in some runs to measure the variation in results. The column "Accuracy" shows in percentage how the system identifies a tweet as belonging to one given topic, according to the evaluation data supplied. Experiment 1 shows the configuration submitted to the TASS 2013 contest.

Tables II and III show information about the distribution of the categories, both for the entry tweets and the results of the classifier used in experiment 1. Note that some tweets belong to more than one category, so for the sake of clarity we have expressed both the occurrence rate, and a normalized occurrence rate. This latter is intended to express the occurrence rate as though the sum of occurrences was 100%.

We present the results by category instead of showing a confusion matrix, because the possibility of finding several categories for one tweet would make the latter large and unintuitive. In Table III the success rate must be interpreted as the proportion of the tagged predictions within the category whose tweet belongs, at least, to that category.

TABLE I

Experiment	Configuration	NER	Accuracy (%)
1	$PR^{0.5}$, PR, PR^2, $HITS^{0.5}$, HITS, HITS, GD	Yes	71.90
2	$PR^{0.5}$, PR, PR^2, $HITS^{0.5}$, HITS, $HITS^2$	Yes	71.62
3	$PR^{0.5}$, PR, PR^2, $HITS^{0.5}$, HITS, $HITS^2$	No	71.38
4	PR	Yes	69.78
5	$PR^{0.5}$	No	69.45
6	$PR^{0.5}$	Yes	71.64
7	$PR^{1/3}$	Yes	71.58
8	$PR^{0.1}$	Yes	69.04
9	HITS	Yes	69.75
10	$HITS^{0.5}$	Yes	71.32
11	$HITS^{1/3}$	Yes	71.35
12	$HITS^{0.1}$	Yes	68.88

TABLE II

Topic	Tweets	Occurrence (%)	Normalized occurrence (%)
movies	596	1.0	0.9
sports	135	0.2	0.2
economy	2549	4.2	3.7
entertainment	5421	8.9	7.8
football	823	1.4	1.2
literature	93	0.2	0.2
music	1498	2.5	2.1
other	28191	46.4	40.5
politics	30067	49.5	43.2
technology	287	0.5	0.4

From the results presented we think that the centrality metric used (PageRank or HITS) does not incur significant difference. On the contrary, the use of a specific normalization may represent a significant improvement (around 2%, for instance, between Experiments 4 and 6). This, together with the good results achieved by using centrality metrics, leads us to believe that choosing an appropriate normalization is of paramount importance for the improvement of results, or in any case, using a metric capable of taking all the factors (size, topology, etc.) into account. We believe that this is an interesting area for future research.

During the execution of the experiments we have detected sensitivity to the available vocabulary. Topics with very few tweets tended to be ignored, such as the case of Technology, because the generated reference graphs are not representative enough. One possible future work could focus on evaluating the sensitivity with larger training sets, and thus determining and measuring how important this effect may be.

TABLE III

Topic	Predictions	Ratio vs. total	Accuracy rate
movies	460	0.77	43.26
sports	67	0.11	47.76
economy	612	1.03	50.16
entertainment	6919	11.66	38.98
football	420	0.71	52.62
literature	60	0.10	25.00
music	1095	1.84	51.60
other	19753	33.29	77.00
politics	29890	50.38	78.27
technology	58	0.10	32.76

In a similar way, this sensitivity could be tested enlarging the NER collection dictionary, so that it can represent in greater detail the topics that the system handles. Maybe given the very limited size of the dictionary used (less than 100 entity names), the impact in the results is not very significant, although consistent (around 0.3%). We should also consider that the NER rate is about 18.3% (occurring rate per tweet) and, as the corpus tweets have been selected, not many different NE recognitions have occurred. Thus we may hypothesize that the impact could be greater within more heterogeneous corpora and bigger dictionaries. This topic is worth to be explored further.

Additionally, the use of the Graph Density in Experiment 1 combines well and was able to improve another 0.3% over the already complex combination of PageRank and HITS in Experiment 2. Nevertheless, it has to be noted that it is not worth increasing unnecessarily the number of characteristics, because as it is shown on Experiment 6, some well chosen metric may be very significant by itself.

The automatic evaluation of the predictive models in Weka is limited because it cannot take more than one prediction per vector, whereas the tweet labelling may include more than one topic per tweet. It is quite possible that a classifier that allows more than one topic per tweet would achieve better results.

Concerning the results for individual categories (Table III), the system appears quite biased towards the main categories (politics and other), as they account for 46.4% and 49.5% respectively of the original tweets. In these cases the system achieves roughly a 78% of correct classification. However, the remaining categories show a rather poor behaviour, many below a mere 50%. Of particular note is the case of entertainment, with a success rate of only 38%, even though it is the third category in the total number of tweets.

We think that an additional experiment with more accurate training could reveal if this behaviour is due to an unbalanced training or to the actual design of the system. Since the number of training texts in some categories (for instance, literature) is rather scarce, we think that a far more complete training set than that currently available would be needed.

REFERENCES

[1] Aseervatham, Sujeevan. 2007. Apprentissage à base de Noyaux Sémantiques pour le traitement de données textuelles. Ph.D. thesis, Université Paris-Nord-Paris XIII.

[2] Blanco, Roi and Christina Lioma. 2012. Graph-based term weighting for information retrieval. Information retrieval, 15(1):54-92.

[3] Brin, Sergey and Lawrence Page. 1998. The anatomy of a large-scale hypertextual web search engine. Comput. Netw. ISDN Syst., 30(1-7):107-117, April.

[4] Dadachev, Boris, Alexander Balinsky, Helen Balinsky, and Steven Simske. 2012. On the helmholtz principle for data mining. In Emerging Security Technologies (EST), 2012 Third International Conference on, pages 99-102. IEEE.

[5] Fernández Anta, Antonio, Luis Núñez Chiroque, Philippe Morere, and Agustín Santos. 2013. Sentiment analysis and topic detection of Spanish tweets: A comparative study of of NLP techniques. Procesamiento del Lenguaje Natural, 50:45-52.

[6] Hall, Mark, Eibe Frank, Geoffrey Holmes, Bernhard Pfahringer, Peter Reutemann, and Ian H. Witten. 2009. The WEKA data mining software: an update. SIGKDD Explorations, 11(1):10-18.

[7] Hassan, Samer, Rada Mihalcea, and Carmen Banea. 2007. Random walk term weighting for improved text classification. International Journal of Semantic Computing, 1(04):421-439.

[8] Kleinberg, Jon M. 1999. Authoritative sources in a hyperlinked environment. J. ACM, 46(5):604-632, September.

[9] Lewis, David D. 1997. Reuters-21578 text categorization test collection.

[10] Mihalcea, R. and P. Tarau. 2004. TextRank: Bringing order into texts. In Proceedings of EMNLP-04and the 2004 Conference on Empirical Methods in Natural Language Processing, July.

[11] Nagao, Makoto and Shinsuke Mori. 1994. A new method of n-gram statistics for large number of n and automatic extraction of words and phrases from large text data of Japanese. In Proceedings of the 15th conference on Computational Linguistics, COLING 1994, Volume 1, pages 611-615. Association for Computational Linguistics.

[12] Padró, Lluís, Samuel Reese, Eneko Agirre, and Aitor Soroa. 2010. Semantic services in freeling 2.1: Wordnet and ukb. In Principles, Construction, and Application of Multilingual Wordnets, pages 99-105, Pushpak Bhattacharyya, Christiane Fellbaum, and Piek Vossen, editors, Mumbai, India, February. Global Wordnet Conference 2010, Narosa Publishing House.

[13] Porta, Jordi and José Luis Sancho. 2013. Word normalization in twitter using finite-state transducers. Proc. of the Tweet Normalization Workshop at SEPLN 2013. IV Congreso Espa nol de Informática.

[14] Salton, Gerard and Michael J McGill. 1983. Introduction to moderm information retrieval.

[15] Shuyo, Nakatani. 2010. Language detection library for java. http://code.google.com/p/language-detection/.

[16] Thakkar, Khushboo S, Rajiv V Dharaskar, and MB Chandak. 2010. Graph-based algorithms for text summarization. In Emerging Trends in Engineering and Technology (ICETET), 2010 3rd International Conference on, pages 516-519. IEEE.

[17] Vilares, David, Miguel A. Alonso, and Carlos Gómez-Rodríguez. 2013. Una aproximación supervisada para la minería de opiniones sobre tuits en español en base a conocimiento lingüístico. Procesamiento del Lenguaje Natural, 51:127-134.

Permissions

All chapters in this book were first published in IJIMAI, by Imai-Software Research Group; hereby published with permission under the Creative Commons Attribution License or equivalent. Every chapter published in this book has been scrutinized by our experts. Their significance has been extensively debated. The topics covered herein carry significant findings which will fuel the growth of the discipline. They may even be implemented as practical applications or may be referred to as a beginning point for another development.

The contributors of this book come from diverse backgrounds, making this book a truly international effort. This book will bring forth new frontiers with its revolutionizing research information and detailed analysis of the nascent developments around the world.

We would like to thank all the contributing authors for lending their expertise to make the book truly unique. They have played a crucial role in the development of this book. Without their invaluable contributions this book wouldn't have been possible. They have made vital efforts to compile up to date information on the varied aspects of this subject to make this book a valuable addition to the collection of many professionals and students.

This book was conceptualized with the vision of imparting up-to-date information and advanced data in this field. To ensure the same, a matchless editorial board was set up. Every individual on the board went through rigorous rounds of assessment to prove their worth. After which they invested a large part of their time researching and compiling the most relevant data for our readers.

The editorial board has been involved in producing this book since its inception. They have spent rigorous hours researching and exploring the diverse topics which have resulted in the successful publishing of this book. They have passed on their knowledge of decades through this book. To expedite this challenging task, the publisher supported the team at every step. A small team of assistant editors was also appointed to further simplify the editing procedure and attain best results for the readers.

Apart from the editorial board, the designing team has also invested a significant amount of their time in understanding the subject and creating the most relevant covers. They scrutinized every image to scout for the most suitable representation of the subject and create an appropriate cover for the book.

The publishing team has been an ardent support to the editorial, designing and production team. Their endless efforts to recruit the best for this project, has resulted in the accomplishment of this book. They are a veteran in the field of academics and their pool of knowledge is as vast as their experience in printing. Their expertise and guidance has proved useful at every step. Their uncompromising quality standards have made this book an exceptional effort. Their encouragement from time to time has been an inspiration for everyone.

The publisher and the editorial board hope that this book will prove to be a valuable piece of knowledge for researchers, students, practitioners and scholars across the globe.

List of Contributors

Teófilo Redondo
ZED Wordwide, Department of Research and Innovation, Madrid, Spain

Guillermo Cueva-Fernandez
University of Oviedo, Department of Computer Science

Jordán Pascual Espada
University of Oviedo, Department of Computer Science

Vicente García-Díaz
University of Oviedo, Department of Computer Science

Martin Gonzalez-Rodriguez
University of Oviedo, Department of Computer Science

A. A. Juan Fuente
Computer Science Department, University of Oviedo, Asturias, Spain

B. López Pérez
Computer Science Department, University of Oviedo, Asturias, Spain

G. Infante Hernández
Laboratory of Software Architecture, University of Oviedo, Asturias, Spain

L. J. Cases Fernández
Laboratory of Software Architecture, University of Oviedo, Asturias, Spain

Fernando Sánchez
University of Málaga, Malaga, Spain

Samuel Benavides
University of Málaga, Malaga, Spain

Fernando Moreno
University of Málaga, Malaga, Spain

Guillermo Garzón
University of Málaga, Malaga, Spain

Maria del Mar Roldan-Garcia
University of Málaga, Malaga, Spain

Ismael Navas-Delgado
University of Málaga, Malaga, Spain

Jose F. Aldana-Montes
University of Málaga, Malaga, Spain

Pushkar Dixit
Faculty of Engineering and Technology Agra College, Agra, India

Nishant Singh
Poornima Institute of Engineering and Technology, Jaipur, India

Jay Prakash Gupta
Infosys Limited, Pune, India

Celia Gutiérrez
Universidad Complutense de Madrid, Spain

Samira Noferesti
Faculty of Electrical and Computer Engineering, Shahid Beheshti University, Iran

Mehrnoush Shamsfard
Faculty of Electrical and Computer Engineering, Shahid Beheshti University, Iran

M Ferreira
Department of Computer Science and Engineering, Instituto Superior Técnico, Technical University of Lisbon, Lisboa, Portugal

M. Casquilho
Centre for Chemical Processes, Department of Chemical Engineering, Instituto Superior Técnico, Technical University of Lisbon, Lisboa, Portugal

Vijay Bhaskar Aemwal
SiemensInformation System, India

Shiv Ram Dubey
GLAU, Mathura, India

Pushkar Dixit
Dept. of Inform. Tech., Dr. M.P.S Group of Institutions College of Business Studies, Agra, India

Nishant Singh
Dept. of Comp. Engg. & Applications, Poornima Group of Colleges, Jaipur, India

Jay Prakash Gupta
Systems Engineer in Infosys Limited, Bangalore, India

Luz Andrea Rodríguez Rojas
University of Oviedo, Asturias, Spain

Juan Manuel Cueva Lovelle
University of Oviedo, Asturias, Spain

Giovanny Mauricio Tarazona Bermúdez
Francisco José de Caldas District University, Bogotá, Colombia

Carlos Enrique Montenegro
Francisco José de Caldas District University, Bogotá, Colombia

Maria Cecília Gomes
CITI/Departamento de Informática, Faculdade de Ciências e Tecnologia, Universidade Nova de Lisboa, Caparica, Portugal

Hervé Paulino
CITI/Departamento de Informática, Faculdade de Ciências e Tecnologia, Universidade Nova de Lisboa, Caparica, Portugal

Adérito Baptista
CITI/Departamento de Informática, Faculdade de Ciências e Tecnologia, Universidade Nova de Lisboa, Caparica, Portugal

Filipe Araújo
CITI/Departamento de Informática, Faculdade de Ciências e Tecnologia, Universidade Nova de Lisboa, Caparica, Portugal

Álvaro Arranz
Zed Worldwide

Manuel Alvar
Zed Worldwide

Zahra Pooranian
Graduate School, Dezful Islamic Azad University, Dezful, Iran

Mohammad Shojafar
Dept.of Information Engineering, Electronic and Telecommunication (DIET), "Sapienza" University of Rome, Rome, Italy

Jemal H. Abawajy
School of Information Technology, Deakin University, Geelong, Australia

Mukesh Singhal
Computer Science & Engineering, University of California, Merced, USA

José del Campo-Ávila
Universidad de Málaga, Andalucía Tech, Departamento de Lenguajes y Ciencias de la Computación, Campus de Teatinos, Málaga, España

Ricardo Conejo
Universidad de Málaga, Andalucía Tech, Departamento de Lenguajes y Ciencias de la Computación, Campus de Teatinos, Málaga, España

Francisco Triguero
Universidad de Málaga, Andalucía Tech, Departamento de Lenguajes y Ciencias de la Computación, Campus de Teatinos, Málaga, España

Rafael Morales-Bueno
Universidad de Málaga, Andalucía Tech, Departamento de Lenguajes y Ciencias de la Computación, Campus de Teatinos, Málaga, España

Fábio Silva
Department of Informatics, University of Minho

Cesar Analide
Department of Informatics, University of Minho

Paulo Novais
Department of Informatics, University of Minho

Francisco Mochón
Universidad Nacional de Educación a Distancia, Madrid, Spain

Oscar Sanjuán
Universidad Carlos III de Madrid, Spain

J. A.Cortés
Systems Engineering Program, Cooperative University of Colombia

J. O.Lozano
Systems Engineering Program, Cooperative University of Colombia

Jorge Bacca
University of Girona, Spain

Silvia Baldiris
University of Girona, Spain

Ramon Fabregat
University of Girona, Spain

Cecilia Avila
University of Girona, Spain

Carlos Alberto Rombaldo Jr
Department of Computing Engineering Digital System of University of São Paulo, São Paulo, Brazil

Solange N. Alves Souza
Department of Computing Engineering Digital System of University of São Paulo, São Paulo, Brazil

Luiz Sergio de Souza
Faculdade de Tecnologia - Carapicuíba, São Paulo, Brazil

Ndeye Massata NDIAYE
Regal team, LIP6, UPMC Paris Jussieu France
Gaston Berger University of Saint-Louis Senegal

Pierre SENS
Regal team, LIP6, UPMC Paris Jussieu France

Ousmane THIARE
Gaston Berger University of Saint-Louis Senegal

Libertad Caicedo Acosta
Universidad Distrital Francisco José de Caldas, Bogotá, Colombia

Camilo Andrés Ospina Acosta
Universidad Distrital Francisco José de Caldas, Bogotá, Colombia

Nancy Yaneth Gelvez García
Universidad Distrital Francisco José de Caldas, Bogotá, Colombia

Oswaldo Alberto Romero Villalobos
Universidad Distrital Francisco José de Caldas, Bogotá, Colombia

Sho Fukuda
Wakayama University, Sakaedani, Wakayama, Japan

Yuuma Yamanaka
Wakayama University, Sakaedani, Wakayama, Japan

Takuya Yoshihiro
Wakayama University, Sakaedani, Wakayama, Japan

Sumit Goyal
National Dairy Research Institute, Karnal, India

Gyanendra Kumar Goyal
National Dairy Research Institute, Karnal, India

Óscar García
R&D Department, Nebusens, S.L

Ricardo S. Alonso
R&D Department, Nebusens, S.L

Dante I. Tapia
R&D Department, Nebusens, S.L

Fabio Guevara
R&D Department, Nebusens, S.L

Fernando de la Prieta
Department of Computer Science and Automation, University of Salamanca

Raúl A. Bravo
Department of Computer Science and Automation, University of Salamanca

Héctor Cordobés
IMDEA Networks Institute, Madrid, Spain

Antonio Fernández Anta
IMDEA Networks Institute, Madrid, Spain

Luis F. Chiroque
IMDEA Networks Institute, Madrid, Spain

Fernando Pérez
U-Tad, Madrid, Spain

Teófilo Redondo
Factory Holding Company 25, Madrid, Spain

Agustín Santos
IMDEA Networks Institute, Madrid, Spain